Handbook on Climate Change and Human Security

Edited by

Michael R. Redclift

*Emeritus Professor of International Environmental Policy,
King's College, University of London, UK*

Marco Grasso

*Senior Lecturer of International Environmental Policy,
Università degli Studi di Milano-Bicocca, Italy*

Edward Elgar
Cheltenham, UK • Northampton, MA, USA

Published by
Edward Elgar Publishing Limited
The Lypiatts
15 Lansdown Road
Cheltenham
Glos GL50 2JA
UK

Edward Elgar Publishing, Inc.
William Pratt House
9 Dewey Court
Northampton
Massachusetts 01060
USA

A catalogue record for this book
is available from the British Library

Library of Congress Control Number: 2013942210

This book is available electronically in the ElgarOnline.com Economics Subject Collection, E-ISBN 978 0 85793 911 1

ISBN 978 0 85793 910 4 (cased)

Typeset by Servis Filmsetting Ltd, Stockport, Cheshire
Printed and bound in Great Britain by T.J. International Ltd, Padstow

Contents

Figures

Tables

Contributors

Karen Bickerstaff Department of Geography, University of Exeter, UK

Hans Günter Brauch United Nations University Institute for Environment (UNU-EHS), Germany

Simon Dalby Balsillie School of International Affairs, Canada

Guy Edwards Center for Environmental Studies, Brown University, USA

Giuseppe Feola Department of Geography and Environmental Science, University of Reading, UK

Des Gasper International Institute of Social Studies (The Hague), Erasmus University Rotterdam, The Netherlands

Nils Petter Gleditsch Peace Research Institute Oslo (PRIO) and Norwegian University of Science and Technology (NTNU), Norway

Marco Grasso Department of Sociology and Social Research, Università degli Studi di Milano-Bicocca, Italy

C. Michael Hall Department of Management, Marketing and Entrepreneurship, University of Canterbury, New Zealand; Department of Geography, University of Oulu, Finland

Emma Hinton Third Sector Research Centre, University of Southampton, UK

Christian D. Klose Think GeoHazards, New York, USA

Michael Mason Department of Geography and Environment, London School of Economics and Political Science, UK

Richard Matthew Center for Unconventional Security Affairs and Department of Planning, Policy and Design, University of California Irvine, USA

Ragnhild Nordås Peace Research Institute Oslo (PRIO), Norway

Mark Nuttall Department of Anthropology, University of Alberta, Canada; Ilisimatusarfik/University of Greenland and Greenland Climate Research Centre, Greenland

Úrsula Oswald Spring Regional Center for Multidisciplinary Research at the National Autonomous University of Mexico, Mexico

Michael R. Redclift King's College, University of London, UK

Elise Remling Stockholm Environment Institute, Sweden

Jesse Ribot Department of Geography, University of Illinois, USA

J. Timmons Roberts Center for Environmental Studies and Department of Sociology, Brown University, USA

Jürgen Scheffran Institute of Geography, University of Hamburg, Germany

David Simon Department of Geography, Royal Holloway, University of London, UK

Sharath Srinivasan Department of Politics and International Studies (POLIS), University of Cambridge, and King's College, Cambridge, UK

Steve Vanderheiden University of Colorado at Boulder, USA; Centre for Applied Philosophy and Public Ethics (CAPPE) at Charles Sturt University, Australia

Elizabeth E. Watson Department of Geography, University of Cambridge, UK

Christian Webersik Department of Development Studies, University of Agder, Norway

Introduction: human security in the age of carbon

Michael R. Redclift and Marco Grasso

Throughout human history individuals and societies have been threatened by environmental change. Nowadays these risks are magnified: there is, in fact, widespread evidence that climate change is increasingly bringing about dramatic impacts on natural and social systems (IPCC 2007) and is seriously endangering the human security of most of the world's population. (Part III of this Handbook examines the repercussions of climate change for human security in some of the world's most sensitive regions.) The earth sciences make it clear that we are in the *Anthropocene* (Crutzen 2002) and that humankind is living in the age of climate change, a global and complex phenomenon that could undermine the stability of natural and social systems and ultimately affect human security (see Scheffran and Remling, this volume). Therefore, in introducing this Handbook, we first need to briefly situate both historically and culturally the relationship between human security and climate change.

The background to the discussion of human security and climate change is provided by earlier debates in which it was suggested we were reaching the limits of resource capacity. During the 1970s resource shortages were seen as a constraint on further economic growth and development. The argument for resource conservation was thus that by conserving resources we were able to facilitate economic growth, subject to natural limits. This was essentially the 'Limits to Growth' position in the early 1970s (Meadows 1972). At the same time existing levels of economic growth were seen as representing a threat to the environment and resources. It was argued that a vicious circle had been created in which economic activity undermined the biosphere resources on which we rely.

The first position – that resource conservation assists necessary economic growth – lost support partly because it was a product of high-energy prices (the oil hikes of the 1970s). As hydrocarbons became relatively cheaper, and the effects of the Green Revolution in expanding food staples to meet population growth began to be acknowledged, it was also clear that the Malthusian position no longer held – that population increased to exceed the resources necessary to feed this growth. And the drive for economic development in the South (circa the Brandt Commission report

1

of 1980) was overtaken by events: at first it was put in jeopardy by the debt crises of the 1980s, the structural adjustment programmes, and post-recovery the deregulation of markets, the retreat of the state and, eventually, higher levels of economic growth in much of the newly developing world, especially the populous economies of Asia.

The genius of the position that came to be referred to as 'sustainable development' was that almost everybody could sign up to it. There were very few dissenting voices (Redclift 1987). The mechanisms which were unleashed via deregulation and the neoliberal ascendancy (the Washington Consensus – a 'consensus', incidentally, in which most people had not been consulted) became the favoured instruments of policy in seeking to achieve sustainable development. These took two forms.

First, attempts to internalize environmental externalities in products and services – or 'ecological modernization'. This was viewed as a competitive strategy by the European Union, in particular, giving Europe a competitive advantage over the United States and any newly developing rivals. Basically, you count the embodied carbon in products, seek to reduce energy and material throughput, and make a 'win/win' gain, by reducing energy costs (hydrocarbons prices were rising) and reducing environmental damage. Trade arrangements also take account of 'embodied carbon'. The more interventionist policies of the European Union facilitated this in the 1990s.

Second, changes occurred with the development of carbon markets, both within industries and, more importantly, between countries. These new markets represented a challenge for entrepreneurship, new market opportunities, and required very little government action. Carbon markets were thus popular among devotees of free-market economics and environmentalism, unlike other interventions such as carbon taxes. Several awkward questions were not posed, however. What might happen when markets fall and the price of carbon drops significantly? What were the wider implications of trading in a *bad* (pollution) rather than a *good*: in institutionalizing the idea of carbon dependency? This latent opposition to carbon trading as a solution remained largely inchoate in the rush to endorse it. Today, in 2013, it is undermined by policy intervention: the European Union issuing fewer carbon permits to ensure that their price rises, rather than falls.

The conversion of governments to a more or less uncritical view of markets was even more evident in the international efforts to 'protect' biodiversity. The biodiversity regime was expressed in the Convention on Biological Diversity (1992) and the Cartagena Protocol on Biosafety (2000). This demonstrated a shift from a focus on the loss of species diversity, and thus the loss of complex ecosystems, to a focus on the preservation of genetic diversity, where the principal gains were in the pharmaceutical industries and agriculture (Paterson 2009). Again the almost imperceptible shift was from nature conservation to nature as commodity. The main opposition to the latter was from groups – mainly *non-governmental organizations* (NGOs) – which argued that marginalized people had rights in nature, which governments and the pharmaceutical industry ignored. However the industry lobby won much of the ideological struggle, insisting that *ex situ* conservation in gene banks should be treated as equivalent to *in situ* conservation in ecosystems.

Finally, the conjunction of newly liberated markets and environmental concern (a necessary contradiction of capitalism seeking a resolution) can with hindsight be seen as a managed senescence, if we continue with the biological metaphors of development.[1] A more mainstream view, however, would be that they addressed system failures, and could even lead to a rejuvenated, if scarcely recognizable, type of materials light capitalism (Lovins et al. 2000).

The hopes that markets and technology would solve the environmental problems associated with accelerated economic growth and the enormous rise in global consumption and carbon emissions were about to be challenged by events.

The financial crisis was fed by the personal greed of many bankers and financial managers, and fuelled by the virtually unregulated production of credit – not because interest rates were low, but because the price attached to housing equity (especially in the United States, the United Kingdom, Spain and Ireland) was unrealistically high. The rise in sub-prime lending and borrowing took place under systems of ineffective governance which emphasized everybody's right to property regardless of collateral and debt levels. Politically it was sold as everybody's right to credit rather than their right to debt. The financial crisis revealed that it was completely unsustainable.

While the policy response paid lip service to the rapidly disappearing Green agenda, it did not support this rhetoric with effective interventions (cf. the almost derisory role of new Green investment in attempts to address the financial crisis). There was now considerable evidence of the effects of the financial downturn on migration, as well as poverty, notably in China, which supported the United States' debt through buying in to its financial packages, and supported raised consumption in the West generally, by lowering the costs of manufactured goods there.

Another process that has gathered speed is that of transnational sourcing of food, minerals and other resources. The internationalization of capital movements and the need to secure resources has led to increased transnational acquisition of land and minerals, on the part of China and some of the Gulf States, principally in Africa. Rather than depend exclusively upon trade relations to meet their domestic resource deficiencies – trade contracts during an economic recession – the advantages of acquisition of land, water sources, food (via virtual water) became evident, especially for their geopolitical reach. Land displacement for crops like soya had already changed international food/land imbalances.

This brief exegesis on the political and economic context of the 'carbon age' suggests that market-based and fossil fuel-centred socio-economic systems, by triggering global climate change, are seriously threatening our planet, together with other forms of resource degradation. At the same time human societies and their environments are bound together in sets of complex relationships in which carbon use and dependence form a central dimension. The recent financial crisis and austerity measures in much of Europe and North America only underline the fragility and co-dependence of human societies and nature. We argue that the concept of human security is the key entry-point for disentangling the entrenched and multi-layered connections between the carbon age and natural and socio-economic systems, and the ambivalent relationship with climate change.

SCOPE AND AIMS OF THE HANDBOOK

The clarification of the relationships between climate change and human security is extremely important in moving from the carbon age, of heavy dependence on hydrocarbons, to a much needed greener global economy. However, beyond the obvious claims of major threats of large-scale disruption to natural and social systems brought about by climate change, the interaction of climate change with human security is not easy to establish (Dalby, this volume). This is due in part to the uncertainty of the dangers posed by climate change, and in part to the still controversial and somewhat blurred notion of human security in the context of, and in connection with, climate change. The general aim of the Handbook is to shed light, at different levels and according to diverse perspectives, on the second issue: the complex and multifarious web of connections between human security and climate change. The primary feature of human security is its focus on the core values of human societies and their continuity over time. This is also the main difference from the traditional concept of security, which, on the contrary, is based on the use of force to prevent threats to autonomy and territorial integrity.

Although the perspective of human security first arose in the 1960s as a response to growing dissatisfaction with the traditional paradigms of security and development, it imposed itself in a structured way only in the early 1990s through the efforts of the United Nations Development Program (UNDP). The 1994 UNDP Report on Human Security stressed the essential properties of the notion of human security: the centrality of people, universality, the interdependency of its components, and its preventive sight, distinguishing seven areas of global concern: economic, food, health, environmental, personal, community and political. Furthermore, the report put forward a two-tiered definition of the concept of human security: '[i]t means, first, safety from such chronic threats as hunger, disease and repression. And second, it means protection from sudden and hurtful disruptions in the patterns of daily life' (UNDP 1994: 23).

There are in fact many possible definitions of human security put forward by the United Nations (UN) and other bodies of the UN system, by governments and by the academic literature. Despite a certain vagueness and controversy surrounding the boundaries, the notion of human security is almost unanimously agreed to concern both needs and rights, to include both individuals and communities, and to prioritize them, to integrate different drivers, and to demonstrate a concern for justice. Furthermore it is generally held that human security depends on a set of interlinked factors (e.g. political freedom, entitlements, economic equality) among which climate change might not be necessarily the most relevant one (Simon, this volume; Scheffran and Remling, this volume). For example, human security can be threatened by global processes acting independently but synergistically, e.g. global environmental change and globalization (Oswald Spring et al., this volume). Moreover, in general, it is recognized that human security is linked to both environmental and societal change, which are themselves mediated by the ecological/geographical and political/institutional context, distribution of rights and resources (Vanderheiden, this volume), and control over and access to assets (Bickerstaff and Hinton, this volume). The relative importance of such conditions and mediating factors, in turn depends on the characteristics of specific regions and communities.

Human security, in fact, is highly scale-dependent (Matthew, this volume) and it is most often observed at community or local level (Hall, this volume). As such, the very genesis of the concept of human security entailed a shift from states to individuals and communities (Gasper, this volume) although national states mediate in many ways in determining the conditions for human security (e.g. Nordås and Gleditsch, this volume; Mason, this volume). For instance, many of the processes affecting human security in the Mediterranean region are occurring at national level (e.g. the Arab Spring) and generate phenomena that call into question the role of national states themselves, for example in relation to immigration (Grasso and Feola, this volume).

Although it is important to distinguish analytically among the different environmental and societal drivers of human security, it is also important to bear in mind that they are closely interlinked, and that threats to human security often arise from the interplay among these drivers in specific contexts (Dalby, this volume). The best example is anthropogenic climate change. In fact, as repeatedly made clear throughout this volume, several environmental changes largely determined by climate dynamics interact with social and institutional ones, and they have the potential jointly to affect human security. Further, it is necessary to note that climate change, owing to its ramified and overarching impacts, can be seen as a threat multiplier (Matthew, this volume; Srinivasan and Watson, this volume) because of its many potential interactions with other factors of human insecurity. For example, climate change is believed to be linked with violent conflict, although the evidence is ambivalent, given that several studies do not find confirmation of such a link (Nordås and Gleditsch, this volume). Climate change is also linked with migration (e.g. Nuttall, this volume), whereby migration often results from a closely intertwined set of factors, among which are also demographic (e.g. overpopulation), economic (e.g. unemployment), and political (e.g. conflict, lack of human rights and freedom) ones (Ribot, this volume).

In general, the prospects for human security under climate change are strongly shaped by the interplay of local and global environmental, socio-economic and political-institutional processes (Webersik and Klose, this volume). Therefore, as we have argued, the broad objective of the current volume is to understand the complex interactions of these societal processes with human security and climate change and the different scales, issues and locations involved. This Handbook brings together the insightful perspectives of prominent scholars, who first explore the bidirectional relationships between climate change and human security, analyse the determinants of human security in climate change, and conduct extensive scrutiny of critical regions, as well as investigate the societal responses to the challenges posed by climate change. It is intended for both academics and decision-makers in the field of human development, climate and development policy and global environmental policy/politics, and aims at bridging the divide between the different groups and domains of scholarship and public policy.

The volume is structured along four major areas. First it seeks to analyse and systematize the intertwined relationships between climate change and human security; further it investigates the determinants of human security in the context of climate change; and then it scrutinizes the spatial dimensions of human security and climate change in some of the most vulnerable regions: Latin America and the Caribbean, the Mediterranean, the Arctic, and Africa. Finally the volume closes by pointing out the most significant societal responses put forward for limiting the adverse impacts of climate change on human security.

OVERVIEW OF THE HANDBOOK

Part I, 'Framing the issue: climate change and human security', explores in depth the evolution of the notion of human security and the emergence of climate change as an issue of human security, and carries out a more detailed investigation of some key conceptual topics involved by human security and climate change. Since climate change remains also a matter of traditional security the focus later turns to this perspective. Finally, in preparing the ground for the remaining parts of the volume, Part I widens the scope of the analysis in order to define the processes involved with human security and their spatial and temporal scales in the context of climate change.

In chapter 1, 'Climate change as an issue of human security', Simon Dalby puts forward a broad conceptual analysis that sets the stage for the entire Handbook, covering much of the most controversial matters of human security and its relationships with climate change. In particular the author opens his chapter by illustrating the switch from a pure environmental security perspective to a more development-oriented one in dealing with climate change. Then he puts forward a history of the idea of security, overviews the extensions of the concept of security through the 1980s and early 1990s as summarized by Emma Rothschild, and explores the UNDP formulation of human security and that of the Commission on Human Security. Dalby goes on to emphasize that the connections between human security and climate change need both to understand people's vulnerability and the causes of such vulnerability, and to engage with the discussions of the international norms of the responsibility to protect which obligates states to protect their populations. In the same vein Dalby argues that human security in climate change, or climate security, besides freedom from want, freedom from fear, and the ability to live a dignified life, rests also on the freedom from hazard impacts and involves intra-generational and intergenerational arguments of justice. Finally the big question raised by Dalby concerns the origins of the authority to implement human security in rapidly changing times. Human security, in fact, challenges traditional assumptions about the locus of political authority and makes it clear that a new political order is urgently needed.

In chapter 2, 'Elements and value-added of a human security approach in the study of climate change' Des Gasper analyses in depth some key-topics of human security in climate change. He first starts with a case, that of Bangladesh, and then analyses some features of human security discourse: (1) whose security? (2) security of what (which values/which sectors)? (3) provision by whom? (4) security as perceived by whom? (5) security against what threats? Gasper proceeds to discuss the application and potential value added of human security in the study of climate change. In particular, a human security approach can help in thinking about climate change by deepening consideration of connectivity and knock-on effects. Human security analysis can also contribute to an increasing sympathetic attention to the difficulties of others, through its consideration both of how 'distant others' live and feel and of the global interconnections. Finally, human security could support the changes that are needed for global sustainability, in respect of how people perceive shared vulnerabilities, shared interests, and shared humanity.

In chapter 3, 'The IPCC, human security, and the climate-conflict nexus' Ragnhild Nordås and Nils Petter Gleditsch define the IPCC as the leading agenda setter for debates about climate change and human security. Interestingly however, the IPCC assessment reports present climate change not only as a challenge to human security, but also in the narrower sense of armed conflict. Nordås and Gleditsch carefully review the Third (TAR) and the Fourth (4AR) Assessment Reports in order to point out how they have dealt with the climate-conflict nexus and to make clear the empirical basis for the claims made. In general they find, especially in the 4AR, that the tone used to relate climate change and conflicts is prudent. Some more attention to this issue is given in the Africa chapter of the 4AR, which contains more references to conflict than any other chapter. Even so, the authors further argue, the IPCC does not fully accomplish its mandate in relation to climate change and conflicts. This is due in part to the state of the climate-conflict literature at the time, but mostly to the unsystematic standard for evaluating evidence and consulting relevant expertise on this particular issue. All in all, the climate change-conflict nexus is neither well-developed nor well-documented, in contrast with the robust scientific analysis of other climate change-related issues. But, hopefully, the IPCC has decided to tackle the security issues more directly in the Fifth Assessment Report, scheduled for publication in 2014. The report from IPCC's Working Group II will, in fact, contain a chapter on human security, including a section on conflict.

In chapter 4, 'Space, time and scales of human security in climate change', Richard Matthew begins by discussing the universal versus particular narratives, then he examines the narratives of climate change and human security and explores the various scales at which climate change and human security can be linked. The chapter puts forward an argument about why these compelling, empirically grounded, and broadly relevant narratives may not generate an effective global response. The chapter concludes by arguing that there is an urgent need to establish global moral principles to guide thinking and particularly responses to climate change. Without this, over the next few decades, a world might emerge of climate change entrepreneurs and climate change victims, a division that may ultimately increase the risk posed by climate change to all of humanity.

Part II, 'The determinants of human security in the climate change context', addresses the environmental and social determinants of human security in the climate change context, and then investigates the notion of vulnerability to climate change and of its constituent parts in relation to human security. Finally it analyses climate impacts and extreme climatic events, disasters, risk, preparedness and management, and their linkages to human security.

In chapter 5 'The environmental determinants of human security in the context of climate change', David Simon first emphasizes that environmental factors continue to impact upon human activities in diverse and important ways, even in high-income countries, and that the human security implications of environmental change are not scale-neutral and therefore demonstrate discontinuities and disjunctures when examined at different scale. The distinct but overlapping elements of environmental change are experienced in ways that reflect existing social, institutional, economic and physical urban structures. The poorest, least well-resourced segments of society will face the most extreme and sustained threats to their human security despite having the smallest ecological footprints and contributing least to the growth in greenhouse gas emissions that underpin environmental change. On the basis of these considerations the chapter puts forward an approach to understanding the nature of environmental change risk that distinguishes it from conventional disaster risk through a focus on the extent, severity and duration of exposure to extreme events and slow-onset changes in environmental conditions. Similarly, simplistic dichotomies of urban versus rural and (semi-) nomadic and nomadic versus sedentary forms of social organization and production are avoided, by means of a dynamic and more relational understanding of human adaptability in the face of changing circumstances. In this light, the chapter surveys many of the key likely dimensions of human security vulnerability in different environmental change contexts, ranging from coastal and inland urban environments in different agro-climatic zones to the progressive erosion of the productive environmental base underpinning remaining hunter-gatherer and pastoralist communities, freshwater and marine fisheries, and the particular complexities of specialized, globally integrated, capital-intensive, export-oriented commercial farming enterprises.

In chapter 6, 'The social dimensions of human security under a changing climate', Jürgen Scheffran and Elise Remling focus on the social dimensions, conditions and determinants of human security and related theoretical concepts. After a thorough analysis of the genesis of the concept of human security, they combine positive and negative security aspects for operationalizing human security as the task of shielding people from threats and empowering people in order to take charge of their lives. Then the authors analyse the linkages between climate vulnerability, adaptation and human security. Climate change, in fact, affects various dimensions of human security in multiple ways. Some of the climate-induced stresses may directly threaten human health and life, such as floods, storms, droughts and heat waves, others gradually undermine the wellbeing of people over an extended period, such as food and water scarcity, diseases, and weakened economic and degraded ecological systems. Once critical thresholds are exceeded, the risks of climate change may turn into existential threats to human security which will depend on the vulnerability of those affected. Scheffran and Remling go on to investigate the impacts of climate change on human needs, capabilities and sustainable livelihoods and also within people's broader social environment, which includes the social interactions and networks in which humans participate as well as institutions and governance structures. They further stress that climate change will particularly increase the burden on those people who are already under stress from other problems, and are lacking essential capabilities and freedoms to take actions that reduce vulnerability and protect human security. Due to low 'social capital' and resilience these groups may be easily overwhelmed by the multidimensional impacts of climate change which further disrupt their societal stability. Against this backdrop the authors conclude that adequate policies for human security under climate change require a long-term planning horizon and an integrative framework to better understand the actual and potential adaptation needs of affected countries and communities from local to global levels.

In chapter 7, 'Vulnerability does not just fall from the sky: toward multi-scale pro-poor climate policy', Jesse Ribot explores causal structures of vulnerability and the relation of vulnerability to climate change. The causes of vulnerability can be traced in the social relations of production, exchange, domination, subordination, governance and subjectivity. They still have to be analysed and understood starting from the instance of crisis in a real place and real time. But, acknowledged anthropogenesis provides a new pathway for attributing social causality, and therefore, blame and liability – and claims for redress and compensation. Ribot further argues that vulnerability is produced by on-the-ground social inequality, unequal access to resources, poverty, poor infrastructure, lack of representation, and inadequate systems of social security, early warning, and planning, and that poverty is the most salient of the conditions that shape climate-related vulnerability. All in all, the chapter frames an approach for analysing the diverse causal structures of vulnerability and identifying policy responses that might reduce vulnerability of poor and marginal populations. In particular, the chapter argues that understanding the multi-scale causal structure of specific vulnerabilities – such as risk of dislocation or economic loss – and the practices that people use to manage these vulnerabilities can point to solutions and potential policy responses. Analysis of the causes of vulnerability can, in fact, be used to identify the multiple scales at which solutions must be developed and can identify the institutions at each scale responsible for producing, and capable of reducing, climate-related risks.

In chapter 8, 'Disasters and human security: natural hazards and political instability in Haiti and the Dominican Republic', Christian Webersik and Christian Klose discuss the impact of natural hazards on human security. In particular, they analyse the cases of two neighbouring countries, Haiti and the Dominican Republic, that occupy the same island, Hispaniola, and that are exposed to a similar level of intensity of annual natural hazards (largely tropical cyclones and earthquakes). However compared to the Dominican Republic, Haiti was more politically unstable over the past decades. The chapter aims at examining the relationship between natural hazards and political instability in Haiti and the Dominican Republic in space and time, using hazard and political data from 1850 to 2007. The chapter demonstrates that natural hazards have no significant impact on political stability in either country. Even though there is no link to political stability, natural hazards affect both countries, especially those populations who are most vulnerable. Low incomes, ineffective government institutions, and environmental conditions reduce the capacity to cope with natural hazards and lower the capacity to adapt to future climate change impacts. Thus, the authors further argue, approaches are needed that respond to local needs through nature protection policies, including education of the public, legal measures to ensure environmental protection and participatory management for forests, fishing, agriculture, and tourism.

As well as the determinants pointed out above, climate change impacts human security of diverse regions in different ways. For instance, some of the most impacted regions will be Latin America and the Caribbean, the Mediterranean, the Arctic, and Africa. Part III of the volume, 'A regional perspective on climate change and human security', overviews the main threats to human security brought about by climate change in these very vulnerable regions.

In chapter 9 'The impact of climate change on human security in Latin America and the Caribbean', Úrsula Oswald Spring, Hans Günter Brauch, Guy Edwards and J. Timmons Roberts focus on the impact of climate change on human security in Latin America and the Caribbean (LAC), particularly on climate change hotspots in Central America, the Caribbean, the Andes, and Amazonia. The chapter addresses four research questions. First, what have been the major conceptual human security debates in LAC countries since 1990? Second, what is the state of knowledge on climate change and its possible human security impacts for LAC? Third, what strategies for climate change adaptation are being implemented in LAC and how are they being financed? And finally, how can projected policies for coping with climate change be interpreted from a human security perspective? To answer these questions, the chapter introduces two global discourses linking climate change impacts and security within LAC and discusses environmental and social vulnerability from a human security perspective. LAC strategies for coping with climate change are assessed and policy debates on financing their adaptation measures are examined.

Following the discussion of the Caribbean and Latin America, in chapter 10 'Human security and climate change in the Mediterranean region', Marco Grasso and Giuseppe Feola investigate human security and its intersections with climate change in the Mediterranean region. The chapter does so by measuring human security at national level, and by critically discussing an ethical approach for improving human security in the Mediterranean. The chapter shows that there exists a significant divide within the Mediterranean region whereby European countries have much higher levels of human security than Middle Eastern and North African ones, whose weakness and vulnerability is largely ascribed to the social, economic and institutional factors producing human security. The chapter further argues that in light of these unbalanced levels of human security and of the determinants that characterize the region, the inclusion of the ethical dimension in a regional approach to human security in the Mediterranean may promote its overall increase. The ethical dimension may mitigate the consequent conflicts among interests, so that the impacts inflicted by climate change on the factors determining human security can be effectively addressed.

In chapter 11, 'Climate change and human security in the Arctic', Mark Nuttall considers climate change and human security in the high latitudes of the circumpolar North. The chapter discusses how the local and regional consequences of ecosystem transformation arising from global climate change have profound and far-reaching implications for human security and human-environment relations in the Arctic. Because of the diversity of human populations across the eight states of the Arctic, the focus is limited to research carried out in Inuit communities in Greenland, Canada and Alaska. However, the chapter points to the need to understand the consequences of climate change in context. The existence of rapid social, economic and demographic change, resource management and resource development, animal-rights and anti-whaling campaigns, trade barriers and conservation policies, all have significant implications for human security in the Arctic. In many cases, climate change merely magnifies existing societal, political, economic, legal, institutional and environmental challenges to human security that northern peoples living in resource dependent communities experience in their everyday lives. The chapter concludes that in the Arctic environment the livelihoods of indigenous peoples will not be changed or influenced by climate change alone, and it is crucial that scientific climate change research does not develop scenarios of possible future states of the Arctic environment, without putting people in the picture and taking rapid social, cultural, economic and demographic changes into account. To this end, approaches to human security also need to proceed from a more nuanced understanding of the complexities of human-environmental relations.

In chapter 12, 'Climate change and human security in Africa', Sharath Srinivasan and Elizabeth Watson start by examining the fundamental issues about climate change in Africa, and its impact upon freedom from want and freedom from fear. The focus is on how emerging science and local complexity are simplified, reinterpreted and deployed within the new discourse of climate change and human security. The chapter then shifts to two case studies that exemplify the politics of representation and of science and politics at different scales. The conflict in Darfur, Sudan, illustrates the ways in which the securitization of climate change activates a particular freedom from fear discourse that gives weight to certain phenomena and causalities over others, directly and indirectly serving selected interests at multiple scales. The case of Marsabit in Kenya examines the ways in which debates about climate change and freedom from want also tend to overlook certain inconvenient sets of processes: often complex and contingent local ones that are ill-suited to the modalities and objectives of external interveners. The chapter finally warns that, unless vigilance is maintained, the discourse on climate change and human security – like many of the environmental discourses in Africa before – could be hijacked to serve certain powerful interests that demand that Africa and Africans need to be *saved* by new and powerful technologies, forms of transnational governance, waves of humanitarian relief and modes of investment.

Part IV of the volume, 'Responses to the threats posed by climate change to human security', focuses on the conditions that make it possible for individuals and communities to increase their human security in response to climate change, as well as to the built environment and critical infrastructures that are expected to increase, or at least to protect, human security against the impacts produced by climate change. The chapters provide both a positive analysis of the likely limited international architectures, policies and instruments for enhancing human security against climate change, and a normative analysis of what international institutions would be needed. Finally the notion of human security when framed as a human rights issue is explored, to point out the related duties that this approach involves, and how this standpoint can strengthen human security in the context of climate change.

In chapter 13, 'Climate change and human security: the individual and community response', Michael Hall explores some of the issues relating to the emerging human security and climate change agenda at the community and individual scale. The chapter argues that the challenge is as much conceptual, with respect to paradigms of behaviour and governance, as it is to the realities of climate change. The chapter is divided into three main sections. The first discusses the ways in which issues of human security and climate change response are framed. Such a discussion is regarded as fundamental to considering capacities to change and interventions to assist and enable such change. The chapter then goes on to examine the multi-level nature of responses to human security and climate change and the way that communities and individuals are embedded within broader social and economic structures. The critical role of trust and values in communities is noted, but it is also stressed that communities should not be romanticized, and it is vital to recognize that they may be ridden with conflict that makes appropriate sustainable solutions extremely hard to achieve. Similarly, the lack of capacity or willingness of individuals to respond to climate change, even if from an outsider's perspective this is the rational thing to do, is noted, with the challenges that this implies for the communication of risk and security. Finally, the chapter returns to the vital role of different paradigms of behaviour and governance and the implications that this has for the nature of interventions and capacity to change behaviour, given the potential of the 'lock in' of communities and individuals within certain socio-technical systems.

Chapter 14, 'Climate change, human security and the built environment', by Karen Bickerstaff and Emma Hinton, explores the role of the built environment through the lens of improving energy efficiency and achieving comfort within the existing building stock – specifically through changing the behaviour of householders. By developing a socio-technical analysis of change – which places stress on the socio-material constitution of everyday life – the chapter not only provides a critical review of dominant models of social transformation, and embedded assumptions about materiality and agency, but also offers an insight into the impacts, and efficacy, of policy efforts to materialize change. In particular future policy should not only focus on removing universal barriers to energy-efficient behaviours (apathy, ignorance, lack of financial interest), but also suggests, by attending to the materiality of innovation, new ways for the promotion of more efficient energy use. Finally, the chapter highlights the need to attend to the role of external players (governments, manufacturers, retailers), in supporting innovation that can encourage a re-evaluation and restructuring of normal routines. This would involve attending both to technologies that can favour pro-environmental behaviours, and to those that offer greater flexibility in the socio-technical distribution of agency and allow people to more actively engage with energy flows, and practices they sustain.

In chapter 15, 'Climate change and human security: the international governance architectures, policies and instruments', Michael Mason first reviews those limited global governance policies and instruments recruited to address the human security implications of climate change. This involves both a survey of the emergence of human security concerns within global climate governance – notably the UN Framework Convention on Climate Change (UNFCCC) – and the recognition by some security governance actors that climate-related harm represents a significant threat to the lives and livelihoods of many people. Reference is made to several governance initiatives to determine whether there is anything other than scattered institutional moves to enhance human security against major climate hazards. The chapter then carries out a normative analysis of whether a more integrated governance to this end is justified and, if so, what it might look like. In this regard the chapter argues that, while the global climate regime holds epistemic and governance authority over the management of 'dangerous' climate change, the effective inclusion of climate concerns in human security decision-making is most likely to be achieved by consolidating the legal coherence and force of the latter – especially in relation to the development of human rights and humanitarian norms on the prevention of climate harm. Such a rights-based architecture would also afford human security governance at least some protection from, and critical engagement with, the power-oriented politics of the international security system.

In chapter 16, 'A human rights-based approach from strengthening human security against climate change', Steve Vanderheiden addresses the question of whether imperatives surrounding climate change might usefully be cast in the language and politics of human rights, which comprise a subset of the moral rights that provide their ethical foundation. In fact, the right not to be harmed as the result of anthropogenic environmental change, especially when such harm involves serious suffering or death, can be regarded as among the most basic of rights. In the context of climate change this right bridges the categories of security and subsistence rights, as extreme weather events can threaten human security directly while climate-related scarcity can threaten subsistence indirectly. The chapter concludes by arguing that although both the seriousness of climate change as a global policy concern and the urgency of action to prevent climate-related suffering suggest the connection to human rights, which are properly reserved for humanity's greatest moral and political challenges, the 'upside' of invoking such rights on behalf of climate change mitigation might best be seen as political rather than philosophical. Among the political benefits of acknowledging these rights, the primary benefits may reside in the recognition and empowerment of current and potential sufferers of climate-related harm, rather than the legal mobilization of recognized political authorities.

From the foregoing discussion it will be evident that this volume seeks to perform two complementary functions. First it seeks to draw, from within its compass, a solid body of scholarship, much of which remains dispersed and, occasionally, located within specialized academic literatures. Second, the volume seeks to take the arguments about the relationship between climate change and human security forward, and in so doing forces us to reflect on both concepts, and the research which underpins them. The editors believe this is a critical step forward, and it is to be hoped that all the original papers published here will resonate with the academic and policy communities and, in turn, lead to a vigorous discussion and a more robust understanding of what is at stake for humanity.

NOTE

1. The expression 'managed senescence' we owe to Graham Woodgate. It continues the biological metaphor of organic growth and development but signals the gradual end of hydrocarbons as a way of achieving this.

REFERENCES

Brandt Commission (1980), *North-South: A Programme for Survival*, London, UK: Pan Books.

Crutzen, P. (2002), 'Geology of mankind', *Nature*, **415** (3), 23.

IPCC (International Panel on Climate Change) (2007), *Climate Change 2007: The Physical Basis of Climate Change. Contribution of Working Group I to the Fourth Assessment Report of the Intergovernmental Panel on Climate Change*, Cambridge, UK: Cambridge University Press.

Lovins, A., Hawken, P. and Lovins, L. Hunter (2000), *Natural Capitalism*, London, UK: Earthscan.

Meadows, D.H., Meadows, D.L., Randers, D. and Behrens, F. (1972), *The Limits to Growth*, London, UK: Pan Books.

Paterson, M. (2009), Global governance for sustainable capitalism? The political economy of global environmental governance', in W.N. Adger and A. Jordan (eds), *Governing Sustainability*, Cambridge, UK: Cambridge University Press.

Redclift, M.R. (1987), *Sustainable Development: Exploring the Contradictions*, London, UK: Routledge.

UNDP (United Nations Development Programme) (1994), *Human Development Report 1994. New Dimensions of Human Security*, New York, USA: Oxford University Press.

PART I

FRAMING THE ISSUE: CLIMATE CHANGE AND HUMAN SECURITY

1. Climate change as an issue of human security
Simon Dalby

INTRODUCTION

Human security has become a major concern in international politics over the last couple of decades and a matter for extensive academic analysis and policy prescription too. Given the recent rising salience of discussions of the disruptions caused by climate change, linking the two themes is an obvious way of thinking about vulnerabilities and policy prescriptions for the future. But as this chapter and others in this section of the Handbook make clear, linking human security and climate change is not an easy matter once one gets past the obvious invocation of general dangers due to large-scale environmental disruption (Webersik 2010). In part this is because of the uncertainties about precisely how climate change is manifesting itself in particular places. Some of the discussion is necessarily speculative, but the difficulty is also linked directly to the multiple dimensions of what is included under the rubric of human security. The specific forms of political and economic arrangements that facilitate human security are also part of the puzzle, and human security itself cannot be simply taken for granted uncritically as a universal norm without thinking through how it is to be provided and by whom.

Climate change involves an extensive set of scientific investigations and numerous attempts to quantify and specify what changes are coming in environmental matters, most notably temperature and rainfall pattern changes. The most obvious difficulties include the implicit assumption that there is a relatively stable natural environment that is reasonably well known and which either already is, or shortly will, be subject to notable perturbations as a result of rising greenhouse gas levels in the atmosphere. Given the highly variable nature of the earth's climate system any notion of a stable baseline is very difficult to establish. That said, it is clear from what meteorological records are available, and numerous satellite and other data sources, trends are towards a warmer world. In particular the huge heat wave in the United States in the summer of 2012 focused attention on the increased frequency that heat records are being broken, a trend that does indeed indicate a warming planet.

But it doesn't follow directly from this that human insecurities are increasing, albeit the casualties from such events are noteworthy tragedies. It doesn't follow directly because to a large degree whether people live or die is a matter of the provision of infrastructure in general and emergency facilities in the particular locations that are affected. There are obvious exceptions in terms of people killed by lightning or trees falling on them in a storm, but the larger point is that technologies, social arrangements and such mundane matters as building designs and functional air condition-ers mediate the impact of "natural" events on "human" vulnerabilities. This is the key point about security: it is a matter of the social provision of living conditions, not a simple matter of natural happenings immediately impacting people (Matthew et al. 2010). If those buildings, and such things as the lack of functioning air conditioners, mean that people die of expo-sure to the elements, their insecurity is not only a matter of meteorological exposure, but a matter of social insecurity interacting with extreme events. Likewise with food security, mostly it is the lack of economic ability to buy food that causes people to starve rather than an absolute shortage of nutrition. Security focuses on these social arrangements even when, as is frequently the case in discussions of climate, direct causation is imputed to "natural" events. It is worth noting that the Intergovernmental Panel on Climate Change (IPCC) in its four major assessment reports to date has not addressed climate in terms of security; clearly how climate becomes a matter for security is not yet established comprehensively or authoritatively.

Much of what follows in this chapter lays out what thinking needs to be done to effectively link human security and climate change (see O'Brien et al. 2013). While there are repeated references in the policy and scholarly literature to environmental threats to human security, the key documents in the 1990s and 2000s that lay out the logic of human security rarely deal with environment systematically, and even less frequently address climate security explicitly. In the case of environment the difficulties lie in the sheer diversity of ways that environment might cause insecurity (Scheffran et al. 2012). In the case of climate until recently much of the discussion presumed that climate change was a matter for the future, or the distant future, not a matter of immediate concern. The United Nations Security Council didn't address climate change as an issue until a couple of years after the principles of responsibility to protect and the larger discussion of human security had already been adopted as UN policy in 2005 and only dealt with the matter in a detailed report in 2009 (United Nations Secretary General 2009). But that said the formulation of human security that has been enshrined recently has a long history, and the contradictions implicit in its original formulation now haunt the discussion of climate change too.

SECURITY

Security is a matter of social arrangements. In Wilhelm von Humboldt's formulation from the late eighteenth century, security is a matter of the legal assurance of freedom (as cited by Neocleous 2000: 9). Such arguments emphasize the emergence in modern societies of security as a principle related to legal and economic matters, a sense of a social guarantee of certain conditions. This is related to the rise of commercial arrangements for the provision of many things, including food, and the legal provision of economic arrangements that facilitate its commercial production and circulation. Liberal societies are dependent on these political arrangements for the provision of economic liberties, a matter of securing the social order which in turn provides the political stability within which commercial societies function.

On the other hand security has also long been invoked as a matter of national security in the sense of protecting at least a notional version of state sovereignty, territorial integrity and the maintenance of social stability within states. The military protection of these conditions relates to the provision of the right to self-defence under international law, and a matter supported by the key provisions of the United Nations charter. During the cold war period however the presence of growing arsenals of nuclear weapons endangered all humanity, not just those of potential adversaries in a nuclear war. Some sense of common insecurity was widely understood as the human situation in these years, and the Palme Commission's formulation of "common security" as an appropriate alternative formulation was in part a response to the dangers of nuclear confrontation (Independent Commission 1982). Understanding the nuclear arsenals as a matter of insecurity rather than security leads to subsequent formulations of security as a matter not only of the collective provision through alliance systems, or more generally through United Nations actions, but a matter that had a larger political constituency among a common humanity.

By the end of the cold war it had become abundantly clear that security understood in terms only of national security was inadequate; shifting the "referent object" of security from states to people seemed a necessity if policies and analytical clarity were to apply to this key political concept. The upshot of all this is that the concept of security, a key formulation in classical liberalism, and a subsequent keystone of state practice, has been extended to encompass many things and apply to numerous aspects of human life. Environmental matters gradually entered this debate in the 1980s. Richard Ullman suggested (1983: 133) that

A more useful (although certainly not conventional) definition might be: a threat to national security is an action or sequence of events that (1) threatens drastically and over a relatively brief span of time to degrade the quality of life for inhabitants of a state, or (2) threatens significantly to narrow the range of policy choices available to the government of a state or to private, nongovernmental entities (persons, groups, corporations) within the state.

In the late 1980s these themes were taken up in a series of policy related statements that urged security thinkers to take environmental matters seriously as a source of threats to many things, although quite how and why environment might be a matter of particular kinds of security was frequently underspecified. It was obvious in the case of stratospheric ozone depletion where the integrity of a key part of the atmospheric system protecting terrestrial life from harmful solar radiation was compromised, and obvious too in the event of a major nuclear war causing both direct casualties and potential massive disruption of the climate system in terms of "nuclear winter" (Sagan and Turco 1990). But how and why North–South wars might occur, or why environmental scarcities in the periphery of the global economy might cause international security problems were not clearly articulated on the basis of any empirical evidence (Homer-Dixon 1999).

These extensions of security through the 1980s and early 1990s were summarized by Emma Rothschild (1995: 55) in terms of four broad themes. First ("downwards"), "(T)he concept of security is extended from the security of nations to the security of groups and individuals". In part this is a response to the recognition that military preparations in the cold war period endangered everyone; states were not providing security for most populations by acting in this manner. Limiting security to national security had become untenable in the face of potential nuclear warfare and related military actions. In the 1980s this had become a matter of discussion in terms of "common security" whereby it became increasingly clear that security could not be ensured by unilateral state action and required cooperative measures in the international community.

Given 1980s concerns about nuclear winter in particular and matters of ozone depletion and biodiversity loss and other loosely defined environmental matters, a reconsideration was required of what is loosely understood under the rubric of "global security". In Rothschild's (1995: 55) terms, in this, a second extension of the concept of security, what she called "upwards" security "is extended from the security of nations to the security of the international system, or of a supranational physical environment". Subsequently these concerns have crystallized into a discussion of "climate security" where the stability of the climate system is what apparently needs to be secured.

Third, Rothschild (1995: 55) argued security required an extension ("horizontally") "from military to political, economic, social, environmental, or 'human' security". Security in these terms is a matter of economics and social services, a broad approach that ultimately has the human as the referent object rather than the state. While states and collective international security are still important in these matters, the provision of security for humans is now the key theme, and the justification for state and international action. Protecting civilians in warfare, and insisting that states have the primary responsibility for doing so as the rationale for their sovereignty shapes the logic of the "responsibility to protect" doctrine in support of what the UN secretary general called in a follow up to the Millennium Summit, a "larger freedom" (United Nations Secretary General 2005).

Fourth and finally Rothschild (1995: 55) argued that,

> the political responsibility for ensuring security (or for invigilating all these "concepts of security") is itself extended: it is diffused in all directions from national states, including upwards to international institutions, downwards to regional or local government, and sideways to non-governmental organizations, to public opinion and the press, and to the abstract forces of nature and the market.

All of which suggests that security, while an overarching desideratum for humanity, and a powerful rhetorical ploy in numerous political debates, is frequently inchoately articulated with little clarity as to where the ultimate responsibility for ensuring it actually lies.

It was from within this larger discussion of security at the end of the cold war that the formulations of human security emerged. But while there is a novelty to the formulation, it is important to remember that the original formulations of security in terms of the provision of social order that facilitates the conduct of commerce and the political order of a liberal society shapes the concept to this day. Security is now also explicitly linked to principles of development, which are now understood as the expansion of economies based on capitalist social arrangements. The inherent contradictions in the concept have persisted, and as this chapter will tease out below, the difficulties in linking human security and climate change in part derive from the origins of the original formulation. Most obviously the difficulties come from specifying who and where climate change will impact, but the tensions in the formulation of security in the first place are also part of the question that needs analysis.

Finally, recent scholarship looking at matters of security has made it clear that security is a contested political term that can be invoked in response to a crisis, or as a matter of heightening the importance of a particular policy issue in public discussion (Buzan et al. 1998). In this sense invoking the language of security is a political act, one that may or may not successfully gain adherence widespread enough to facilitate action. Invoking security is usually allied to requirements for emergency action, a state where normal political relations are suspended and exceptional action justified an argument that has been made repeatedly about environmental matters (Floyd 2010). Where this becomes tied into larger invocations of political order and the principles of the responsibility to protect, human security becomes part of the larger calculus of international politics, as the case of the military intervention in the 2011 civil war in Libya suggests. Where climate change fits into such discussions adds interesting nuances, especially if the responsibility to protect is applied to cross-national actions.

HUMAN SECURITY

Human security was initially widely promulgated in the 1994 UNDP report, in terms similar to Ullman's specification of an extended definition, as "first, safety from such chronic threats as hunger, disease and repression. And second, it means protection from sudden and hurtful disruptions in the patterns of daily life – whether in homes, in jobs or in communities" (United Nations Development Program 1994: 23). The human security agenda has since the beginning been concerned with both larger structural issues in the global political economy as well as short-term emergency situations that render people vulnerable. Both are important in considering human security in the context of climate change, and both are interconnected in important practical ways, but in ways that vary widely across the global economy.

The original 1994 UNDP report suggested that human security had at least four essential characteristics. First, it is a universal concern relevant to people everywhere. Second, the components of security are interdependent. Third, human security is easier to ensure through early prevention. Fourth, and perhaps the crucial innovation in this formulation, is the shift of the referent object of security from states to people. This in turn raises issues of rights, justice, vulnerability, power relations, and empowerment. Security is now about much more than maintaining the political control of elites within states or the related matter of preparing for wars with other states.

While many causes of insecurity on the UNDP list of threats to human security are local threats, global threats to human security were said to include at least six categories, caused, it is important to note in contrast to earlier concerns with warfare, more by the actions of millions of people than by deliberate aggression by specific states. These six are (1) unchecked population growth, (2) disparities in economic opportunities, (3) excessive international migration, (4) environmental degradation, (5) drug production and trafficking, and (6) international terrorism.

Notable was the inclusion of environment as an issue but notable too was the failure to spell out how environment actually caused insecurity, beyond the obvious statements about pollution and resource shortages and the assumptions, spelled out in the earlier 1987 World Commission on Environment and Development, that suggested that resource shortages would lead to conflict. Climate change wasn't a major priority in these early formulations, but in the decades since it has increasingly had an impact on all facets of the human security agenda, although most obviously the first four listed above. Now it is intertwined with virtually all aspects of the agenda, and repeated, mostly exaggerated, warnings about potential armed conflict as a result of climate change have been linked to terrorism too. Fears of climate induced migration causing political disruptions in metropolitan states faced with large numbers of foreigners attempting to gain entry have also raised the spectre of using military action to prevent boundary crossings (Smith 2007). This has been done in some key situations, and the militarization of border controls to control immigration has proceeded apace in the last decade, frequently under the mantle of the war on terror (Jones 2012).

In the years since the UNDP formulation numerous additions and qualifications to the core notions of human security have been added to the discussion. Human security has been discussed in terms of a series of formulations that reflect von Humboldt's original theme in terms of the legal assurance of freedom. Freedom from want and freedom from fear are key formulations focusing on poverty and violence specifically as problems that require state provision to prevent them occurring. Governments have, so the formulation of the responsibility to protect suggests, an obligation to provide basic needs for their population (Kaldor 2007). Failure to provide these things can, should the international community choose to invoke the principles, allow the violation of state sovereignty and military intervention by external forces to provide what a state government has failed to produce. Sovereignty is thus, at least in some cases, contingent on the provision of human security.

CODIFIYING HUMAN SECURITY

The larger discussion of human security in what is as close to a formal codification of the discussion as exists, in the 2003 report of the commission on human security on "Human Security Now" starts precisely with the interconnected nature of the present world. Clearly it argues, right from the beginning of its discussion, that this interconnected condition is the starting point for considering human security in its multiple aspects. Its opening sentences suggest that:

> Today's global flows of goods, services, finance, people and images spotlight the many inter-linkages in the security of all people. We share a planet, a biosphere, a technological arsenal, a social fabric. The security of one person, one community, one nation rests on the decisions of many others – sometimes fortuitously, sometimes precariously. (Commission on Human Security 2003: 2)

While this might all seem entirely common sense, it is important to note the emphasis on interconnectedness in the provision of security; obligations stretch across frontiers just as the causes of insecurity do. This is, as will be argued later in this chapter, doubly so in the case of climate security, where the interconnections between people and places are key to the discussions of international obligations. The Commission on Human Security (2003: 4) defines human security as being:

> to protect the vital core of all human lives in ways that enhance human freedoms and human fulfillment. Human security means protecting fundamental freedoms – freedoms that are the essence of life. It means protecting people from critical (severe) and pervasive (widespread) threats and situations. It means using processes that build on people's strengths and aspirations. It means creating political, social, environmental, economic, military and cultural systems that together give people the building blocks of survival, livelihood and dignity.

This is obviously a very wide ranging definition but one that once again echoes the traditional liberal formulation of the concept of security, suggesting that people need protection in some cases from their own states as well as protection from other states. The ability to engage as self-reliant entrepreneurs and citizens is the key to this definition of human security. The provision of the essentials of life is the key theme, a matter of what social theorists these days, following Michel Foucault's (2008) ruminations on these themes in the 1970s, call "biopolitics". While it is obvious that the systems that provide the "building blocks" will be widely varied given the numerous differing contexts within which people live, it is clear that at the heart of human security are claims to universal principles that limit the actions of political elites in some crucial ways, and simultaneously shape their obligations to their charges. But implicit too is the key theme that failure to provide for safety and security implies an obligation to intervene across international borders.

Human security is related to matters of human rights, and the aspirations of poor people to the benefits of economic development too. Peace is obviously key here; violent conflict undermines "building blocks" for most people, although not obviously those who profit from conflict by supplying weapons or reaping windfall profits from economic disruptions related to the conflict. The commission quotes Secretary General Kofi Annan's (2000) suggestion that human security directly links together other United Nations agendas:

> Human security in its broadest sense embraces far more than the absence of violent conflict. It encompasses human rights, good governance, access to education and health care and ensuring that each individual has opportunities and choices to fulfill his or her own potential. Every step in this direction is also a step towards reducing poverty, achieving economic growth and preventing conflict. Freedom from want, freedom from fear and the freedom of future generations to inherit a healthy natural environment – these are the interrelated building blocks of human, and therefore national, security.

Here, although not specified as such but included as part of a healthy natural environment, the issue of climate change emerges as a challenge to one of the crucial things that human security is supposed to provide. In the section that follows this discussion of building blocks, "menaces" to human security are itemized, but again in a way that implies but doesn't actually mention climate change. "State security has meant protecting territorial boundaries with – and from – uniformed troops. Human security also includes protection of citizens from environmental pollution, transnational terrorism, massive population movements, such infectious diseases as HIV/AIDS and long-term conditions of oppression and deprivation" (Commission on Human Security 2003: 6). In all this freedom from want suggests the provision of essential food, water and shelter, while freedom from fear suggests the absence of political violence and oppression. Climate change is now frequently discussed as a threat precisely because it may disrupt food supplies and through storms in particular directly endanger shelter and health too (Webersik 2010). Likewise there is very considerable literature that expresses concern that such disruptions will induce violent conflict, although the social science research suggests that much of this is alarmist (Kahl 2006); nonetheless security agencies are now frequently using such arguments in their analyses of potential future conflict.

While environmental themes recur throughout the rest of this key report, it was written prior to the emergence of climate as a central theme in the environmental security discussion. Climate change relates to both immediate short term hazards from storms, droughts and disruptions as well as the longer-term trajectories that may dramatically change how the climate system functions at a global scale (Bogardi 2004). Nonetheless both the dangers of immediate hazards and the longer-term dangers of environmental degradation are clearly present as causes of human insecurity that need to be dealt with by government policies that deal with both the immediate dangers and long term conditions for human flourishing.

CLIMATE SECURITY

Teasing out the connections between climate change and human security requires working through where people are vulnerable, and the causes of their vulnerability due to political and economic institutions as well as the geographical locations where these social processes play out. But now too it requires an engagement with the discussions of the international norms of the responsibility to protect, which obligates states to protect their populations. Meteorological hazards matter too, both immediately in terms of hurricanes, storms, floods, and heat waves, and in the longer-term effects of shifting precipitation patterns and droughts. International obligations to help in the face of climate disruption are increasingly a part of this discussion as are concerns about inter-generational equity. Climate justice matters in terms of obligations both to vulnerable people now and to the potential disruptions that may cause difficulties for people in the future.

But how these issues play out requires thinking simultaneously about the large-scale changes to the climate system and the transformation of the human condition as we increasingly become an urban species. Urbanization has changed the context of vulnerabilities and the practical technologies of human existence in ways that now make humans vulnerable in complex artificial contexts (Graham 2010). In so far as human security shifts the focus to anticipatory action and away from reactive violence after a problem has arrived, it is now about institution building, infrastructure planning and emergency services. All this requires a careful focus on the precise modalities of vulnerability, and this is now part of the extended human security agenda in terms of claims concerning rights to be free of hazard impact.

Finally all these things have to be considered in terms of the simple and fundamental fact that in the face of global environmental change the past cannot be taken as a reliable guide to the future in terms of weather patterns and practical matters such as crop yields or the design criteria for infrastructure. With the increasing frequency of extreme events, planning everything from bridge construction to food supply cannot assume that what was normal over the last century is a reliable guide to the future

(Pascal 2010). Anticipating the unexpected is becoming a key part of the action required of human security planners, because failure to anticipate may require all sorts of emergency interventions to rescue people facing disasters and dislocations. Key to human security in the UNDP formulation is anticipation and preparation, in contrast to the traditional invocation of security in terms of emergency action and the use of violence to restore control after an event.

The widespread discussion of risk, and how to manage it feeds into this discussion too. Preparing for contingencies, using financial instruments to attempt to spread risk widely, whether through traditional methods of insurance or more complicated financial instruments such as catastrophe bonds, is now part of the larger calculus of security. In Rothschild's terms the responsibility for security is now widely dispersed. But much of this discussion relates only to the formal sectors of government and the economy. Marginal and poor populations, and those primarily dependent on subsistence and non-commercial economic arrangements are frequently ignored in these calculations (Grove 2010). A detailed discussion of how these matters have changed the formulations of security in theoretical terms is beyond the ambit of this chapter, but in terms of social theory the focus on biopolitics, and of making particular forms of economic action flourish, suggests that security is not about protecting non-commercial modes of human life. Most obviously, in terms of what Naomi Klein (2007) calls "disaster capitalism" it seems that in current circumstances states and businesses frequently take advantage of disasters to accelerate the conversion of places and economic opportunities from subsistence and non-commercial modes into those incorporated within commercial societies. What is being secured here is the processes of economic development understood in terms of markets and finance, not other modes of life. Given the dominant discussions of development in these terms this is to be expected. But the contradiction is noteworthy, especially so when the commercial sector is powered by carbon fuels.

All this is made more complicated by the rise of a series of economic and political practices that are very loosely labelled "neo-liberalism" in recent years (Harvey 2005). Privatization, the extension of international commercial activities, the primacy of the financial system, and the use of state power to extend the reach of capitalism all emphasize the supposed superiority of corporate economic practices to provide growth, and implicitly human welfare. These practices have, under the guise of development, accelerated the migration from rural occupations to urban areas, and from subsistence modes of economic activity to purely commercial transactions. But unlike the urbanization of the nineteenth century when people moved to European and later growing American cities to work in the new factories of the industrial revolution, populations in the global South are now frequently entering cities as part of an informal economic system that is not based on industrial jobs. As such the economic systems of the informal sector are key to the human security of many people in the burgeoning cities of Africa, Latin America and much of Asia.

When it comes to matters of climate change the contradiction here becomes ever more obvious because it is the fossil fuelled mode of economic development that is being secured by contemporary neo-liberalism that is the primary cause of greenhouse gas emissions which are destabilizing the climate, and in turn making many people more vulnerable to the rapid disruptions to ways of life that both Ullman's theoretical formulation, and the more practical formulation of abrupt changes in the UNDP 1994 document, suggest are key to insecurity. Focusing on these contradictions is now a key matter for any discussion which links climate change to human security with any analytical clarity, especially so when the international connections between economic activities and human insecurity are considered (Dalby 2009). Who has the responsibility to provide what kind of protection where, focuses attention on the contradictions between security understood as a universal human norm, and the logics of national security of supposedly autonomous states. If following Hans Günter Brauch's (2005) argument, discussed below, states ought to provide a freedom from hazard impact too, then the question of the locus of political authority, and the obligation to anticipate future dangers takes another turn that directly links to the matter of climate change.

THE RESPONSIBILITY TO PROTECT

The principle of the responsibility to protect emerged from the discussions of interventions, and the failure of the UN to intervene in some of the humanitarian disasters of the 1990s. An apparently tardy military response in the Balkans and a failure to effectively deal with clear warnings about imminent genocide in Rwanda suggested that some general principles were needed to guide policy makers and the UN system as to when sovereignty was no longer applicable. With the cold war standoff no longer thwarting collective action in response to political violence, numerous advocates of humanitarian action worked on establishing the principles that might govern future interventions.

Much of this was codified in the International Commission on Intervention and State Sovereignty (ICISS) report on the "responsibility to protect" in 2001, principles that were subsequently adopted by the UN in 2005 in ways that effectively make sovereignty contingent on states' provision of basic safety to their populations (Elden 2009). There are of course no guarantees that failure to provide for safety will lead to intervention, and at least in the case of the military actions against the Quaddafi regime in Libya in 2011, such principles can be used for political purposes such as supporting one side in a civil war. Nonetheless formally this broad ranging initiative in the UN system has so far focused on how to deal with matters of genocide, war crimes, crimes against humanity and ethnic cleansing. If a situation is adjudged to constitute one of these four activities, then the logic is that the international system ought to act in ways that might include military intervention if less drastic measures fail to protect the populations of the states affected.

None of the four themes directly address matters of environment generally or climate change in particular. While these do not yet extend to failures of states to protect people in the face of storms and other disasters, the discussion of possible interventions to protect people in Myanmar/ Burma in the aftermath of tropical storm Nargis, did raise the question of interventions on these grounds while emphasizing the point that some states choose to prioritize state security over human security in a crisis to the considerable cost of their populations (Seekins 2009). How such thinking might address climate change, as opposed to disaster response, depends on what causal connections emerge between climate and the political processes said to be set in motion by environmental change. But as the empirical work on climate change and conflict has made clear repeatedly, there are no simple connections and causations between the two (Theisen et al. 2011). Institutions are more important than environmental changes in explaining when violence erupts. In extreme cases of environmental scarcity it is worth emphasizing that people usually starve rather than fight. Given that this is usually in part because of a failure of governance institutions to provide for their populations, famine is obviously a matter of a failure of human security. Michael Watts (1983) termed such matters a case of "silent violence" in his key investigation of famine in Nigeria to emphasize the point that while conflict wasn't overt, people died as a result of the failures of governing powers to act effectively in providing what might now be called human security.

All of this is made more complicated by the argument that migration is a logical response to environmental change. Many species adapt to changing circumstances by movement to more congenial climes, and there is no good argument that humans shouldn't do the same. Apart, that is, from the obvious point that the widely accepted political geography principle of rule now is that people are understood to be citizens of a particular state. Consequently their human security ought in theory to be provided in situ rather than by migration. While most migration related to environmental matters is within states, clearly some people are in motion in part because of climate changes and resultant agricultural difficulties. But separating this out from economic processes, especially the rapid enclosures and extension of commercial modes of farming in many parts of the world as well as the massive current migration to cities, is very difficult to do (White 2011).

Nonetheless despite the absence of formal legal categories of "environmental refugee" or "climate refugee" there continues to be discussion of migration in these terms as part of the larger discussion of climate change and security. The implicit, and sometimes explicit, argument is that failure to prevent environmental change is forcing people to move, and given the failure to provide for their needs within the state of origin, the international obligations to help people on the move require that such people designated in terms of forced migration be given analogous treatment to political refugees. This argument has persisted in various forms since the 1980s, highlighting the implicit geography of much of the discussion of human security.

It raises quite directly the question as to whether there ought to be a freedom to migrate in the face of climate disruption. This becomes especially interesting where societies close to sea level are inundated by rising sea levels. Given that most climate change is caused either directly or indirectly by the fossil fuel consumption and resource extraction projects of the OECD states then it would seem to follow logically that they have an obligation to provide for the human security of those people whose states are going under the rising seas. Migration would allow the human security of displaced peoples to be provided. But given the widespread invocation of territorial integrity as a basic principle of national security and the portrayal of migrants as potential threats in Northern countries, the contradiction between national and human security is palpable once the international dimensions of the consequences of climate change are worked into the discussion of climate security.

FREEDOM FROM HAZARD IMPACTS

Insofar as climate change causes more severe hazards then obviously in the short term human security is directly implicated. While there will remain difficulties attributing individual weather events to climate change, clearly the likelihood and frequency of extreme events is increasing and states concerned with their population's welfare should be making preparations. Such considerations have suggested that a fourth "pillar" of human security, following on from the first three (freedom from want, freedom from fear, and the ability to live a dignified life), be formulated in terms of freedom from hazard impacts. Putting the matter directly it is clear that a policy agenda of "freedom from hazard impacts" is an obvious extension of the freedom from want and freedom from fear parts of the human security agenda (Brauch 2005). Given the importance of preparations to deal with hazards this extension of the agenda would be appropriate regardless of whether climate change is enhancing the frequency and severity of hazards, but the logic of this extension becomes all the more obvious if climate change is taken seriously.

While the obvious objection to such a formulation is that climate or environmental hazards are of a different order to political violence, it is clear from both the research literature in environmental security and the extensive research on disasters that environmental factors are mediated by social ones to produce human consequences. The provision of shelters in storms determines survival rates much more than the intensity of the storm. Public provision of basic protections, as in the successful Cuban experiment with hurricanes, makes it very clear that it is human actions or inactions that turn hazards into disasters (Sims and Vogelman 2002). Hence freedom from hazard impacts focuses on the impacts rather than the hazards in suggesting preparations in advance to deal with storms, droughts, floods and other disruptions.

Some states have taken measures to ensure their infrastructure against extreme events, although the flooding of New Orleans in the aftermath of Hurricane Katrina suggests that frequently such things are low on the list of political priorities. However some of these state measures take the form of financial instruments concerning insurance and catastrophe bonds that are designed in theory to provide funds for rebuilding should the worst happen. But these measures are not obviously going to rebuild the informal settlements and slums that many of the most vulnerable people in harm's way need, even if they do provide funds for rebuilding highways or electrical grids. Neither are these measures the anticipatory ones that are seen as key to the prevention of harm which is part of the human security agenda.

What is protected, at least in so far as the funds from catastrophe bonds appear in the aftermath of a disaster, is the infrastructure for the formal commercial part of the economy. Beyond that in many cases populations are on their own to rebuild their lives and economies after a disaster. While disaster aid, interventions by militaries or aid agencies undoubtedly help, the focus on the international economy at the expense of local survival is a telling reflection on how the economic priorities of the present work. Climate change is not a separate matter from the global economic situation and the inequities in that system are mirrored in the likely vulnerabilities of people in the face of global change (Roberts and Parks 2008). The implicit assumption that economic growth will enrich whole populations seems increasingly unlikely as the consequences of the global financial turmoil take their toll in many places. Nonetheless, it is also important to note that despite this turmoil and a temporary slowdown in greenhouse gas emissions in 2009, the global economy has continued to grow, albeit highly unevenly. The consequences of this are that it is increasingly unlikely that the growth of greenhouse gases will be curtailed in time to prevent major disruptions to the climate system with all the unknowable long-term consequences for human life.

CLIMATE SECURITY AND JUSTICE

In the larger pattern of climate change the discussion focuses in particular on matters of agricultural production and assumptions that droughts or extreme weather events will make numerous people food insecure due to climate disruptions. Subsistence farmers are especially vulnerable to climate disruptions if their crops fail to mature; they have few other options to feed themselves if their economic circumstances do not provide other sources of income that allow food to be purchased. More generally debates about food supplies focus on overall production figures for food crops and matters of price where the ability to purchase food, in particular by urban populations, is a key factor in food security. Here the assumption is usually that markets will provide, but that price might prohibit access by poor people; hoarding and speculative pricing are persistent problems in famine situations. The discussion of entitlements and access to food make it clear that climate alone isn't the source of insecurity, albeit it is obviously an aggravating factor. In so far as human security is about the provision of basic necessities for all, crucial questions about food security are at the heart of discussions of how climate affects human security. But as the literature on famine makes clear, food is not just a matter of supplies, but very much about economic arrangements to ensure that what is available is distributed. Climate makes dealing with these issues ever more pressing. It relates to justice between urban and rural dwellers as well as divisions between rich and poor at various scales. Indeed many of the calls for improved security meld well with larger calls to environmental justice (Stoett 2012).

In so far as human security is about the anticipation of potential problems limiting human potential in the future, then the longer-term issues of climate change are especially important if actions today are pre-cluding options for future generations. Sustainable development, popularized in the 1987 World Commission on Environment and Development report *Our Common Future*, in terms of ensuring the needs of present generations without precluding possibilities for future generations, clearly specifies matters in terms of inter-generational justice. While it is hard to know exactly what future generations will need, a stable climate system would seem to be key. Other matters such as an ocean system that is not acidified beyond the range that most aquatic life can handle, a wide range of biodiversity, and a lack of toxic pollution in ecosystems would also seem to be essential. All these issues are contemporary problems which, despite many technical fixes to pollution issues, are increasingly serious if assumptions are made that the planet ought to be handed to future generations in a condition something analogous to the state that facilitated the emergence of civilization in the first place. That is the explicit premise of the United Nations Framework Convention on Climate Change predicated on the necessity of avoiding dangerous anthropogenic climate change.

Now with accelerating climate change the question becomes what legacy current decisions are leaving for future generations, and whether the current generation is precluding prospects for future ones (Vanderheiden 2008). If current modes of economic activity are destroying prospects for future generations then clearly human security in the future is being compromised. But all this depends on projections of what is coming which are dependent in turn on predictions of what kind of economic activity will happen in coming decades. Thus human security is inextricably linked into economic decisions that are being taken now, only most obviously related to the matter of "carbon lock-in" whereby power-plants and other infrastructure and production facilities commissioned now lock society into certain fuel use patterns for the duration of their lifetime.

EMPOWERMENT AND THE PROVISION OF HUMAN SECURITY

Finally the big question hanging over all this discussion is the matter of where the authority to implement human security in rapidly changing times will come from. As the discussion at the beginning of this chapter made clear the extension of security beyond traditional themes of national security means that numerous agencies are involved; human security implies that more than states are involved in its provision. The failure of states in the international system to seriously tackle the rapid growth of greenhouse gases over the last few decades has spawned numerous attempts to think about innovations in business and politics that might tackle matters more effectively. Activists in many places are directly tackling coal plants and pressuring authorities to look to alternative sources of energy. While activists may be sceptical about business initiatives on climate change, the emergence of significant carbon trading schemes suggests that here too serious efforts at de-carbonization are underway in at least some markets (Newell and Paterson 2010). Reports from China suggest that political elites there too are paying attention to the necessity to rethink economic strategies in the face of looming climate disruptions. In the United States despite the political logjam that is Washington, numerous state and city initiatives are tackling carbon emissions in ways that seem to suggest that possibly the US has passed its point of highest carbon emissions.

No central authority is taking effective decisions that relate to the future level of greenhouse gases but nonetheless many initiatives that have consequences relating to human security are doing so (Hoffman 2011). Cities building infrastructure to deal with future flooding events, municipalities supporting solar and wind power installations, and business efforts to reduce travel and hence jet-fuel consumption will have an impact, even if as yet they are not fundamentally changing the trends towards higher greenhouse gas levels. All this suggests that the extension of the responsibility for security provision is as Emma Rothschild (1995) argued, coming into play in matters that relate to climate change. It is hard to summarize this concisely; that is just the point. The provision of human security isn't something that can or ought to be left to states whose record so far on many things including climate change hasn't been impressive. Human security challenges traditional assumptions about the locus of political authority and does so by focusing on those frequently neglected in the calculations of state political elites (Oswald Spring 2008).

Given the emphasis in the liberal formulations of security on the provision of circumstances that allow for the operation of commercial society, and the clear implication that liberal assumptions about economic rationality have led to the situation where humanity is dramatically altering its circumstances, the difficulties in accepting this notion of security as the benchmark condition for thinking and acting now are great. While much of the literature on climate security assumes that what has caused the problem will provide the solution, the juxtaposition of climate and security suggests that this assumption, and with it the formulation of security that has underpinned much of modernity, is no longer a valid formulation for the long-term future of human civilization. This contradiction is no longer avoidable, but how security might be rethought in these new circumstances isn't clear. Focusing on human security makes it clear that simply maintaining the political order that has produced climate change in the first place is an entirely inadequate formulation of security, despite the obvious temptations that it presents to political elites anxious to retain their power in the face of events that are in danger of spinning out of their control.

REFERENCES

Annan, Kofi (2000), 'Secretary-General salutes international workshop on human security in Mongolia', New York, USA: United Nations, Press Release SG/SM/7382.

Bogardi, Janos J. (2004), 'Hazards, risks and vulnerabilities in a changing environment: the unexpected onslaught on human security?' *Global Environmental Change*, **14**, 361–365.

Brauch, Hans Günter (2005), *Environment and Human Security: Towards Freedom from Hazard Impacts*, Bonn, Germany: United Nations University Institute for Environment and Human Security Intersections.

Buzan, Barry, Ole Wæver and Jaap de Wilde (1998), *Security: A New Framework for Analysis*, Boulder, USA: Lynne Rienner.

Commission on Human Security (2003), *Human Security Now*, New York, USA: United Nations Publications.

Dalby, Simon (2009), *Security and Environmental Change*, Cambridge, UK: Polity.

Elden, Stuart (2009), *Terror and Territory*, Minneapolis, USA: University of Minnesota Press.

Floyd, Rita (2010), *Security and the Environment: Securitization Theory and US Environmental Security Policy*, Cambridge, UK: Cambridge University Press.

Foucault, Michel (2008), *The Birth of Biopolitics*, New York, USA: Palgrave Macmillan.

Graham, Steve (ed.) (2010), *Disrupted Cities: When Infrastructure Fails*, London, UK: Routledge.

Grove, Kevin (2010), 'Insuring "Our Common Future"? Dangerous climate change and the biopolitics of environmental security', *Geopolitics*, **15** (3), 536–563.

Harvey, David (2005), *A Brief History of NeoLiberalism*, Oxford, UK: Oxford University Press.

Hoffman, Matthew (2011), *Climate Governance at the Crossroads: Experimenting with a Global Response after Kyoto*, Oxford, UK: Oxford University Press.

Homer-Dixon, T. (1999), *Environment, Scarcity and Violence*, Princeton, USA: Princeton University Press.

Independent Commission on Disarmament and Security Issues (1982), *Common Security: A Blueprint for Survival*, New York, USA: Simon and Schuster.

Jones, Reece (2012), *Border Walls: Security and the War on Terror in the United States, India, and Israel*, London, UK: Zed.

Kahl, C. (2006), *States, Scarcity and Civil Strife in the Developing World*, Princeton, USA: Princeton University Press.

Kaldor, Mary (2007), *Human Security: Reflections on Globalization and Intervention*, Cambridge, UK: Polity.

Klein, Naomi (2007), *Shock Doctrine: The Rise of Disaster Capitalism*, Toronto, Canada: Alfred Knopf.

Matthew, Richard A., Jon Barnett, Bryan McDonald and Karen L. O'Brian (eds) (2010), *Global Environmental Change and Human Security*, Cambridge, USA: MIT Press.

Neocleous, Mark (2000), 'Against security', *Radical Philosophy*, **100**, 7–15.

Newell, Peter and Matthew Paterson (2010), *Climate Capitalism: Global Warming and the Transformation of the Global Economy*, Cambridge, UK: Cambridge University Press.

O'Brien, Karen, Johanna Wolf and Linda Sygna (eds) (2013), *The Changing Environment for Human Security: New Agendas for Research, Policy, and Action*, London, UK: Earthscan.

Oswald Spring, Úrsula (2008), *Gender and Disasters. Human, Gender and Environmental Security: A HUGE Challenge*, Bonn, Germany: United Nations University Institute for Environment and Human Security, Intersections.

Pascal, Cleo (2010), *Global Warring: How Environmental, Economic and Political Crises Will Redraw the World Map*, Toronto, Canada: Key Porter.

Roberts, J.T. and B.C. Parks (2008), *A Climate of Injustice: Global Inequality, North-South Politics and Climate Policy*, Cambridge, USA: MIT Press.

Rothschild, Emma (1995), 'What is security?', *Daedalus*, **124** (3), 53–98.

Sagan, Carl and Richard Turco (1990), *A Path Where No Man Thought: Nuclear Winter and the End of the Arms Race*, New York, USA: Random House.

Scheffran, J., M. Broszka, H.G. Brauch, P.M. Link and J. Schilling (eds) (2012), *Climate Change, Human Security and Violent Conflict: Challenges for Societal Stability*, Berlin, Germany: Springer Verlag.

Seekins, Donald (2009), 'State, society and natural disaster: Cyclone Nargis in Myanmar (Burma)', *Asian Journal of Social Science*, **37** (5), 717–737.

Sims, Holly and Kevin Vogelmann (2002), 'Popular mobilization and disaster management in Cuba', *Public Administration and Development*, **22** (5), 389–400.

Smith, Paul J. (2007), 'Climate change, mass migration and the military response', *Orbis*, **51** (4), 617–633.

Stoett, Peter (2012), 'What are we really looking for? From ecoviolence to environmental injustice', in M.A. Schnurr and L. Swatuk (eds), *Natural Resources and Social Conflict: Towards Critical Environmental Security*, London, UK: Palgrave Macmillan, 15–32.

Theisen, Ole Magnus, Helge Holtermann and Halvard Buhaug (2011), 'Climate wars? Assessing the claim that drought breeds conflict', *International Security*, **36** (3), 79–106.

Ullman, Richard (1983), 'Rethinking security', *International Security*, **8** (1), 129–153.

United Nations Development Program (1994), *Human Development Report*, New York, USA: Oxford University Press.

United Nations Secretary General (2005), *In Larger Freedom: Towards Development, Security and Human Rights for All*, New York, USA: United Nations.

United Nations Secretary General (2009), *Climate Change and its Possible Security Implications*, New York, USA: United Nations.

Vanderheiden, Steve (2008), *Atmospheric Justice: A Political Theory of Climate Change*, Oxford, UK: Oxford University Press.

Watts, Michael (1983), *Silent Violence: Food, Famine and Peasantry in Northern Nigeria*, Berkeley, USA: University of California Press.

Webersik, Christian (2010), *Climate Change and Security: A Gathering Storm of Global Challenges*, Santa Barbara, USA: Praeger.

White, Greg (2011), *Climate Change and Migration: Security and Borders in a Warming World*, Oxford, UK: Oxford University Press.

World Commission on Environment and Development (1987), *Our Common Future*, Oxford, UK: Oxford University Press.

2. Elements and value-added of a human security approach in the study of climate change
Des Gasper

WHAT CAN HUMAN SECURITY ANALYSIS ADD IN THE STUDY OF CLIMATE CHANGE?

To present climate change as an issue of human security means to focus on the impacts and implications in the lives of ordinary people, not only in the agendas of armies, states or national economies. It means, for example, looking at life expectancies and patterns of nutrition, morbidity and mortality, not only at whether stressed populations might explode into armed conflict—the extreme 'Darfur' scenario. That sort of scenario and such preoccupations often reflect traditional state-centred and military-focused concerns more than person-focused ones. That poor people in most cases seem to lack the organization, cohesion or resources to initiate armed conflict does not mean that their plight is not an issue of—human—security. As part of a humanist perspective, the human security approach means looking at more also than aggregates of monetized economic variables, but rather at the contents, objective and subjective, of the lives of all of the people—at their 'do-ings' and 'be-ings', the constraints that they face, the real opportunities that they have, or lack, and the meanings that they experience—not only at the parts and the persons that are counted in money terms.

Security language is commonly tacitly oriented to prioritizing and safeguarding the interests of powerful groups within nation-states. 'Human security' is in origin a counter-concept, that attempts to turn the frequently predominant implicit association of 'security' with the security of the state, to a focus instead on the security of human persons, seen as real individuals in the round, and not only as bodies and statistics. The attempted turn involves, first, looking at the security of distinct human individuals, in the circumstances of their particular lives at the intersection of many different forces. Why focus on this set, human persons? Why not ignore some of them? Or why not also include the security of bacteria, algae or clouds? Evidently, human security thinking is a humanist approach, sister to the perspectives of human rights and human development; better put, it is a sister face of a humanist perspective. Further, with all human individuals taken as the set of primary objects of value concern, there is implicitly or typically an interest in the security of the human species, including future generations. According to some authors, there should also be a concern with the security in some respects of human groups, cultures and meaning-systems, since persons are group members (see e.g. Roe 2010). Securing one thing (such as a group) may, however, sometimes be in competition with securing others (such as individual persons). Different types of human security approach reflect different stances on how to conceptualize human identity and human interests, and how to weigh competing considerations. In effect there are different viewpoints about the nature of being human.

While there has been enormous discussion of the meanings of 'security', which is the term whose meaning a 'human security' approach attempts to turn away from a state-centred presumption, there has been too little attention in much of the work on human security to thinking about the meaning of 'human' and thus to what supposedly is to be secured, in addition to bare life. Yet, since 'security' is fundamentally a prioritizing term, rather than an entity in itself, the main focus should be on what is being secured and why. So human security analysis must be led to engage with the central issues of what it is to be human, and what are and/or should be priorities. It connects here to earlier and ongoing work in basic human needs analysis (e.g. Burton 1990) and to traditions of exploration in humanist philosophy.

Being human involves more than having sufficient food and other basic items in order to survive. Humans are embodied, and, not least, gendered; they are not only mortal but have a life-cycle, of growth, maturity and decline. The long period of growth involves enormous enculturation, acquisition of attitudes, knowledge and skills that are not genetically encoded but instead socially constructed, communicated, inculcated and adapted, and that build on our latent capacities, including those for reasoning and choice. Humans desire to employ their capacities, in some degree, to exercise agency. These capacities are a key focus in recent human security theory, in the concept of 'securitability': the ability of people to establish and maintain their own security (UNDP 2003). Perhaps related to the long periods of dependence in the life-cycle, human identity and functioning are intensely social and relational; people do not become human in isolation. Nor are they human in the abstract: our physical, felt presence is in a particular time, place and social setting. Humans have complex systems of emotions which mediate their dependencies and help to mobilize and orient their agency. Security and insecurity are matters both of objective vulnerability and of emotional state and subjective perception (Leaning

2008; Leaning et al. 2004). Put differently, vulnerability is not only physical but emotional and, more broadly, existential.

Overall, the intersecting multiplicity of their life circumstances—their physical location, personal capacities, health, gender, social status and enculturation, assets and options, habits and perceptions, relationships and exposure—determines people's security and insecurity. So for example: 'In the 2007 Cyclone Sidr [in Bangladesh] some people died who declined to go to shelters because that would leave their animals unattended, since some who went to shelters during a previous warning were robbed at home' (Saferworld 2008: iii).

Studying the determinants of real people's insecurity requires holistic thinking at the level of individual lives—a narrative approach, that studies people's historically determined situation and constrained agency—and holistic thinking at the level of grander systems, about the socio-economic drivers of some people's marginalization in their own countries and the drivers of economic, environmental and climate change at a global scale (Gasper 2010, 2013a). This combination of, first, empathetic detailed attention to specific local vulnerabilities and, second, bold but cool attention to global interconnections, including to linkages that frequently generate costs—ecological, medical, psychological, cultural, economic costs—for other people, costs that are ignored in narrowly national, disciplinary, or commercial calculations, was well articulated by Denis Goulet, for example in *The Cruel Choice* (Goulet 1971; Gasper 2011). His work on 'development ethics' consciously proceeded in the spirit articulated in the 1940s by the French movement 'Économie et Humanisme': that we should consider 'le développement de tout l'homme et de tous les hommes'. *The Cruel Choice* was one of the major precursors of the wave of human security thinking that has unfolded since around 1990; its second chapter, for example, was entitled 'Vulnerability: the key to understanding and promoting development'.

Goulet made clear that the required combination of types of analysis is unlikely to happen in the absence of a certain perceptual and emotional base. It is not a free-floating technical instrument. As mentioned earlier, why, for example, should we not otherwise leave out some (even most) humans from our field of attention and concern? In cases where we do not fear them (or their successors) or hope to gain from them, conventional security analyses leave them out. Dalby remarks in his chapter in this book that the conventional formulation of security corresponds to the Westphalian era of competing nation-states and concerns the stabilization of the operating environments required for smooth running of commercial society within and across national boundaries; it excludes other concerns.

Human security analysis does consider closely the nexuses between deprivation and insecurity on the one hand and violence and crime and migration on the other. But its interest in people everywhere extends beyond self-interest and fear about what they might do. It shares the humanist values expressed by the human rights tradition, and so does not look only at other people and their difficulties if they are feared, but fundamentally because they are respected and valued as fellow humans. Human rights are an essential element in a human security approach. Arguably though, a human security approach adds something that has not been well developed in mainstream human rights work: an ontological perspective of interconnectedness (Gasper 2012, 2013a). The combination of this with a human rights style concern for the dignity of every person is central, for the motivation for tracing the connections across national, disciplinary and organizational boundaries rests in part on sympathetic interest in fellow humans' lives around the globe.

For climate change analyses this means close attention to identifying how particular people, in particular life-niches, are affected and how they might be able to respond: for example, 'rural communities in Bangladesh, farmers in Ethiopia and slum dwellers in Haiti' (UNDP 2007:3), in the words of the Human Development Report 2007/8 on climate change. Citing human rights criteria, this Report proceeded partly in the spirit of a human security approach, repeatedly invoking the interests of particular vulnerable groups, including children, 'our children and grandchildren' (p. 2), and 'future generations'. In contrast, the counterpart report on climate change from the World Bank, the World Development Report 2010 (World Bank 2010; WDR) barely mentions children and never once uses the terms 'grandchildren' or 'future generations' (Gasper et al. 2013a). It focuses on abstract aggregates, especially money aggregates. Sadly, the successor Human Development Report (HDR) 2011 on environmental sustainability almost completely abandons the substantive human language and corresponding focus of the 2007/8 HDR (see Gasper et al. 2013b). It retreats instead to the virtually useless hyper-aggregated category of a global Human Development Index (UNDP 2011: 2), which conceals the implications of climate change for the poorest groups. An aggregated index like the HDI sums achievements for all inhabitants of a country or region—in this case the world—and conceals the scale of impacts on poor people, many of whom may fare far worse than the national or regional average, and even more so in the case of a global average. A human security approach aims to do the opposite, and to reveal and to respond to these people's situations.

We will start with a case, that of Bangladesh; then essay a specification of key dimensions and components in a human security approach; and discuss its application and potential value-added in the study of climate change, including through use of a scenarios methodology for identifying, appreciating and responding to future threats and opportunities.

BANGLADESH: WHICH THREATS TO WHICH VALUES AND WHICH PERSONS?

... the level of insecurity relating to 'freedom from fear' is perceived as being relatively low compared to 'freedom from want'. Bangladeshis consider issues such as poverty, employment, food security and health to be much greater concerns than crime. (Saferworld 2008: 104)

In a survey of 2000 households in Bangladesh, undertaken by the Research Division of BRAC, the largest development NGO in Bangladesh and perhaps the world, many more people identified 'freedom from want' issues as big problems—poverty (69 percent), unemployment (65 percent), provision of utilities (56 percent) and vulnerability to natural disasters (51 percent)—than wished to highlight issues that are conventionally placed under a 'freedom from fear' heading (including crime, extortion, and availability of firearms). Crime was rated as one of the five main concerns by only about one third of the sample.

With 50 percent of Bangladeshis living below the poverty line, poverty and unemployment are the greatest concerns for most people. Poverty underlies many other problems. It is a major cause of food insecurity, since many people lack the resources (including land and agricultural products) either to grow their own food or to buy it from others. Limited resources make it harder to access basic services such as healthcare, sanitation and education. Poverty and unemployment are also seen as being the two most important drivers of crime and injustice. (Saferworld 2008: ii)

Similarly, when the interviews moved on from the priority values that are considered in jeopardy, to which sources of threat made people feel insecure:

[n]atural disasters were the most common concern (53 percent), followed closely by a lack of healthcare (48 percent). The third most popular answer was a perceived increase in crime (28 percent), and fourth was drug abuse (23 percent). Worries about natural disasters are a much greater concern in rural areas [58 percent in rural areas, versus 37 percent in urban areas], ... where the two most frequent responses are crime and drug abuse. (Saferworld 2008: ii)

In the period 1900–2008, sixty-six recorded cyclones caused on average over 9000 deaths (p. 21). In 1970, apart from Cyclone Bhola's direct impact the resultant surge inland of sea waters killed approaching half a million people in total. 'Natural disasters' frequently have a large human component, in terms first of their causation and second of the determinants of the extent and distribution of their impacts. Causation of the two main such threats in Bangladesh, cyclones and floods, certainly has a human component, though the causes lie largely outside Bangladesh; and in addition the structure of the humanly (re)constructed social and physical landscape is a major determinant of the impacts when these threats strike.

Interestingly, even when people were asked about the providers of security in relation to *freedom from fear*, the leading responses were not the conventional security apparatus:

> The five most popular answers were education institutions (43 percent), NGOs/ micro-finance institutions (34 percent), [well above] police stations (28 percent), Union Parishads/municipalities (28 percent) and hospitals/healthcare facilities (20 percent). These results show that Bangladeshis understand human security to be about much more than crime and justice. (Saferworld 2008: vii)

'Natural disasters' were thus already the greatest felt threat. Climate change is likely to seriously increase their frequency and magnitude. '. . . [B]y some estimates, a one metre sea-level rise will submerge about one-third of the total area of Bangladesh, thereby uprooting 25–30 million of our people . . . They are most likely to become refugees of climate change', declared the head of government (Ahmed 2007). Even much smaller rises, and extreme weather events, could uproot large numbers of people:

> . . . urbanisation (or 'slumisation') would be likely to increase even more rapidly as desperate people moved towards the cities in search of work and food. This would further over-stretch already insufficient infrastructure and governance mechanisms . . . Secondly, there would likely also be much more emigration, in particular economic migration to India. This would be a very sensitive and possibly explosive political issue for both states . . . (Saferworld 2008: 24)

Despite this, 'little serious analysis is available of possible scenarios for the effects of climate change on human security in Bangladesh', at least as of a few years back, declared the Saferworld report (2008: xvi). This might reflect the sheer scale of the challenges faced, which exceed the capacity of business-as-usual politics, particularly the viciously divided party politics of Bangladesh. But in addition, the neglect reflects that the people at risk are predominantly the poorer groups, those on the lands most at risk of flooding, those with the least robust houses, those who feel they cannot risk leaving their animals unattended during a flood or cyclone.

DIMENSIONS AND CHARACTER OF A HUMAN SECURITY APPROACH

Simon Dalby, in the preceding chapter in this book, reviews Rothschild's (1995) list of four shifts or extensions proposed in human security and other 'non-traditional security' analyses: shifts away from the emphases adopted in state-centred and/or violence-centred security analyses. We noted earlier the first two shifts, which both concern the conception of *whose* security: (1) downwards from the state to the human person, and (2) upwards from the state to the human species. Rothschild presents the latter in terms of the international system and the physical environment, but arguably in both those cases the ultimate object of security is the human species. Conceivably, security of all individuals equates to security of the species, but an explicit emphasis on the human species highlights the concern for individuals who are as yet unborn.

We have touched also on the other two shifts, concerning: (3) security of which values, and which associated sectors (for example, food, water, physical and mental health); and (4) security through action by which providers. The extensions beyond the national state in this last case are both international and intra-national, including attention also to persons' self-provision and securitability. 'Securitability' is a concept coined for the Latvia Human Development Report 2003: 'the ability to avoid insecure situations and to retain a [psychological] sense of security when such situations do occur, as well as the ability to reestablish one's security and sense of security when these have been compromised' (UNDP 2003: 15).

The list can itself be usefully extended. Another dimension which Rothschild explored was: (5) Security as perceived by whom—the issue of so-called objective (expert-perceived) security versus subjective (self-perceived) security. Next, Oscar Gomez amongst others highlights three supplementary questions: (6) Security against which threats, in the sense not merely of 'the threat of serious damage to priority value V' but with regard to which specific types of event or force threaten to do that damage. Only because there are threats (at least felt threats) does security become an issue at all, but the threats that are perceived, considered relevant, and prioritized for attention vary considerably. (7) How much security—this point elaborates the third, concerning which values are to be secured, by asking: secured to what extent, what are the targets for the level of assurance? Last: (8) Secured by what instruments—this point elaborates the fourth, concerning who are the providers, with fuller attention now to the specific means that will be employed (Gomez et al. 2013). Table 2.1 encapsulates the checklist of questions and the corresponding agenda.

So, associated with the re-orientation of the answer to 'whose security?'

Table 2.1   Some features of human security discourse

Questions/Issues	Some emphases/features in human security discourse
1. Whose security?—I	Human persons (and communities) not only bigger systems. Downwards shift of focus, from the state.
2. Whose security?—II	Human species, not only national systems. Upwards shift of focus from the state, including to the physical environment that sustains human society.
3. Security of what (which values / which sectors)?	In terms of sectors: focus on basic sectors. The conventional list of Freedoms ('freedom from want', 'freedom from fear', 'freedom from indignity'). Meanings of human ➔ what priorities? ➔ bodily safety, health, . . . but also typically beyond bare survival. Securing of human rights. Securing of present levels?—the concern here is with stability in only some respects.
4. Provision by whom?	Securitability; empowerment and not only treatment. The idea of a 'constellation of providers' (UNDP 2003).
5. Security as perceived by whom?	Both 'objective' (expert-perceived) and 'subjective' (self-perceived).
6. Security against what threats?	Not only against threats of physical violence. Ontology of interconnection and intersectionality. Nexuses. Thresholds as sometimes danger points.
7. How much security?	Within sectors: ensure attainment of at least basic levels. To ensure capability (the potential to be secure), or to ensure functioning (actual security)? Principle of harm avoidance, and precautionary principle.
8. By what instruments?	Human rights. Prevention not only palliation. Use also diverse and unorthodox instruments. Promoting securitability.

is a system of linked features (Gasper 2005, 2010, 2013b). We saw earlier that underlying this agenda are a humanist normative concern, an ontology of interconnectedness (see, e.g., Thomashow 2002), and a resulting perception of human lives as marked by a combination of capability and vulnerability. Diverse types of threat to priority values for particular persons and groups can arise out of their particular locations in intersected, interconnected systems of systems: climatic, ecological and epidemiological, economic, social and political.

We need to consider the perceptions which underlie choices in these various dimensions. In particular we should reflect on the significance of the term 'human', and correspondingly on the arguments in support of the priority claims implied by a 'security' label. One cannot reasonably merely assert priority via use of security language; one should have convincing reasons for it, including ideas about the nature of persons—the referent in human security claims—and about nexuses of interconnection, threats, thresholds and potential damage or even collapse. Let us look, for a majority of the questions, at some examples.

Regarding Question 1: Security for Whom?—I: Persons or 'The Economy'? Persons or Citizens?

The World Development Report on climate change, written for the 2009 Copenhagen conference, gives much attention to economic calculations of the costs of climate change, and the net returns to programs for mitigation and adaptation. It estimates modest potential GDP losses from global warming: 'a global average GDP loss of about 1 percent' (World Bank 2010: 5). The impact is more in low-income countries but even there it is not more than a year's foregone growth. The estimates were based on the conservative and now superseded IPCC 2007 Report's climate projections, but that is not the issue here. Our concern is the methodology of calculation, namely estimation of the possible aggregate direct monetary costs of damage. This ignores other human costs and the indirect impacts of changes in disease patterns and of social and political unrest, conflict and migration, such as suggested for example by the Darfur crisis. It aggregates the losses of some people, who may be poor and driven into destitution, with the gains of others, who may be already affluent. And it values effects according to market prices that put very little weight on the lost housing, the lost livestock, and even the lost life of the marginal low-income rural resident in Darfur or Bangladesh. Correspondingly, after extended discussion of various economic modelling exercises along these lines, seeking to determine whether a scenario of 2, 2.5, 3 or 3.5 degrees rise in average global temperatures offers the best economic rate of return, the

Report deals shockingly briefly with the expected deaths from business-as-usual economic expansion. 'And over 3 million additional people could die from malnutrition each year', declares a single sentence at the end of a paragraph in the Overview (World Bank 2010: 5). As Dalby observes in the previous chapter in this book, security as interpreted by ruling establishments has usually been about 'making particular forms of economic action flourish, . . . not about protecting non-commercial modes of human life.'

In contrast, the Human Development Report 2007/8, also fully devoted to climate change, focused more on human impacts than on GDP: 'In terms of aggregate world GDP, these short term effects may not be large. But for some of the world's poorest people, the consequences could be apocalyptic' (UNDP 2007: v). The Report makes clear how their lives can be broken and stunted. Overall, the World Development Report's considerably lesser sense of urgency in response to climate change reflects its preoccupation with monetary magnitudes and thus with the interests of people who already have more resources to protect themselves from possible future stresses (Gasper et al. 2013a: 32).

A second vital contrast is between a human rights based concept of human security, which covers all affected persons, and an interpretation of 'citizen security' which excludes non-citizens. The concept of 'citizen security' has risen in prominence in and from literature in Spanish, including in Spain's post-Franco Constitution and in post-dictatorship Latin America of the past generation. It has been defined as 'the personal, objective and subjective condition of being free from violence or from the threat of intentional violence or dispossession by others' (UNDP 2005: 14–15). The concept is prominent in various other National and Regional Human Development Reports (e.g. UNDP 2010, 2012) and in the World Development Report 2011 on conflict (World Bank 2011). While the label's reference is close to the 'personal security' component in the 1994 HDR's seven-fold list, its use of 'citizen' rather than 'personal' may partly grow out of the tradition of thinking about security as stabilization of the operating environments for commercial society within a nation-state system and about people as 'citizen consumers within the boundaries of nation states' (Dalby 2009: 160). Yet commercial society has increasingly drawn in large groups of non-citizens to very many countries, where their human rights often are at risk (Edwards and Ferstman 2011; Truong and Gasper 2011); and fossil-fuel based economic development within the nation-state system now endangers the livelihoods of non-citizens worldwide and in generations to come.

Regarding Question 2: Security for whom?—II: The human species and future generations?

If one does not focus on persons, then the issue of future generations may not arise: instead all discussion is in terms of abstractions and aggregates—the nation, the economy, the national economy, the state. If one does focus on persons then one will eventually note that many of the people affected by environmental change have no current voice, including because most of them are not yet born or are not yet adults. So the Human Development Report 2007/8 talks specifically and frequently of 'humanity', the 'human community', 'the world's poor', 'our children', 'our children and grandchildren', 'future generations', 'the world's poor and future generations'.

The World Development Report 2010 avoids all such terms (see Table 2 in Gasper et al. 2013a). Its 11,000 word Overview never mentions even grandchildren or future generations. In comparison to the HDR 2007/8, the WDR devotes more time to discussion of an insurance rationale for investment in climate change mitigation and adaptation; insurance feels like safely conservative technical economic language. But the Report presents the case for insurance with an unusual amount of tentativeness and hedging (Gasper et al. 2013a: 35). For in this context an insurance argument introduces an assumption of moral community, at some level or levels. Climate change insurance is largely for the benefit of future generations, worldwide, and thus mostly for people who live abroad. Even if the argument is that rich countries should invest now in the efforts of developing countries in climate change adaptation and mitigation, because this investment prevents future 'catastrophic risks' for all (World Bank 2010: 9), there is a presumption of solidarity of the present-day rich North with at least the future generations in the North. These ethical choices are hidden behind the World Bank's economic calculations; and unlike the Human Development Report Office it is not ready to bring them out for discussion.

Regarding Question 3: Securing Which Values?

The conventional slogans-cum-labels in human security discourse for referring to priority values have been 'freedom from want', 'freedom from fear', and subsequently also 'freedom from indignity' (CHS 2003; IIHR n.d.). 'Freedom from hazard impact' (Brauch 2005) is sometimes mentioned as if it was a new category to add to this set, but it is better seen as a highlighting of one type of threat which might endanger the basic freedoms. Not every hazard impact—for example the damage in a cyclone to a billionaire's private vacation island castle—should be listed as a human security concern. The priority values are partly the basis of broad international consensuses, reflected in the major conventions on human rights and the Millennium Declaration, and are partly subject to further selection and specification through context-specific debate and nego- tiation. Thus the GECHS (Global Environmental Change and Human Security) research program defined human security as the capacity of indi- viduals and communities to respond to threats to their social, human and environmental rights, and the UN Trust Fund for Human Security takes a similar line. They recognize that achievement of rights cannot be achieved instantly and all at once, and that interpretation and prioritization must occur *in situ*.

Some authors, often from a conventional security studies background, wish to exclude 'freedom from want' issues, let alone 'freedom from indignity', from the concept of human security, and in effect to limit it to bodily safety (see e.g. MacFarlane and Khong 2006). This is despite the frequently demonstrated greater importance—both objective and (as we saw for Bangladesh) subjective—of 'freedom from want', for promoting human life and for quality-adjusted life-years. In the context- specific prioritizations and choices of focus required in National Human Development Reports (NHDRs) on human security, a wide range of priority values have emerged for special consideration in the reports con- cerned (see Table 2.2). Some reports attempt a comprehensive mapping of major threats to basic values (e.g., the prominent Arab Countries report of 2009 and Latvia report of 2003); a second group can be called 'citizen security' or citizen safety reports and focus on crime and physical violence; a third group focuses on state-building, as an essential basis for human security that demands priority in some cases; and the fourth group of reports chooses one or two other lead challenges felt in the particular country at the particular time (e.g., the 1998 Chile study of social moderni- zation, and the 2012 African regional report on food security and human development). Various other NHDRs have addressed climate change (e.g. Kazakhstan 2008, Moldova 2009/10, Cambodia 2011) and could also be viewed as part of this fourth group.

Barnett et al. (2010: 20) underline that this sort of prioritization, using the name of 'security' to prioritize individual and community needs, rights, and values at risk from environmental change, also engages diverse policy communities, including those concerned with socio-economic develop- ment, sustainable development, human rights, and foreign policy. Thus the meaning of 'human security' is not left to the traditional purveyors of security and is instead continually negotiated in ways that are far less likely to justify the strengthening of the state at the expense of human security.

Table 2.2 *Categorization of some National Regional Human Development Reports that have explicitly adopted a human security theme*

Comprehensive focus		Narrower focus	
Investigation of the context-specific range of primary threats to primary values, done without restriction in terms of how to organize security provision	Focus on one priority set of threatened values: 'citizen security', often with main attention given to use of conventional security instruments	Focus on a priority threatened means: the State	Focus on one or two selected, context-specific threatened values or primary threats, without restriction in terms of how to organize security provision
Comprehensive mapping reports	*'Citizen security' reports*	*State-building reports*	*Lead-challenge driven reports*
Arab Countries (2009)	Caribbean (2012)	Afghanistan (2004)	Africa (2012)
Benin (2011)	Costa Rica (2005)	Democratic	Chile (1998)
Djibouti (2012/13)	Philippines (2005)	Republic	Macedonia (2001)
Kenya (2006)	Bangladesh (2002)	of Congo (2008)	Mali (2009)
Latvia (2003)		Occupied	Senegal (2010)
Thailand (2009)		Palestinian	Uruguay (2012/13)
		Territories (2009/10)	

Source: Based on Gomez et al. 2013.

Some discussions of human security—for example after the Asian financial crisis of 1997–98—have emphasized stability, in the sense of preventing major or drastic downturns. However stability of everything would mean stabilizing things which are already in excess—such as the lifestyles that drive global climate change and generate insecurity for poor people—and stability of some things which are in grossly deficient supply. The idea of human security refers instead to attainment or attainability of priority values or basic needs, and not to stabilizing the status quo.

The agenda of clarifying and focusing on basic human values and what threatens or is felt to threaten them has, by an interesting paradox, meant that human security analysis with its apparently restricted focus has sometimes penetrated more deeply than have more open-ended discussions of human development. Mahbub ul Haq, lead author of the global Human Development Reports of 1993 and 1994 that launched the main stream of human security analysis, stressed that by 'human' is meant more than merely the solitary individual, the unit in casualty counts; and correspondingly, exploration of the content of 'human' is required, not only of meanings of 'security' (Lama 2010).

Part of being human is that people have and need meaning systems, including identities. Denis Goulet argued that each person and every society wants to be treated by others for their own sake and on their own terms, as having intrinsic worth regardless of their usefulness to others (Goulet 1975). Human security has essential dimensions that concern felt security, including how to make sense of one's world, as explored for example in the Chile national report (UNDP 1998) and the UNESCO regional report on Western Europe (Burgess et al. 2007). People's wish and ability to contribute in choosing for themselves and their families has implications for who are the appropriate responders to insecurities; the 'providers', in bureaucratic parlance, should include the people themselves.

Regarding Question 4: Which Providers?

Security and insecurity are determined not just by the level of exposure to threats, but by the degree of vulnerability to damage in the face of a given threat, and the extent of resilience, the ability to recover after damage. Hence the policy agenda of human security covers all three areas: exposure, sensitivity and resilience. Many contributors are needed, acting in complementary ways, including to strengthen people's own securitability. Entitlements analysis, which was elaborated by Amartya Sen and others to understand famines and food insecurity amongst poor and marginal groups (Sen 1981; Dreze and Sen 1991), was then extended to consider more broadly issues of welfare and social security (Dreze and Sen 1989). Such analysis helps to reveal the wide range of factors that affect exposure and vulnerability, and also the wide range of opportunities for counteraction and for strengthening resilience.

Next, the contributors to damage, and hence also potential contributors to remedial action, are international as well as intra-national, not only national, especially when we speak of environmental insecurity. In the case of climate change, the Kyoto system's exclusively national framing of issues and of responsibilities has arguably contributed to the global policy deadlock. It has allowed inattention to high-emitters in low-income countries, which has reduced the readiness of high-emitter countries to act (Harris 2010).

Third, human security analyses' focus on issues of basic human needs, including needs for self-expression and self-determination, is matched by a frequent anthropological style of attention to real individuals' whole lives and own efforts, which leads to emphases on human capability and not only human vulnerability. Empowerment and protection have thus long been equally stressed (CHS 2003; Dreze and Sen 1989, 2002; Gasper 2008). The GECHS research program (1999–2009) came to a definition of human security as where 'individuals and communities have the options necessary to end, mitigate or [sufficiently] adapt to threats to their human, social and environmental rights; have the capacity and freedom to exercise these options; and actively participate in pursuing these options'. Similarly, for Barnett and Adger (2007: 140): 'Human security is taken here to mean the condition where people and communities have the capacity to manage stresses to their needs, rights, and values'; in other words, where they possess adequate securitability. Since people can and should contribute towards their own security, a concern for security is not in itself a language of paternalism; it includes helping people to enjoy the conditions needed so that they can to a large extent act and adapt independently.

Regarding Question 6: What Are the Threats?

Threats are situation-specific, so there is no fixed universal list of threats. Some scholars try to insist that intentional physical violence is the overwhelming priority threat and the only justified focus for human security studies; MacFarlane and Khong (2006) try to limit the focus even further, to organized intentional physical violence. These are arbitrary attempts to monopolize the concept of human security, that diverge from commonly felt priorities (as we saw for Bangladesh) and from the typical realities of the causation of human insecurity, suffering or deaths (Gasper 2010). Dangerous global environmental change is underway caused by fossil-fuel intensive lifestyles, which can realistically be called major threats to human security (Dalby 2009; Urry 2011).

While threats vary, some themes recur. First, many threats have major indirect effects, knock-on effects, after they strike. The vulnerability of coastal mega-cities to extreme weather events, sea-level rise and so forth means the indirect vulnerability of their vast nationwide and worldwide hinterlands (O'Brien and Leichenko 2007: section 5). Although the 2008 Bangladesh human security study found that the felt threats to 'freedom from want' were much greater than those to 'freedom from fear', it warned of important linkages. Not least: 'This research finds that political insecurity lies at the heart of many human security challenges in Bangladesh

[. . . for] it makes it less likely that other insecurities will be adequately addressed by the state' (Saferworld 2008: ix).

Second, threats combine. Old people are particularly vulnerable in the face of climate change (O'Brien and Leichenko 2007: section 6), especially when this combines with changes in family structure that leave more of them living alone and with changes in public finances or public policy that bring reduced or absent state support. Human security thinking looks at such intersections of diverse forces in persons' and groups' lives. Leichenko and O'Brien's *Environmental Change and Globalization* (2008) shows how abstracted disciplinary discourses can miss these vital combinations and interactions. Much human security analysis adopts, in contrast, a supra-disciplinary perspective that looks at particular people in particular physical and social locations and at the intersecting forces in their lives, which generates case-specific insights:

> Examining the dynamics of vulnerability among rural households in Zambia, Malawi, and South Africa, Ziervogel et al. [2006] found that different households experienced different stresses to be the most important. . . . An emphasis on human capabilities and human security draws attention to the differential consequences of climate change for individuals and communities resulting first and foremost from disparities in human development. (O'Brien and Leichenko 2007: 10, 14)

Third, human security analysis highlights a particular type of danger: thresholds beyond which negative effects significantly intensify, sometimes even leading to system collapse; for example, in the case of persons, leading to death or highly increased chances of death, as when small children or old people are exposed to exceptional stresses. Fourth, threats also bring opportunities; attention to these and to securitability and resilience is vital to avoid reducing human security thinking to only a depressing roll-call of hazards.

Many of the cases concerned lend themselves to scenarios thinking rather than to deterministic projections or sets of statistically securely grounded possible pathways. Trying to think intelligently about possible futures is essential, and calls for enhanced skills in listening and narrative imagination that can identify some of the implications of intersections, thresholds, and knock-on effects.

THINKING ABOUT FUTURE THREATS AND RESPONSES

Gwynne Dyer remarks that IPCC type climate models and projections have stayed 'well clear of any attempt to describe the political, demographic and strategic impacts of the changes they foresee' (Dyer 2010: 3). Most of the projections by international organizations such as the World Bank have assumed, in effect, that there will be no surprises—which would be the real surprise, for reasons we will come to. They thus fail to become convincing scenarios. Future forms of so-called 'adaptation' will include far more than the peaceful building of dykes. Adaptation, responses to climate change—including to changed variance, more frequent extreme events, particular local climatic and environmental quirks—will include many types of human reaction. If climate scientists and social scientists do not study these possibilities enough, military planners worldwide appear to be busy doing so, as for example in two major American studies already from 2007: *National Security and the Threat of Climate Change* from the Center for Naval Analysis (CNA 2007) and *The Age of Consequences*, from the Center for Strategic and International Studies in Washington DC (Campbell et al. 2007).

The sorts of scenarios that they generate suggest: first, the 'magnification of [the] physical effects by likely political and social responses' (Dyer 2010: 16). Second, 'nonlinear climate change [occasional rapid shifts] will produce nonlinear political events' (Fuerth 2007: 72). And third, unlike 'the kind of approach that is often taken in public policy, which is that you only need to do THIS, and the problem will be solved now and forever', we should instead 'Expect that any solutions you apply are likely to disturb the system, leading to an infinite series of surprises' (interview with Leon Fuerth; at Dyer 2010: 21). This third insight matches use of a narrative approach in futures studies: stories will keep on unfolding, with periodic surprises.

Reviewing the historical record of human responses to environmental crises, the historian J.R. McNeill notes how troubles beget troubles. Disasters fuel mutual suspicions and religious zeal. People under pressure often get nasty. More elegantly stated: 'Restraint and civility can quickly perish when confronted with imperious necessity. This much has been obvious to observers since Thucydides's analysis of the Corcyran Revolution . . . [Political] reaction to shocks often [in history] took the form of scapegoating minorities and foreigners' (McNeill 2007: 29). In a sister chapter in *The Age of Consequences*, Gulledge warns against the myth that 'climate change will be smooth and gradual. The history of climate reveals that climate change occurs in fits and starts, with abrupt and sometimes dramatic changes rather than gradually over time' (Gulledge 2007: 37). So, as highlighted also by Fuerth, the social impacts and forms of 'adaptation' too could work out differently than suggested by the smooth curves in the scientific reports.

The Age of Consequences builds three climate scenarios. To start with the least worrying: Scenario 1 traces the impacts of the IPCC's 2007 main projection, through to 2040. 'It is a scenario in which people and nations are threatened by massive food and water shortages, devastating natural disasters, and deadly disease outbreaks. It is also inevitable' (Podesta and Ogden 2007: 55). The various component crises 'are all the more dangerous because they are interwoven and self-perpetuating' (p. 56). Scenario 2 adds the early impacts of the dangerous feedback effects that were explicitly not included by the IPCC 2007 Report on the grounds that it was not able to model them—for example the danger that the carbon embedded in Siberian permafrost ('more than currently resides in the atmosphere'; Woolsey 2007: 83)—will be released as methane due to warmer temperatures, greatly accelerating the greenhouse effect. The scenario envisages warming of 2.6 degrees C by 2040. 'Agriculture becomes essentially nonviable in the dry subtropics' (Fuerth 2007: 71), where the average warming could be greater and the frequency of extreme weather events would greatly increase. Human systems worldwide will come under major stress, and 'massive nonlinear events in the global environment will give rise to massive nonlinear societal events' (p. 76). Scenario 3 follows this story through to a world in 2100 that is 5.6 degrees C warmer and where the sea level has already risen two metres. One metre may be enough to flood one-third of Bangladesh, we saw. The study traces the possible diverse human impacts of these climate scenarios, as people and organizations react and interact, leading to an infinite series of surprises.

Dyer illustrates such an approach in richer detail, while never claiming that scenarios are predictions. Pakistan figures prominently: a country of 180 million people, with nuclear weapons. It is already the world's university for armed Islamic militants. A fast growing population of unemployed young men can continue to provide recruits like the perpetrators of the 2008 Mumbai massacres. The country's environmental vulnerability was illustrated in the extraordinary 2010 floods, due to exceptional rains apparently related to a La Niña event in the Pacific. In addition, the country has the largest contiguous irrigation system in the world, a system that relies on river waters from the Himalayas. The shrinking of the Himalayan glaciers is envisaged to affect Pakistan's winter water supply and to lead to intense tensions with India, from where several of Pakistan's major rivers come.

Bangladesh, a country with a population almost equal to Pakistan's, is even more environmentally vulnerable. Given a combination of historic, current, and impending damage and grievance, security planners worldwide (e.g. Campbell et al. 2007) ponder scenarios in which the country becomes a second university for armed militants. Dyer's scenario five concerns a Bangladesh subjected to ever more frequent and destructive cyclones. Its government finally threatens to 'upload a million tonnes of powdered sulfates into the stratosphere—in order to cut incoming sunlight and drop the global temperature unilaterally—if there were not swift global agreement on doing it by less noxious means' (Dyer 2010: 161–62).

The story form has a number of advantages for considering human trajectories. It provides descriptions that are not only more vivid but in many respects more realistic. Documents like the Stern Review (Stern 2007) and the 2010 World Development Report employ a too limited cast of characters: for example, in some analyses just 'developed', 'developing', and 'emerging' countries. They lack specifics, and, related to that, they miss some of the resulting dynamics. Narratives and scenarios bring us to the concrete particularities, the actual strange combinations, contiguities and coincidences that can and do occur. Second, because stories can better respect complexity they are better in giving understanding. A standard piece of advice in interpretive policy analysis is the Goldberg Rule: don't ask people what's the problem, ask them what's the story. In that way one can get deeper, including in identifying the real problems (Forester 1999). Or one can ask the sister questions, 'How has this issue come into your life?', or 'What did you do when that happened?' (Forester 2009). People respond not with speculation but with revealing narratives.

Third, stories are sometimes better for prediction. They show human reactions that are both intelligently calculated and emotionally driven; and they do not shy away from considering the interactions between environmental, economic, social and political impacts that are beyond our ability to formally model. The Stern Review when costing possible impacts in rich countries did not include the feedback effects from economic crisis in 'poorer countries who are more vulnerable to climate change . . . with increasing pressures for large-scale migration and political instability' (Stern 2007: 139). The Review recognized a whole series of such omissions (pp. 169–73) but had no methodology for dealing with them. In contrast, stories help us to think about the diverse potentials of complex interactions, and reveal risks, possibilities and opportunities that can otherwise be overlooked. Scenario exercises help people to perceive connections and possibilities that their mental frames, routines and authority structures may normally screen out (Gasper 2013a). In stories we can consider the highly improbable combinations that could occur and that, if they do occur, would change everything. While any particular such combination is highly improbable, the chance of occurrence of some such world-changing improbable combination is much higher. Yet social science has too little interest in such 'Black Swan' events, argues Nassim Taleb (2010), since it cannot model them. So, fourthly, stories may be better for promoting preparedness even where we cannot predict.

Fifthly, stories can have strengths in promoting ameliorative action, though they also have dangers. In interpretive policy analysis, telling one's story, showing one's reasons, has the potential to establish a party as a recognized actor in the eyes of the other parties, and also to provide information and mutual awareness that open up previously unseen possibilities in the mutual relationships (Forester 1999). Faced with clearly conflicting values espoused by different parties, there is little point in addressing such a conflict head-on. Instead it becomes essential to explore the worldview, history and humanity of each of the parties, to create sufficient mutual understanding and acceptance and find enough pragmatic handles to be able to move forward. The stories of one's interlocutors reveal that they are more multi-featured persons than in one's stereotypes. Thus 'when we face value- and identity-based disputes, we need to mine stories, not sharpen debates' (Forester 2009: 71). Telling personal stories is particularly important for global-scale issues (Schaffer and Smith 2004), given 'the power of personal narratives to displace stereotypes and expectations' (Forester 2009: 126).

Sixthly, stories and even scenarios can motivate us better. They feel more real and so have advantages in capturing attention, being remembered, and connecting to action. They engage our emotions, which reinforces all those advantages. They bring us closer to the lives and minds of other people, and show us the human implications of abstracted projections and generalized trends (Gasper 2013a). Through stories we are emotionally educated, and made both more knowledgeable and more sensitive in relation to others (Forester 1999: Ch. 2). Abstract talk cannot do most of this work.

Much of the climate literature warns however that doomsday scenarios can generate resignation or disbelief and rejection, and strengthen individualist responses, including the seeking of self-esteem through money, image and status. Stories need to move us beyond focusing only on problems: past, present or future. So exercises in scenarios planning typically seek to identify plausible desirable future paths too. Scenarios work shows how some of the benefits from inclusive story-making at micro-levels can be extended to much greater scales of operation (e.g., Raskin et al. 2002). More study is needed on which types of narrative and scenario may be helpful for which contexts and tasks.

Dyer's climate-wars scenarios lead him to conclude that a form of cosmopolitan egalitarianism would be the only arrangement that could ensure long-term global peace and survival. The only sustainable basis for a 'global deal' will be not a calculation of 'what is the most we need to concede' but a principle that conveys equality of esteem, a principle such as the one propounded by the Global Commons Institute: that 'everybody on the planet is entitled to the same basic personal allocation of greenhouse gas emission rights, and that those who exceed that allocation must compensate those who use less than their allocated amount' (Dyer 2010: 72).

Morally myopic thinking, such as in 'deals' based purely on self-interest, can tend to induce explanatory myopia too. A focus only on one's own interests can be associated with a shortage of attention to, understanding of, and flow of reliable information from other people. It may bring an underrecognition of connections that bind even the strong to the weak in a globalized world. Giddens (2009: 213ff.) notes thus the failure of the Bush-Cheney 'realist' foreign policy regime of brusque use of military and economic power to enforce its own interests. In Friedman's (2009) view too, this testosterone-driven *folie de grandeur* proved to be the opposite of realism: it reinforced rather than reduced American reliance on imported oil, and boosted often anti-American autocrats and dictators in oil- and gas-exporting states plus Islamism worldwide (Friedman's 1st Law of Petropolitics). Storytelling helps to make this web of connections clearer than regressions alone ever could.

CONCLUSION

A human security approach can help us in thinking about climate change, in various ways. It deepens consideration of connectivity and knock-on effects. It guides us to reflect on the situations of human actors who react to changing pressures and opportunities and are not simply punchbags for greater powers. Damage done to others will spread. Human security analysis may then help to widen the awareness by important elite groups of these global interconnections, and to counteract narrowly self-absorbed stances (Campbell et al. 2007; Moran 2011; Gasper 2012). Ideally, human security analysis can contribute also to increasing sympathetic attention to the difficulties of others, through its consideration of how 'distant others' live and feel and of the global interconnections. It may support the changes that are needed for global sustainability in respect of how people perceive shared vulnerabilities, shared interests, and shared humanity (The Earth Charter; Gasper 2009).

Karen O'Brien and Robin Leichenko's closing summary from their background paper for the 2007/8 Human Development Report provides a good short synthesis:

> climate change has for the most part been perceived, debated, and addressed as an environmental issue, rather than as a human security issue. A human security

approach to climate change can be considered a people-oriented approach that emphasizes both equity issues and the growing connectivities among people and places. It focuses on the management of threats to the environmental, social and human rights of individuals and communities, while at the same time enhancing the capacity to respond to both change and uncertainty. Responding to climate change from a human security perspective requires that both mitigation and adaptation strategies consider the equity and connectivity dimensions of human security. (O'Brien and Leichenko 2007: 32)

Notwithstanding the importance of connectedness and knock-on effects, the greatest value-added of human security analysis rests in its equity agenda: less in the warning of damage to the rich than in the insistence on focusing on the rights of the poor. Some authors suggest that the uncertainties about the precise forms and effects of climate change in particular places can make it very difficult to clearly link climate change and human security. Precision and a high degree of certainty are requirements in tort law, for proving that actions by agent A are responsible for damages D to victim V. But tort law is not the only relevant mechanism for responding to the global interconnections in which myriads of emitters endanger and harm myriads of vulnerable people dispersed around the world. We have enough precision and certainty in our knowledge of the general connections to justify establishing a system of global social insurance and protection (Penz 2010; Gasper 2012). Further, the uncertainties about exact causes of particular instances of damage—the periodically devastated livelihoods in delta and low-lying areas, for example—make human insecurity more than just an issue of calculable risk. It involves the question of ethical ordering of the responses to such uncertainties and the determination of who should bear or share the costs and the (in the general, informal sense) risks. Human security analysis helps us to look at who and which values are being secured, how much attention is given to basic aspects of the lives of vulnerable people, and whose interests are referred to when dealing with risks and uncertainties.

The foremost insecurity is that of vulnerable marginal populations in much of Africa, Asia and Latin America, as well as on islands and in low-lying areas elsewhere (as Hurricane Katrina illustrated). Possible knock-ons to wider destabilization of political and economic systems are more speculative. Whatever the uncertainties regarding the latter there is considerable certainty regarding the former insecurity. The IPCC's recent report on extreme weather shows vividly how the combination of increased average temperatures and increased weather variability will dramatically increase the frequency of very hot weather, as well as more modestly increase the frequency of hot weather, and also affect the frequency of cold and very cold weather (see diagrams SPM.3; IPCC 2012: 5). We know how 'higher temperatures may mean that previously unsuitable areas will become suitable for transmission of vector borne diseases such as malaria' (O'Brien and Leichenko 2007: 10). And the UN advisory group on the ethics of science and technology underlines how 'even a 2 degree rise in average temperatures [a figure that will likely be greatly exceeded during this century] will have catastrophic effects for populations living on small islands, large river deltas, or other low-lying areas' (COMEST 2010: 36).

Barnett et al. when reporting at the end of the Global Environmental Change and Human Security (GECHS) research programme concentrate at first on the possible link from environmental change to armed conflict, although this is not the most important impact in human security terms (Barnett et al. 2010: 10–15); indeed the only specific issue to which they give its own section in the book is possible armed conflict. The significance of the nexus between environmental change, including climate change in particular, and violent conflict, lies however not in the presence of a universal high linkage, which does not exist, but in the possibility of conflagration in a few of the numerous potential tinderboxes. And even more important, as cited by Nordas and Gleditsch in their chapter in this book: 'Gleick (1998) also argues that, "for the most part, inequities [in regard to water access] will lead to poverty, shortened lives, and misery, but perhaps not to direct conflict" (p. 113)'. Poverty, shortened lives, and misery; the significance of climate change for human insecurity, as opposed to state insecurity, is evident.

The equity and connectivity dimensions in human security thinking are both vital and are interlinked, not accidental partners. Sensitivity to border-crossing connections comes partly through sensitivity to the lives that are affected. Human security thinking combines a normative ontology of the value of human persons, as in human rights work, and an explanatory ontology of interconnectedness (Gasper 2012, 2013a). Much rights thinking has been emphatically individualist and stressed people's dignity alone. But the human rights commitment needs to be embedded in a richer and more realistic understanding of the human species, its vulnerabilities and capabilities, including both its dependence on its habitat and its power now to dangerously destabilize it. A human security framework helps to sustain such an understanding.

REFERENCES

Ahmed, F. (2007), Statement by His Excellency Dr. Fakhruddin Ahmed, Honorable Chief Adviser of the Government of the People's Republic of Bangladesh at the High-level Event on Climate Change, New York, 24 September. Cited in Saferworld 2008, p. 23.

Barnett, Jon, and W. Neil Adger (2007), 'Climate change, human security and violent conflict', *Political Geography*, **26**, 639–655.

Barnett, Jon, Richard A. Matthew, and Karen L. O'Brien (2010), 'Global environmental change and human security: an introduction', in Richard A. Matthew, Jon Barnett, Bryan McDonald and Karen L. O'Brien (eds), *Global Environmental Change and Human Security*, Cambridge, USA: MIT Press, pp. 3–32.

Brauch, Hans Günter (2005), *Environment and Human Security: Towards Freedom from Hazard Impacts*, Bonn, Germany: United Nations University Institute for Environment and Human Security.

Burgess, J. Peter, et al. (2007), *Promoting Human Security: Ethical, Normative and Educational Frameworks in Western Europe*, Paris, France: UNESCO.

Burton, J.W. (1990), *Conflict: Basic Human Needs*, New York, USA: St. Martin's Press.

Campbell, Kurt, et al. (2007), *The Age of Consequences: The Foreign Policy and National Security Implications of Global Climate Change*, Washington, DC, USA: Center for Strategic and International Studies/Center for a New American Security.

CHS (Commission on Human Security) (2003), *Human Security Now*, New York, USA: UN Secretary-General's Commission on Human Security.

CNA (Center for Naval Analysis) (2007), *National Security and the Threat of Climate Change*, Alexandria, USA: CNA Corporation.

COMEST (2010), *The Ethical Implications of Global Climate Change*, World Commission on the Ethics of Scientific Knowledge and Technology (COMEST), Paris, France: UNESCO.

Dalby, Simon (2009), *Security and Environmental Change*, Cambridge, UK: Polity.

Dreze, Jean, and Amartya Sen (1989), *Hunger and Public Action*, Oxford, UK: Clarendon.

Dreze, Jean, and Amartya Sen (eds) (1991), *The Political Economy of Hunger* (3 vols.), Oxford, UK: Clarendon.

Dreze, Jean, and Amartya Sen (2002), *India: Development and Participation*, Oxford, UK: Oxford University Press.

Dyer, Gwynne (2010), *Climate Wars*, Oxford, UK: Oneworld Publications.

Earth Charter, The Earth Charter. Available at: http://www.earthcharterinaction.org/content/pages/read-the-charter.html

Edwards, Alice, and Carla Ferstman (eds) (2011), *Human Security and Non-Citizens: Law, Policy and International Affairs*, Cambridge, UK: Cambridge University Press.

Forester, John (1999), *The Deliberative Practitioner*, Cambridge, USA: MIT Press.

Forester, John (2009), *Dealing with Differences*, New York, USA: Oxford University Press.

Friedman, Thomas (2009), *Hot, Flat and Crowded – Release 2.0*, New York, USA: Picador.

Fuerth, Leon (2007), 'Security implications of Climate Scenario 2', in Campbell et al., pp. 71–79.

Gasper, Des (2005), 'Securing humanity–Situating "human security" as concept and discourse', *J. of Human Development*, **6** (2), 221–245.

Gasper, Des (2008), 'From "Hume's Law" to policy analysis for human development – Sen after Dewey, Myrdal, Streeten, Stretton and Haq', *Review of Political Economy*, **20** (2), 233–256.

Gasper, Des (2009), 'Global ethics and human security', in G. Honor Fagan and Ronaldo Munck (eds), *Globalization and Security: An Encyclopedia*, Vol. 1, Westport, USA: Greenwood, pp. 155–171.

Gasper, Des (2010), 'The idea of human security', in K. O'Brien, A.L. St. Clair and B. Kristoffersen (eds), *Climate Change, Ethics and Human Security*, Cambridge, UK: Cambridge University Press, pp. 23–46.

Gasper, Des (2011), 'Denis Goulet', in D. Chatterjee (ed.), *Encyclopedia of Global Justice*, Volume 1, Dordrecht, Netherlands: Springer, pp. 457–459.

Gasper, Des (2012), 'Climate change – the need for a human rights agenda within a framework of shared human security', *Social Research: An International Quarterly of the Social Sciences*, **79** (4), 983–1014.

Gasper, Des (2013a), 'Climate change and the language of human security', *Ethics, Policy and Environment*, **16** (1), 56–78.

Gasper, Des (2013b), 'From definitions to investigating a discourse', in M. Martin and T. Owen (eds), *The Routledge Handbook of Human Security*, Abingdon, UK: Routledge, 28–42.

Gasper, Des, Ana Victoria Portocarrero, and Asuncion St. Clair (2013a), 'The framing of climate change and development: a comparative analysis of the Human Development Report 2007/8 and the World Development Report 2010', *Global Environmental Change*, **23**, 28–39.

Gasper, Des, Ana Victoria Portocarrero, and Asuncion St. Clair (2013b), 'An analysis of the Human Development Report 2011 "Sustainability and Equity: A Better Future for All"', *S. African J. on Human Rights*, **29** (1), 91–124.

Giddens, Anthony (2009), *The Politics of Climate Change*, Cambridge, UK: Polity.

Gomez, Oscar A., Des Gasper, and Yoichi Mine (2013), *Good Practices in Addressing Human Security through National Human Development Reports*, Report to Human Development Report Office, New York, USA: UNDP.

Goulet, Denis (1971), *The Cruel Choice*, New York, USA: Atheneum.

Goulet, Denis (1975), 'The high price of social change – on Peter Berger's "Pyramids of Sacrifice"', *Christianity and Crisis*, **35** (16), 231–237.

Gulledge, Jay (2007), 'Three plausible scenarios of future climate change', in Campbell et al., pp. 35–53.

Harris, Paul (2010), *World Ethics and Climate Change*, Edinburgh, UK: Edinburgh University Press.

IIHR, n.d. 'What is human security?' San Jose, Costa Rica: Inter-American Institute of Human Rights. Available at: http://www.iidh.ed.cr/multic/default_12.aspx?contenido id=ea75e2b1-9265-4296-9d8c-3391de83fb42&Portal=IIDHSeguridadEN.

IPCC (2007), *Fourth Assessment Report*, Intergovernmental Panel on Climate Change. Cambridge, UK: Cambridge University Press.

IPCC (2012), *Managing The Risks Of Extreme Events And Disasters To Advance Climate Change Adaptation: Summary For Policymakers*, Cambridge, UK: Cambridge University Press.

Lama, Mahendra (2010), *Human Security in India*, Dhaka: The University Press Ltd.

Leaning, Jennifer (2008), 'Human security', in M. Green (ed.), *Risking Human Security: Attachment and Public Life*, London: Karnac Books pp. 25–50.

Leaning J., S. Arie and E. Stites (2004), 'Human security in crisis and transition', *Praxis: The Fletcher Journal of International Development*, **19**, 5–30.

Leichenko, Robin, and Karen O'Brien (2008), *Environmental Change and Globalization: Double Exposures*, New York, USA: Oxford University Press.

MacFarlane, Neil and Yuen Foong Khong (2006), *Human Security and the UN – A Critical History*, Bloomington, USA: University of Indiana Press.

McNeill, J.R. (2007), 'Can history help us with global warming?', in Campbell et al., pp. 23–33.

Moran, Daniel (ed.) (2011), *Climate Change and National Security: A Country-level Analysis*, Washington, DC, USA: Georgetown University Press.

O'Brien, Karin, and Robin Leichenko (2007), *Human Security, Vulnerability and Sustainable Adaptation*, Occasional Paper 2007/9, New York, USA: Human Development Report Office, UNDP.

Penz, Peter (2010), 'International ethical responsibilities to "climate change refugees"', in J. McAdam (ed.), *Climate Change and Displacement – Multidisciplinary Perspectives*, Oxford, UK: Hart Publishing, pp. 151–174.

Podesta, J., and P. Ogden (2007), 'Security implications of Climate Scenario 1', in Campbell et al., pp. 55–69.

Raskin, Paul, Banuri, T., Gallopín, G., Gutman, P., Hammond, A., Kates, R., and Swart, R. (2002), *Great Transition – The Promise and Lure of the Times Ahead*, Boston, USA: Stockholm Environment Institute. http://www.gtinitiative.org/documents/Great_Transitions.pdf

Roe, P. (2010), 'Societal security', in A. Collins (ed.), *Contemporary Security Studies*, Oxford, UK: Oxford University Press, pp. 202–217.

Rothschild, Emma (1995), 'What is security?', *Daedalus*, **124** (3), 53–98.
Saferworld (2008), *Human Security in Bangladesh*, London, UK: Saferworld.
Schaffer, Kay, and Sidonie Smith (2004), *Human Rights and Narrated Lives: The Ethics of Recognition*, New York, USA: Palgrave.
Sen, Amartya (1981), *Poverty and Famines*, Oxford, UK: Clarendon.
Stern, Nicholas (2007), *The Economics of Climate Change – The Stern Review*, Cambridge, UK: Cambridge University Press.
Taleb, Nassim (2010), *The Black Swan – The Impact of the Highly Improbable*, revised edition, London, UK: Penguin.
Thomashow, Mitchell (2002), *Bringing the Biosphere Home*, Cambridge, USA: MIT Press.
Truong, Thanh-Dam, and Des Gasper (eds) (2011), *Transnational Migration and Human Security*, Heidelberg, Germany: Springer.
UNDP (1998), *Chile National Human Development Report: Paradoxes of Modernity – Human Security*.
UNDP (2003), *Latvia Human Development Report: 2002–2003: Human Security*. Riga, Latvia: United Nations Development Program.
UNDP (2005), *Overcoming Fear: Citizen (In)security and Human Development in Costa Rica*. San Jose, Costa Rica: United Nations Development Program.
UNDP (2007), *2007/2008 Human Development Report: Fighting Climate Change: Human Solidarity in a Divided World*, New York, USA: United Nations Development Program.
UNDP (2009), *Abrir espacios para la seguridad ciudadana y el desarrollo humano*, Central American Human Development Report 2009/10.
UNDP (2011), *Human Development Report 2011 – Sustainability and Equity: A Better Future for All*, New York, USA: United Nations Development Program.
UNDP (2012) *Human Development and the Shift to Better Citizen Security*, Caribbean Human Development Report 2012, New York, USA: United Nations Development Program.
Urry, John (2011), *Climate Change and Society*, Cambridge, UK: Polity.
Woolsey, R.J. (2007), 'Security implications of Climate Scenario 3', in Campbell et al., pp. 81–91.
World Bank (2010), *World Development Report 2010: Development and Climate Change*, Washington DC, USA: World Bank.
World Bank (2011), *World Development Report 2011: Conflict, Security and Development*, Washington DC, USA: World Bank.
Ziervogel, G., Taylor, A., Thomalla, F., Takama, T., and C. Quinn (2006), *Adapting to Climate, Water and Health Stresses: Insights from Sekhukhune, South Africa*, Stockholm: Stockholm Environment Institute.

3. The IPCC, human security, and the climate-conflict nexus
Ragnhild Nordås and Nils Petter Gleditsch[1]

THE SECURITIZATION OF THE CLIMATE CHANGE DEBATE

Climate change has increasingly been framed as a major security concern in media and public discourse over the last decade. The security scenario gathered particularly strong momentum in 2007 with the UN Security Council's debate on the security implications of climate change and the awarding of the Nobel Peace Prize to the Intergovernmental Panel on Climate Change (IPCC) and Al Gore.

In his speech on the occasion of the award of the Nobel Peace Prize for 2007 to Al Gore and the Inter-governmental Panel on Climate Change (IPCC), the then chair of the Norwegian Nobel Committee adopted a polemical stand against 'those who doubt that there is any connection between the environment and the climate on the one hand and war and conflict on the other', and told the audience that global warming not only has negative consequences for 'human security' in a wide sense, but that it 'can also fuel violence and conflict within and between states'. He provided two examples: First, the 'melt-down in the Arctic is giving a sharper edge to the new series of sovereignty claims' in the North. And, secondly, 'in Darfur and in large sectors of the Sahel belt . . . we have already had the first "climate war"', with 'nomads and peasants, Arabs and Africans, Christians and Muslims' clashing repeatedly as a result of desertification' (Mjøs, 2007). Upon accepting the prize on behalf of the IPCC, its Chair acknowledged 'the threats to stability and human security inherent in the impacts of a changing climate' and cited the threats from 'dramatic population migration, conflict, and war over water and other resources as well as a realignment of power among nations', in addition to the possibility of 'rising tensions between rich and poor nations, health problems caused particularly by water shortages, and crop failures as well as concerns over nuclear proliferation'. Finally, he argued that non-climate stresses, such as conflict, can increase the vulnerability to climate change and reduce the capacity to adapt to it (Pachauri, 2007).[2,3]

The climate-conflict link has been promoted by pundits, journalists, and media commentators, but also repeated by influential world leaders such as UN Secretary-General Ban-Ki Moon. When President Obama first presented his energy and environment policies on 26 January 2009, he stated that 'urgent dangers to our national and economic security are compounded by the long-term threat of climate change, which, if left unchecked, could result in violent conflict . . .'[4]

On the other hand, systematic studies of the climate-conflict relationship have so far failed to uncover any robust relationship (for reviews of the literature, see Bernauer et al., 2012; Gleditsch, 2012; Gleditsch et al., 2007; Nordås and Gleditsch, 2007a, b; Scheffran et al., 2012). Why this discrepancy? Normally, one would turn to the assessment reports of the IPCC for authoritative guidance. For reasons that will become clear, that does not settle the issue. In this chapter, we review how the IPCC reports from 2001 and 2007 have dealt with the climate-conflict nexus,[5] what this has done to the framing of the climate-conflict nexus, and evaluate the basis for their conclusions.

The next section introduces the IPCC and the principles guiding their process of producing climate change reports. We then outline the methodology used in this chapter, and our main finding regarding the discrepancy between the knowledge base in peer-reviewed literature and the sources and evidence used in the 2001 and 2007 IPCC reports. We then outline why establishing empirically well-founded baseline knowledge about the potential for climate change-induced conflict is critical. The potential for 'conventional wisdom' based on the diffusion of hearsay and unfounded claims is clearly evident – and potentially detrimental in several ways. We therefore suggest avenues for research to fill some of the knowledge gaps, and conclude that the signal that the IPCC is now going to engage more of the conflict and peace research communities in particular, is a promising step towards a more knowledge-based IPCC evaluation of the climate change-conflict scenario.

THE IPCC AS THE AGENDA SETTER

The IPCC is the main agenda setter for the climate change debate. The panel was created because policymakers need an objective source of information not only about climate change itself, but about the socio-economic consequences and options for adaptation and mitigation. The IPCC describes its own reports as: (1) up-to-date descriptions of the knowns and unknowns of the climate system and related factors, (2) based on the knowledge of the international expert communities, (3) produced by an open and peer-reviewed professional process, and (4) based upon scientific publications whose findings are summarized in terms useful to decision makers (IPCC, 2001 [Technical summary in Synthesis report]: 22). It is added, however, that while the assessed information is policy relevant, the panel does not establish or advocate public policy.

Review is an essential part of the IPCC process according to its main guiding principles.[6] As an intergovernmental body, the IPCC submits its documents to a review by experts but also by governments.[7] In this and other ways, the review process in the IPCC differs from the peer review practiced by academic journals. However, the IPCC's reference to sources cited by the reports as peer-reviewed seems to conform to standard definitions in academic circles.

The IPCC review process generally takes place in three stages: (1) expert review, (2) government/expert review, (3) government review of the *Summaries for Policymakers, Overview Chapters* and/or the *Synthesis Report*. The first part in the process is the selection of lead authors for sections and chapters.[8] These authors prepare a first draft of a report which is sent out for expert review. Although the IPCC aims at producing reports based first and foremost on peer-reviewed sources,[9] a wide variety of different source materials can be included. A guiding principle is that 'contributions should be supported as far as possible with references from the peer reviewed and internationally available literature'. The pool of reviewers is selected by the relevant Working Group/Task Force Bureau and, in addition, experts from lists provided by governments and participating organizations. The review circulation should include experts who have significant expertise or publications in particular areas covered by the report they are reviewing.

Three principles govern the review process.[10] First, that 'the best possible scientific and technical advice should be included so that the IPCC Reports represent the latest scientific, technical and socio-economic findings and are as comprehensive as possible'. Second, the peer-review process in itself aims at 'a wide circulation process, ensuring representation of independent experts (i.e. experts not involved in the preparation of that particular chapter)', from as many countries as possible, including developing countries. Third, the review process should be 'objective, open and transparent'. IPCC claims to ensure this in part is by use of information found in sources that have *not* been published or peer-reviewed (e.g., industry journals, internal organizational publications, non-peer reviewed reports or working papers of research institutions, proceedings of workshops etc.).[11] On the other hand, however, the openness of the review process (in the interpretation above) is in contrast to the principle of double-blind review, which is frequently practiced in the social sciences, less so in the natural sciences (Gleditsch, 2002).

The IPCC prides itself on high academic standards in its evaluation of scientific evidence related to climate change, and also on the inclusiveness of its data collection. However, the climate change-conflict nexus is neither well-developed nor well-documented. This stands in remarkable contrast to the scientific knowledge of the physical processes related to climate change, a knowledge that the IPCC has been instrumental in collecting, assessing, and disseminating. Indeed, contrary to the evidence that constitutes the basis for the evaluation of the physical processes related to climate change, the evidence and sources used to buttress the claims made about a climate-conflict link, are of a decidedly more uncertain quality. A wide-ranging examination of the IPCC by the InterAcademy Council (IAC, 2010: 18) acknowledged that some governments, particularly in developing countries, had not always nominated the best experts, that the author selection process suffered from a lack of transparency, and that the regional chapters did not always make use of experts from outside the region. It also cited a study that found that while 84 percent of the sources for IPCC's Working Group 1 on the physical science basis derived from peer-reviewed sources, it was only 59 percent for Working Group 2 on the vulnerability of socio-economic and natural systems to climate change (IAC, 2010: 16).

METHODOLOGY

In order to evaluate the extent to which the IPCC has made a connection between climate change and (violent/armed) conflict, and to assess the basis for their claims, we have conducted, as described below, an extensive search in the two last IPCC reports (from 2001 and 2007). Previously, in an analysis of 'historical documents' on climate change (including the IPCC reports up to 2007), Detraz and Betsill (2009) distinguished between three general discourses linking environment and security, environmental conflict (focusing on militarized conflict), environmental security (including a broader set of human consequences of environmental degradation), and ecological security (examining how human activities threaten the environment). They found that as yet there was no shift in the direction of the narrower concept, but cautioned against such a shift in the future. They found only a dozen references to 'conflict' in these documents. However, it seems that in the IPCC reports they searched only the summaries for policymakers. As we shall show below, this gives a somewhat simplified picture of these reports.

To investigate the claims made about climate-conflict links, we have searched for the following terms in all substantive parts of the IPCC reports:

'armed'
'conflict'
'violen'[to catch 'violence' as well as 'violent']
'war'

We searched for any occurrence of any of these terms, e.g. so that 'conflict' would also give a hit for 'conflicts', 'conflict-ridden', 'conflicting', 'war' for 'wars', 'warlike' etc. More limited searches using a wider list of search terms (including 'riot', 'uprising', 'insurrection', 'revolution', 'genocide', 'massacre') did not yield additional hits, and from reading the various sections of the reports, we have found no indication that a wider search would have yielded additional material. In our search we have not reported hits that are simply 'noise'.[12]

Occurrences of the search terms were classified into *irrelevant* hits, *low relevance* hits (where conflict refers to conflict of interest without violence; or where the reference is to consequences of conflict), *secondary relevance* (where climate change is believed to increase the duration or severity of an on-going conflict)[13], *high relevance* (where it is argued that violent conflict has been caused by resource scarcity or other environmental problems in the past or can be caused by similar problems in the future and the type of resource scarcity is one that may be exacerbated by climate change), and *very high relevance* (where it is argued directly that climate change will cause violent conflict).

In our assessment of how the climate-conflict nexus is portrayed in the 2001 and 2007 IPCC reports, we focus here on assessing the quality of the sources in cases of high relevance or very high relevance. This is done by assessing the quality of the research by studying the evidence, as well as whether the research is peer-reviewed. In addition, we trace sources cited. Sometimes, the chain of cites continues reproducing conventional wisdom or hearsay or ends up in a circular pattern. We have generally limited our search to the first two 'generations' of sources, except in cases where we want to point to circular referencing and reproduction of unsubstantiated claims.

THE THIRD ASSESSMENT REPORT (TAR)

The word 'conflict' appears in its various forms more than one hundred times in IPCC's TAR (2001), whereas 'war' appears seven times (excluding hits in the article, paper, or book titles in literature references). 'Armed' occurs one time, in a secondary relevance context. 'Violence' did not occur anywhere. Many of the 'conflict' occurrences are clearly irrelevant or of low or secondary relevance to the climate-conflict nexus. We focus our attention on references that are of relevance (high and very high) to the climate-conflict nexus.

The TAR is structured into three parts. We focus here on part 2 on *Impacts, Adaptation and Vulnerability* and part 3 *on Mitigation*. In the second part, the terms 'conflict' or 'war' occur several times. Some instances of 'conflict' refer to 'conflict of interest' or 'conflicting evidence' rather than violent conflict. Some statements take the potential for armed conflict as a result of climate change for granted in discussing related issues, and again other comments are made without any empirical backing or cited references. Of the relevant references, however, the most prominent put forward state how climate change can imply (a) a potential for 'water wars', (b) that climate change-induced migration can generate conflicts (due to e.g. sea-level rise and environmental stress), and (c) that we could be seeing resource wars in the wake of climatic shifts. We discuss each in turn next.

Water Wars

The IPCC (2001: 84) states that 'negative trends in water availability have the potential to induce conflict between different users', citing a report by Kennedy et al. (1998). This report notes that 'water wars' have been 'relatively rare' (p. 31), provides no examples of such wars, and presents no evidence for a greater likelihood of water wars in the future (Kennedy et al., 1998: 30–31).

A similar formulation occurs later ('a change in water availability has the potential to induce conflict between different users', IPCC, 2001: 225). Here, the IPCC cites Biswas (1994) and Dellapenna (1999), which are more relevant than the Kennedy et al. report. In Biswas (1994), a chapter by Aaron Wolf notes an ongoing or impending conflict (of interest) over water quantity or quality between co-riparians on the Euphrates, the Tigris, and the Nile (Wolf, 1994: 5). These may escalate to the level of threat, but on most accounts do not involve direct violent attacks. Wolf also mentions that threats related to water have been made by Israeli officials in the 1960s and 1970s (Wolf, 1994: 5; Gleditsch and Nordås, 2009). However, the threats of violence never materialized. Wolf also notes that 'on the Jordan, the resulting tensions helped lead to a cycle of conflict, which, exacerbated by other disputes, ended in war in 1967' (p. 24). However, the calculus that water cannot be 'won' with violence seems to have prevailed. Wolf also emphasizes water as a potential vehicle for *cooperation* in the Middle East, as water issues have brought states together for peace talks and negotiations (Wolf, 1994: 37). In other words, negotiations over water resources may encourage dialogue on other more contentious issues.

Some of the claims made by Wolf (1994) were largely refuted by himself a few years later in a number of new publications. For instance, Wolf (1998) clearly states that 'no war has ever been fought over water' (p. 251), and that there were only seven crises related to water in the twentieth century, and in three of them no shots were fired. Wolf (1998) points to weaknesses in case studies of interstate water conflicts which lead to an overrating of the link between water and war and often a 'complete lack of evidence' (Wolf, 1998: 254) that water was a causal factor behind violence. This source is not cited in the IPCC's (2001) chapter on water.

The second source, Dellapenna (1999), states that 'considerable evidence suggests that *cooperative* solutions to water scarcity problems are more likely than prolonged conflict' (emphasis added) (p. 1312). He remarks that 'the widespread and frequent disputes over water have generally exhibited a rather remarkable feature: They do not lead to war' (ibid.). Because water is such a vital resource, it is too risky to fight over, which serves to deter the parties from violence. Dellapenna (1999: 1312) points to the relationship of Pakistan and India. Despite their many conflicts since 1948, India and Pakistan have negotiated and implemented a complex treaty on sharing the waters of the Indus River system and 'during periods of hostility they did not target water facilities nor interfere in the cooperative water management arrangements'.

In terms of water resources, IPCC (2001: 225) also states that 'where there are disputes, the threat of climate change is likely to exacerbate, rather than ameliorate, matters because of uncertainty about the amount of future resources that it engenders', citing Gleick (1998). Gleick lists four conditions that could influence the likelihood that water will be the object of military action: water scarcity, the extent to which the supply is shared between groups, the relative power of those groups, and the ease of access to other sources of water. In the case of the 1967 war in the Middle East, and associated border disputes, he defines water as 'an important factor' (p. 109). The second case in point is a conflict over water allocation within India in 1992 after a court decided to allocate more water originating in Karnataka to downstream Tamil Nadu. Over 50 people were killed in the ensuing riots (Gleick, 1993). The remaining examples do not involve casualties.

Gleick (1998) also argues that, 'for the most part, inequities will lead to poverty, shortened lives, and misery, but perhaps *not* to direct conflict' (p. 113) (emphasis added). But in some cases, he adds, they will exacerbate existing disputes (local, regional, or international), create refugees, and decrease the ability of nations and societies to resist military aggression.

This statement, which is the main point repeated by the IPCC, does not come with any further evidence, clear case examples, or an assessment of the likelihood that disputes will escalate over such issues.

TAR's chapter on Africa (Ch. 10) states that 'growing water scarcity, increasing population, degradation of shared freshwater ecosystems, and competing demands for shrinking natural resources [. . .] have the potential for creating bilateral and multilateral conflicts' (IPCC, 2001: 495). Thus, a failure to put into place robust legal frameworks to ensure equity in access to and accountability for water supply and water quality management between states sharing rivers, 'could lead to water resources-related conflict' (IPCC, 2001: 499–500). No sources are cited for this claim. The same is the case for a statement about areas with a dry climate in arid regions of Africa where, with climate change, 'the constraint will be in finding limits to water extraction that do not adversely impact communities downstream and result in conflicts' (p. 521). However, the nature of these conflicts is not specified, nor does TAR discuss adaptation mechanisms.

IPCC also suggests, based on Caponera (1996), that regional coop-eration protocols in shared river basins minimize adverse impacts and potential for conflicts. International river and lake basins 'by their very nature . . . carry the potential for conflicts' according to Caponera (1996: 105), so IPCC (2001: 516) suggests that legal regulation of water sharing between nations of Africa region must be developed sooner rather than later to avoid situations of 'water-related political tensions like those in the Middle East and north Africa'.

IPCC (2001: 566) also expresses concern that 'radical changes in water management strategies and substantial investments will be required in Asia' in order to 'secure sustainable development and avoid potential intersectoral and international water conflicts' (ibid.). TAR does not clarify whether there is a risk of *violent* conflict, and no sources are provided.

Ironically, the IPCC ignores academic empirical work that suggests that shared water resources in international rivers may be associated with an increased risk of conflict (Toset et al., 2000).[14]

Resource Wars

According to the IPCC (2001: 580), climate change could lead either to cooperation or to conflict over the world's major resources. This claim is not backed by any references, and there is no evaluation of the relative likelihoods or predicted prevalence associated with either trajectory. TAR states that 'competing demands for shrinking natural resources . . . have the potential for creating bilateral and multilateral conflicts' (p. 498), referring to a report by Gleick (1992), which is focused mainly on water and conflict.

The IPCC lists examples of possible large-scale singularities (events) that are triggered by climate causes, one being 'destabilization of international order by environmental refugees and emergence of conflicts as a result of multiple climate change impacts' (IPCC, 2001: 950). The causal process is described stemming from a development where 'climate change – alone or in combination with other environmental pressures – may exacerbate resource scarcities in developing countries' (ibid.). This could have severe social effects, and 'may cause several types of conflict, including scarcity disputes between countries, clashes between ethnic groups, and civil strife and insurgency, each with potentially serious repercussions for the security interests of the developed world' (ibid.). This hypothesis, that violent conflict should be a likely outcome of climate change via resource scarcity, is based on five sources: Homer-Dixon (1991), Myers (1993), Schellnhuber and Sprinz (1995), Biermann et al. (1998), and Homer-Dixon and Blitt (1998). We discuss each in turn.

Homer-Dixon (1991), an academic article in a leading international peer-reviewed journal, defines (acute) conflict as 'involving a substantial probability of violence' (Homer-Dixon, 1991: 77). The article does not provide an empirical assessment of the likelihood or prevalence of conflict in the wake of environmental (or climatic) change. It suggests possible linkages, but primarily sets out a research agenda for future research on environmental change and acute conflict. Homer-Dixon and Blitt (1998), however, goes somewhat further empirically. Specifically, the role of environmental scarcity is explored in five conflict contexts – Chiapas, Gaza, South Africa, Pakistan, and Rwanda. However, no clear link between resource scarcity and war is established. Indeed, the book opens with a statement that research shows that 'the answer lies somewhere in between environment being very important and unimportant'. Furthermore, this study states that global warming and ozone depletion are 'unlikely to be immediate causes of violence' (Homer-Dixon and Blitt, 1998: 2). Several studies have criticized the studies by Homer-Dixon (1991, 1994; Homer-Dixon and Blitt, 1998) and questioned the reliability of the conclusions, particularly due to problems of selection of cases on the dependent variable (e.g. Levy, 1995; Gleditsch, 1998). Such critical studies were not cited by the IPCC in 2001.

Myers (1993: 199f) discusses resource scarcity particularly in the form of water scarcity and drops in food production, and links it to potential conflict through mass migration. Environmental refugees will flee famine due to climate-induced reductions in grain production, which will cause conflict when migrants try to resettle in a new area (pp. 200f). Myers

(1993) does not mention war as a likely outcome, but points out how some countries have experienced social conflicts due to strain on the economic and institutional capacity of some states that deal with mass influxes of migrants.

Schellnhuber and Sprinz (1994) is a general survey of environmental security, which does not report any new research. Biermann et al. (1998), however, reports a relationship from shared rivers to conflict among riparian states and also find that states with acute water scarcity tend to have more internal conflict. However, their analysis is problematic, *inter alia* because it does not include any control variables.

A critical reading of the sources that lead the IPCC to conclude that violence might be the outcome of climate change-induced resource scarcity and forced environmental migration, suggests that the climate-conflict link is somewhat overstated. Moreover, the substantial uncertainty of the conclusions was not sufficiently acknowledged by the IPCC.

Climate change is frequently argued to be costly (e.g. *Stern Review*, 2006), and subsequent competition over resources is argued to be a key risk factor. In the case of international fisheries, for instance, the TAR refers to McKelvey (1997), who reports that cooperative harvesting agreements have often 'degenerated into mutually destructive fish wars' (IPCC, 2001: 760). However, no lethal violence has been used in such 'fish wars'. Commenting on associating costs with climate changes, IPCC (2001: 125f) highlights that particularly in terms of social effects, measures often do not adequately recognize nonmarket costs. One such cost is assumed to be conflict over scarce resources. IPCC refers to Schneider et al. (2000) which assumes *a priori* or indirectly that climate change has the 'cost' of conflicts over environmentally dependent resources (among other things); however, no substantiation for this assumed relationship is provided. Stephen Schneider was part of the core writing group of the IPCC and co-author of 'Uncertainties in the IPCC Third Assessment Report: Recommendations to Lead Authors for More Consistent Assessment and Reporting' in 2000, as well as founder and editor of the interdisciplinary journal *Climatic Change*. Hence, if he took a climate-conflict link for granted, this could possibly be a common assumption within the IPCC.

Climate Migrants and War

Another link between climate and conflict made by the IPCC is related to refugees. The IPCC states in the overview chapter (Ch. 1) that 'migration of populations affected by extreme events or average changes in the distribution of resources might increase the risks of political instabilities and conflicts' (IPCC, 2001: 85). This statement is based on three cited sources: Myers (1993), Kennedy et al. (1998), and Rahman (1999). None of these sources present any convincing evidence for a link between climate change and armed conflict, and the probability and relative weight of the migration factor is not reflected in the TAR.

Myers (1993) devotes considerable space to a discussion of 'environmental refugees'. His estimates of the scope of this (potential) problem range from 100 million environmental refugees due to global warming effects (p. 10), to 150–200 million (p. 191). The book is devoted to assessing the potential scope of environmental refugees due to climate change, rather than assessing the implications for conflict. He refers in passing to 'economic and social dislocations' and 'cultural and ethnic problems' (p. 201) and argues that this 'would prove a deeply destabilizing factor in international relations', that could 'undermine security' (Myers, 1993: 202f). He provides many examples of places where people might have to move as a result of sea-level rise, but very few examples of where environmental refugees have created instability in the past. The work by Kennedy et al. (1998) is quite substantial, but cautiously formulated, reflecting the tentative knowledge of the social consequences of environmental change. Rahman (1999) is a chapter titled 'Climate change and violent conflict' in an edited volume. However, the chapter is neither about climate change nor about violent conflict.

Finally, the IPCC (2001: 719) discusses climate change and health in Ch. 14 on Latin America, citing Patz (1998). Patz' statement is that 'Environmental refugees could present the most serious health consequences of climate change. Risks that stem from overcrowding include virtually absent sanitation; scarcity of shelter, food, and safe water; and heightened tensions – potentially leading to social conflicts'. The original wording in Patz (1998: 51) even refers to forced environmental migration 'leading potentially to war'. However, this is merely an aside in the article, citing Myers and Kent (1995). The rest of the article does not mention conflict. Myers and Kent (1995) elaborates in greater detail on the potential for environmental migration, but touches the possibility of conflict only in passing, citing Ayres and Walter (1991) for the point that refugee camps and shantytowns could become breeding grounds for civil disorder and 'even violence of various sorts'; and Cline (1992), a book on the economics of climate change, for the view that 'peoples have fought wars to avoid being forced to leave their homelands' (p. 119), without examples or citations. Yet, the authors clarify that among the indirect effects on conflict, 'the "greenhouse effect" contribution is likely to be relatively minor compared to other factors' (p. 119).

Mitigation: Futurist Scenarios

The third main part of the TAR, the *Mitigation* report, includes some relevant hits on the word 'conflict' based on futures research. The focus on this research is perhaps rather surprising, given the IPCC's focus on evidence-based reporting. Specifically, the futures research focuses on scenarios in terms of possible links from greenhouse gas emissions to violent conflict. In Ch. 2 on Mitigation, futurist scenarios are reported by IPCC to range from peace to 'many wars' and even 'world war' (IPCC 2001, Mitigation, Ch. 2: 138). The most prevalent scenario, however (based according to IPCC on 76 scenarios), is that conflict will be rising. Out of the 76 scenarios, 36 propose a rise, 26 a decline, whereas 14 scenarios envision no change compared to the current situation as a result of rising emissions.

The futures research cited is based on thought experiments about scenarios for the future, with little if any documentation about methods used to get to conclusions, and no empirical backing of the claims but rather discussions of theoretical possibilities. Many of the views are highly normative rather than predictive, and contrast sharply with IPCC's self-proclaimed attention to academic excellence, openness, and accountability.

One possible exception, Ramphal (1992), refers to international conflict over water, but he only repeats well-known points from the water wars literature. He mentions how distribution of water in the Euphrates and Tigris river basins has often caused friction among Syria, Iraq, and Turkey, and that leaders in the region have warned about water wars.[15] He also expresses fear (without providing a clear justification) that unresolved water issues will likely pose danger to global security although water disputes tend to be local or regional. Two other sources cited by the IPCC for futurist views are McRae (1994) and Mercer (1998), but 'climate change' does not occur in the index of either book; global warming and conflict is barely mentioned. Overall, a long list of ideas about what the future might hold is reported, but a link between climate change and conflict is not one of them.

Indeed, many of the sources cited by IPCC do not seem to deal with climate change and security specifically, several of them are near impossible to acquire (even with access to world-leading library systems), and the basis for IPCC statements is often unclear. This makes a full evaluation of the validity of the claims cumbersome and at times near impossible. In sum, we have not found any convincing evidence of a climate change-violent conflict link in the futures literature cited, nor that this link is emphasized. The relatively prominent space given to this literature is therefore questionable.

THE FOURTH ASSESSMENT REPORT (AR4)

The Fourth Assessment Report (IPCC, 2007) also contains references to a possible climate-conflict link, although this has been moderated somewhat from the 2001 report. The 2007 Synthesis report contains only two references to the search terms, one is 'low relevance', and the other one is of 'secondary relevance'. There are no longer any references to conflict in the categories we have labeled 'high relevance' or 'very high relevance'. Only in the index to WG2 are there references to 'conflicts and war' – all of these are in the chapter on Africa. The impression therefore is that there are few references to conflict, and that they mostly fall in the categories we have labeled 'secondary relevance'.

The Regular Chapters

In the report by Working Group 2, *Impacts, Adaptation and Vulnerability*, there are 45 references to 'conflict' in several chapters (in one case 'armed conflict'), but only three to 'violence/violent' (all of them of secondary relevance), and none to 'War'. This includes Chs. 1–8 and 10–20, but for the moment we keep the Africa chapter (Ch. 9) apart. In these 19 chapters, five hits are of secondary relevance, whereas four are of very high relevance. The rest are of low relevance or irrelevant.

Of the hits that are of very high relevance, the first (in Ch. 5, p. 299) is a statement that 'Unilateral adaptation measures to water shortage related to climate change can lead to competition for water resources and, potentially to conflict and backlash for development.' Given the use of the word 'potentially' (rather than 'probably') we might have excluded this altogether. However, even if we accept it, it doesn't take us very far. The reference is to Dalby (2004), a brief survey in a non-peer-reviewed newsletter of the different approaches to environmental security. However, this survey is not really about water conflict and a subsequent reference to Honduras and two of its neighbors shows that the IPCC authors must have intended to refer to the next article in the newsletter, Lopez (2004). This is in fact a two-page article that explains why, despite the potential for conflict, the three countries (El Salvador, Honduras, and Guatemala) have achieved transborder *cooperation*.

The second reference (IPCC, 2007, Ch. 7: 365) argues that 'An argument can also be made that rising ethnic conflicts can be linked to competition over natural resources that are increasingly scarce as a result of climate change' but goes on to argue that many other intervening and contributing causes of conflict need to be taken into account. This suggests caution in the prediction of such conflicts as a result of climate change. The reference here is Fairhead (2004), which is a chapter in an edited volume focusing on sustainability in Africa, not on conflict. Again, the references to conflict in the AR4 seem to be from sources that are only indirectly or marginally on conflict – and by researchers who do not specialize in conflict research.

The third reference (Ch. 10: 488) cites the considerable impact that climate change is likely to have on 'human sustenance and livelihoods', which in turn 'can lead to instability and conflict', citing Barnett (2003). Barnett's work is a careful evaluation of possible causal paths leading from climate change to conflict, but the link to conflict refers to a possibility rather than a probability.

Finally, the fourth reference is found in Ch. 17 (p. 733). Here, Schneider et al. (2000) are cited saying that 'coercion to migrate, conflict over resources, cultural diversity, and loss of cultural heritage sites' might be non-market costs of climate change. There is no further discussion of conflict in this article. The article cites Nordhaus (1994) and Roughgarden and Schneider (1999). Nordhaus interviewed 22 natural and social scientists about their views on the consequences of climate change. The natural scientists tended to estimate larger damages, but placed them in the non-market category, and they cannot be calculated very precisely. There is no discussion of the costs of conflict in Roughgarden and Schneider (1999).

In short, therefore, these references offer little or no support for a climate change to armed conflict scenario.

The Africa Chapter

The Africa chapter contains 11 references to conflict, more than any other chapter.[16] Most of these are classified as of secondary relevance, but there are three search terms hits of 'very high relevance'.

The first of these (p. 443), notes that Africa has seen many armed conflicts in the recent past and that 'climate change may become a contributing factor to conflicts in the future, particularly those concerning resource scarcity, for example, scarcity of water' (Ashton, 2002; Fiki and Lee, 2004). These sources provide only qualified support for a climate-conflict nexus. Ashton provides a detailed overview of the potential for water conflict in Africa, which he sees as 'inevitable', 'unless we can jointly take preventive actions' (p. 240). He also concedes that 'territorial sovereignty issues have been implicated in virtually every dispute or conflict that has taken place over, or near to, water', that 'in most cases, these disputes have been linked to disagreements over the precise positions of territorial boundaries', and that 'despite many predictions to the contrary, "true" water wars have happened very rarely, if at all' (p. 239). Climate change is mentioned only once, and then peripherally: 'There is also compelling, though as yet unproven, evidence that projected trends in global climate change will worsen this situation' (p. 236). Fiki and Lee (2004) discuss local conflicts between pastoralists and farmers in two drought-prone communities in north-eastern Nigeria, some of which have led to violence. Most of these conflicts occurred in resource-rich rather than resource-poor environments and were usually over ownership rights. The article discusses how traditional modes of conflict resolution can mitigate such conflicts. The impact of climate change is not discussed.

The second case is Figure 9.5 (p. 451), a map of Africa that portrays 'current and possible future impacts and vulnerabilities associated with climate variability and climate change'. The map includes a few 'conflict zones' but these are clearly zones of present conflict, so this source might more appropriately have been classified as a case of secondary relevance.

In the third case (p. 454), climate change is seen possibly contributing to migration of agriculturalists into marginal lands, which in turn can trigger conflict. However, there are no references provided to the point about conflict.

THE AUTHORS

The two IPCC reports of 2001 and 2007 list a large number of authors, but very few of them are names that one would recognize from the literature on armed conflict. As a first cut, we took the twenty top-ranked authors in the study of 'Armed conflict' in ISI's Essential Science Indicators from November 2006 (http://esi-topics.com/armed-conflict/authors/b1a.html). None of them is listed as an author or cited in the 2001 or 2007 reports. We also searched for references to the top twenty armed conflict journals, from the same study (http://esi-topics.com/armed-conflict/journals/e1a.html). None of them was cited in connection with the search terms, with one exception: One article in *Journal of Conflict Resolution* is cited (Ron, 2005). The article deals with conflict over natural resources, but climate change is not relevant. This is in stark contrast to the field of environmental studies, where a number of well-known journals are cited.

When we look at the authorship of the key chapters mentioning a climate-conflict link we also get the same impression. Although the authors are generally well qualified, holding PhDs in various natural sciences, there are in most instances no authors with a social science qualification. For instance, in the chapter on Africa in the 2007 report, of the main authors of the chapter, four authors have a background in physical geography; three authors have PhDs in plant sciences – agronomy, plant genetics, forestry, and vegetable crops; one works in a geology department; and another holds a PhD in medical entomology. It is adamantly clear that the range of expertise is severely limited, and that the authors as a group are not well placed to make qualified assessments of conflict risks.

The chapter on water and hydrology in the 2001 IPCC report (Ch. 4), gives the same impression. Here, the authors generally do have advanced academic degrees, but not within the social sciences or political science. The majority of the authors are from engineering, and there are representatives from economics, meteorological sciences, and hydrogeology. Even in this chapter, where a water-conflict link is discussed, there are no authorities on conflict among the authors, nor do the authors have any formal qualifications that should make them well placed to summarize the literature on that particular issue.

RECENT HEADWAY AND PROSPECTS FOR THE FUTURE

Considerable headway has been made in the literature on a potential climate change-conflict link since the AR4. However, a number of issues still remain underexplored. Both cross-national comparative and statistical studies and case studies of areas of stated particular concern have been published in peer-reviewed journals focusing on conflict (e.g. Benjaminsen et al., 2012; Bernauer and Siegfried, 2012; Gleditsch, 2012; Slettebak, 2012; Theisen et al., 2013). This work has also included a wider set of violent conflicts, and a careful disaggregation of key variables of interest to produce more specific tests of proposed relationships (e.g., Hendrix and Salehyan, 2012; Theisen, 2012). The literature is therefore moving towards uncovering how climate change might affect human security, looking, for example, at the complex of interrelated threats such as civil war, and also genocide and the displacement of populations (Human Security Report, 2005), but keeping the definition of conflict to a narrow interpretation of human security, focusing on protection of individuals from violence. In so far as a broader understanding of human security is included in the analyses (including such factors as hunger, disease and natural disasters), it is as a possible intermediary factor between climate and violent conflict.

Some recent work has also contributed to a tighter coupling of climate change models and conflict models. Studies by Hendrix and Glaser (2007), Raleigh and Urdal (2007), and Buhaug (2010) are pioneering efforts in this regard. However, there is considerable promise in further cross-disciplinary collaboration focused on integrating the knowledge of both the natural and social sciences.

Given the new empirical research on the climate-conflict nexus, the IPCC should be in a position to provide a more comprehensive and reliable review of the relevant research. In 2007, the IPCC could have decided to leave out the security implications altogether and leave that to more qualified observers. Indeed, one of Pachauri's final remarks in his Nobel Lecture could be interpreted to this effect, but is perhaps better understood as a plea for more research: 'How climate change will affect peace is for others to determine, but we have provided scientific assessment of what could become a basis for conflict' (Pachauri, 2007). Be that as it may, the IPCC has decided to tackle the security issues more directly in the Fifth Assessment Report (AR5), scheduled for publication in 2013–14. The report from IPCC's Working Group II will contain a chapter on human security, including a section on conflict. The list of authors for this chapter indicates that the team has considerably stronger competence on human security in a broad sense than on conflict. Indeed, not a single prominent conflict scholar is part of the team. Some have, however, been invited to critique drafts of the human security chapter. It remains to be seen how much influence they will have over the final outcome. One potential hazard is that 'human security' may be given an excessively broad definition.[17] In that case, even successful measures of adaptation to climate change, such as urbanization or industrialization, can be interpreted as threats to human security, because they imply social change that may upset traditional life styles.

CONCLUSION

Global climate change will have profound implications for the quality of life of hundreds of millions of people. A high degree of agreement has been reached on the likely consequences of climate change on the natural environment. The IPCC has been instrumental in gathering and disseminating knowledge about these natural processes, building on hundreds of (mainly) peer-reviewed research contributions. The basis of knowledge is considerably weaker regarding the social consequences of climate change and particularly on the issue examined here: the question of how climate change will affect security and (potentially) violent conflict.

Claims about a causal link from climate change to conflict seem often to be cited more or less uncritically from one source to the next, with insufficient weeding of second- and third-rate sources and without any real accumulation of knowledge. The overall impression from the IPCC reports of 2001 and 2007 is therefore that the link between climate change and conflict is ambiguous and when it is stated it is weakly substantiated.

AR5, to be published in 2013–14, provides an opportunity for the IPCC to improve its record in this area.

NOTES

1. Our work has been supported by the Research Council of Norway. This work draws in part on Nordås and Gleditsch (2007b) and Buhaug, Gleditsch and Theisen (2010). Unless another date is specified, all Internet sources were last accessed on 12 November 2012.
2. Pachauri also cited various historical examples of the collapse of states due to environmental degradation, and the historical links between temperature fluctuations and warfare in Eastern China over the last millennium. Zhang et al. (2007a: 403, 2007b) did indeed find an association between warfare and the decline of agriculture, but the peaks in warfare occurred during *cooling* phases.
3. Co-winner of the Nobel Peace Prize, Al Gore (2007) generally stayed away from questions of war and peace in his lecture, although he noted in passing that 'Climate refugees have migrated into areas already inhabited by people with different cultures, religions, and traditions, increasing the potential for conflict.'
4. See *Washington Post*, 26 January 2009: 'Obama announces new energy, environmental policies': http://www.washingtonpost.com/wp-dyn/content/article/2009/01/26/AR2009012601157.html.
5. We use the shorthand terms 'climate-conflict nexus', 'climate-conflict' and 'climate change-armed conflict nexus' interchangeably. In all three versions we refer to the notion that human-induced climatic changes will lead (directly or by traceable causal chains) to armed and organized violent conflict either within or between states.
6. See the main guiding principles of the IPCC online at: www.ipcc.ch/pdf/ipcc-principles/ipcc-principles.pdf.
7. Expert reviewers may be nominated by governments, national and international organizations, Working Group/Task Force Bureau, Lead Authors and Contributing Authors.
8. Coordinating Lead Authors and Lead Authors are selected by the relevant Working Group/Task Force Bureau. See: www.ipcc.ch/pdf/ipcc-principles/ipcc-principles-appendix-a.pdf.
9. See: www.ipcc.ch/pdf/ipcc-principles/ipcc-principles-appendix-a.pdf.
10. For description of the review process see: www.ipcc.ch/pdf/ipcc-principles/ipcc-principles-appendix-a.pdf.
11. The IPCC specifies that this option may be chosen for instance to obtain case study materials from private sector sources for assessment of adaptation and mitigation options. See: www.ipcc.ch/pdf/ipcc-principles/ipcc-principles-appendix-a.pdf.
12. This includes hits on 'farmed', 'harmed', or 'warmed' (for 'armed') and on 'warm', 'warming', 'warning', etc. (for 'war').
13. We have not examined the secondary relevance cases in the same detail since (like most of the conflict literature) we focus on the onset of conflict. However, a cursory examination of these cases reveals many of the same problems as in the high relevance cases.
14. This finding was supported later by Furlong et al. (2006) and Gleditsch et al. (2006), although the most recent work with new data questions whether the effects of the shared river basin variable can be distinguished from the effects of contiguity (Brochmann and Gleditsch, 2013).
15. Such as President Sadat commenting that water was the only matter which could take Egypt to war again (stated in 1979), and King Hussein of Jordan who reportedly said in 1990 that a dispute over water could lead to a war between Jordan and Israel (Ramphal, 1992: 47).

16. Bringing the total number of references to 56 in the Fourth Assessment Report, still less than half of the number in the TAR from 2001.
17. For a discussion of broad and narrow interpretations of human security, see the Human Security Report (2005: viii).

REFERENCES

Ashton, Peter J. (2002), 'Avoiding conflicts over Africa's water resources', *Ambio*, **31** (3), 236–242.

Ayres, Robert U. and Jorg Walter (1991), 'The greenhouse effect: damages, cost, and abatement', *Environmental and Resource Economics*, **1** (3), 237–270.

Barnett, Jon (2003), 'Security and climate change', *Global Environmental Change*, **13** (1), 7–17.

Benjaminsen, Tor A., Koffi Alinon, Halvard Buhaug and Jill Tove Buseth (2012), 'Does climate change drive land-use conflicts in the Sahel?', *Journal of Peace Research*, **49** (1), 97–111.

Bernauer, Thomas and Tobias Siegfried (2012), 'Climate change and international water conflict in central Asia', *Journal of Peace Research*, **49** (1), 227–239.

Bernauer, Thomas, Tobias Böhmelt and Vally Koubi (2012), 'Environmental changes and violent conflict', *Environmental Research Letters*, **7** (1), 1–8.

Biermann, Frank, Gerhard Petschel-Held and Christoph Rohloff (1998), 'Umweltzerstörung als Konfliktursache? Theoretische Kozeptualisierung und empirische Analyse des Zusammenhangen von "Umwelt" und "Sicherheit"' [Environmental Degradation as a Cause of Conflict? Theoretical Conceptualization and Empirical Analysis of the Relationship between 'Environment' and 'Security'], *Zeitschrift für Internationale Beziehungen*, **5** (2), 273–308.

Biswas, Asit K. (ed.) (1994), *International Waters of the Middle East: From Euphrates-Tigris to Nile*, Oxford, UK: Oxford University Press.

Brochmann, Marit and Nils Petter Gleditsch (2013), 'Shared rivers and conflict – a reconsideration', *Political Geography*, **31** (8), 519–527.

Buhaug, Halvard, Nils Petter Gleditsch and Ole Magnus Theisen (2010), 'Implications of climate change for armed conflict', ch. 3 in Robin Mearns and Andy Norton (eds), *Social Dimensions of Climate Change: Equity and Vulnerability in a Warming World*. New Frontiers of Social Policy, Washington, DC, USA: World Bank, pp. 75–101.

Buhaug, Halvard (2010), 'Climate not to blame for African civil wars', *PNAS*, **107** (38), 16477–16482.

Buhaug, Halvard, Nils Petter Gleditsch and Ole Magnus Theisen (2010), 'Implications of climate change for armed conflict', ch. 3 in Robin Mearns and Andy Norton (eds), *Social Dimensions of Climate Change: Equity and Vulnerability in a Warming World*, New Frontiers of Social Policy, Washington, DC, USA: World Bank, pp. 75–101.

Caponera, Dante A. (1996), 'Conflicts over international river basins in Africa, the Middle East, and Asia', *Review of European Community and International Environmental Law*, **5** (2), 97–106.

Cline, William R. (1992), *The Economics of Global Warming*, Washington, DC, USA: Institute of International Economics.

Dalby, Simon (2004), 'Conflict, cooperation and global environment change: advancing the agenda', *Update*, **4** (3), 1–3.

Dellapenna, Joseph W. (1999), 'Adapting the law of water management to global climate change and other hydropolitical stresses', *Journal of the American Water Resources Association*, **35** (6), 1301–1326.

Detraz, Nicole and Michele Betsill (2009), 'Climate change and environmental security: for whom the discourse shifts', *International Studies Perspectives*, **10** (3), 303–320.

Fairhead, James (2004), 'Achieving sustainability in Africa', in Richard Black and Howard

White (eds), *Targeting Development: Critical Perspectives on the Millennium Goals*, London, UK: Routledge, pp. 292–306.

Fiki, Charles and Bill Lee (2004), 'Conflict generation, conflict management and self-organizing capabilities in drought-prone rural communities in north eastern Nigeria: a case study', *Journal of Social Development in Africa*, **19** (2), 25–48.

Furlong, Kathryn, Nils Petter Gleditsch and Håvard Hegre (2006), 'Geographic opportunity and neomalthusian willingness: boundaries, shared rivers, and conflict', *International Interactions*, **32** (1), 79–108.

Gleditsch, Nils Petter (1998), 'Armed conflict and the environment: a critique of the literature', *Journal of Peace Research*, **35** (3), 381–400.

Gleditsch, Nils Petter (2002), 'Double-blind but more transparent', *Journal of Peace Research*, **39** (3), 259–262.

Gleditsch, Nils Petter (2012), 'Whither the weather? Climate change and conflict', *Journal of Peace Research*, **49** (1), 3–9.

Gleditsch, Nils Petter and Ragnhild Nordås (2009), 'Climate change and conflict: a critical overview', *Die Friedens-Warte*, **84** (2), 11–28.

Gleditsch, Nils Petter, Kathryn Furlong, Håvard Hegre, Bethany Lacina and Taylor Owen (2006), 'Conflicts over shared rivers: resource scarcity or fuzzy boundaries?', *Political Geography*, **25** (4), 361–382.

Gleditsch, Nils Petter, Ragnhild Nordås and Idean Salehyan (2007), *Climate Change, Migration, and Conflict*, Coping with Crisis Working Paper Series, New York, USA: International Peace Academy.

Gleick, Peter H. (1992), *Water and Conflict*, Occasional Papers Series on the Project on Environmental Change and Acute Conflict, Toronto, Canada: Security Studies Programme, American Academy of Arts and Sciences, University of Toronto.

Gleick, Peter H. (1993), 'Water and conflict: fresh water resources and international security', *International Security*, **18** (1), 79–112.

Gleick, Peter H. (1998), *The World's Water: The Biennial Report on Freshwater Resources*, Washington, DC, USA: Island Press.

Gore, Al (2007), 'Nobel Lecture', acceptance speech delivered at the Nobel Peace Prize award ceremony, Oslo, 10 December, http://nobelpeaceprize.org/en_GB/laureates/laureates-2007/gore-lecture/.

Hendrix, Cullen and Sarah M. Glaser (2007), 'Trends and triggers: climate, climate change and civil conflict in Sub-Saharan Africa', *Political Geography*, **26** (6), 695–715.

Hendrix, Cullen S. and Idean Salehyan (2012), 'Climate change, rainfall, and social conflict in Africa', *Journal of Peace Research*, **49** (1), 35–50.

Homer-Dixon, Thomas (1991), 'On the threshold: environmental changes as causes of acute conflict', *International Security*, **16** (2), 76–116.

Homer-Dixon, Thomas (1994), 'Environmental scarcities and violent conflict: evidence from cases', *International Security*, **19** (1), 5–40.

Homer-Dixon, Thomas and Jessica Blitt (1998), *Ecoviolence: Links Among Environment, Population and Security*, Lanham, USA: Rowman and Littlefield.

Human Security Report (2005), *Human Security Report 2005. War and Peace in the 21st Century*, New York, USA: Oxford University Press for the Human Security Centre.

IAC (2010), *Climate Change Assessments, Review of the Processes and Procedures of the IPCC*, Amsterdam, The Netherlands: InterAcademy Council, http://reviewipcc.interacademycouncil.net/.

IPCC (2001), *Third Assessment Report. Climate Change 2001*, Geneva, Switzerland: Intergovernmental Panel on Climate Change and Cambridge, UK: Cambridge University Press.

IPCC (2007), *Fourth Assessment Report. Climate Change 2007*, Geneva, Switzerland: Intergovernmental Panel on Climate Change and Cambridge, UK: Cambridge University Press.

Kennedy, Donald D., with David Holloway, Erika Weinthal, Walter Falcon, Paul Ehrlich, Roz Naylor, Michael May, Steven Schneider, Stephen Fetter and Jor-San Choi

(1998), *Environmental Quality and Regional Conflict*, Washington, DC, USA: Carnegie Commission on Preventing Deadly Conflict.

Levy, Marc A. (1995), 'Is the environment a national security issue?', *International Security*, **20** (2), 35–62.

Lopez, Alexander (2004), 'The Lempa River Basin: transborder cooperation in an international river basin with high potential for conflict', *Update*, **4** (3), 10–11.

McKelvey, Robert (1997), 'Game theoretic insights into the international management of fisheries', *Natural Resource Modeling*, **10** (2), 129–171.

McRae, Hamish (1994), *The World in 2020: Power, Culture and Prosperity*, London, UK: HarperCollins.

Mercer, David (1998), *Future Revolutions: A Comprehensive Guide to the Third Millennium*, London, UK: Orion Business.

Mjøs, Ole Danbolt (2007), 'Speech given by the Chair of the Norwegian Nobel Committee, Oslo, 10 December. URL: http://www.nobelprize.org/nobel_prizes/peace/laureates/2007/presentation-speech.html.

Myers, Norman (1993), *Ultimate Security: The Environmental Basis of Political Stability*, New York, USA: Norton.

Myers, Norman, with Jennifer Kent (1995), *Environmental Exodus: An Emergent Crisis in the Global Arena*, Washington, DC, USA: Climate Institute.

Nordhaus, William (1994), 'Expert opinion on climatic change', *American Scientist*, **82** (1), 45–52.

Nordås, Ragnhild and Nils Petter Gleditsch, guest editors (2007a), 'Climate Change and Conflict', Special Issue, *Political Geography*, **26** (6), August.

Nordås, Ragnhild and Nils Petter Gleditsch (2007b), 'Climate change and conflict', *Political Geography*, **26** (6), 627–638.

Pachauri, Rajendra K. (2007), 'Nobel Lecture', on the occasion of the Nobel Peace Prize awarded to the Intergovernmental Panel on Climate Change, Oslo, 10 December. URL: http://www.nobelprize.org/nobel_prizes/peace/laureates/2007/ipcc-lecture_en.html.

Patz, Jonathan A. (1998), 'Climate change and health: new research challenges', *Health and Environment Digest*, **12** (7), 49–53.

Rahman, A. Atiq (1999), 'Climate change and violent conflicts', in Mohamed Suliman (ed.), *Ecology, Politics, and Violent Conflicts*, London, UK: Zed, pp. 181–210.

Raleigh, Clionadh and Henrik Urdal (2007), 'Climate change, environmental degradation and armed conflict', *Political Geography*, **26** (6), 674–694.

Ramphal, Shridath (1992), *Our Country, The Planet: Forging a Partnership for Survival*, Washington, DC, USA: Island Press.

Ron, James (2005), 'Paradigm in distress? Primary commodities and civil war', *Journal of Conflict Resolution*, **49** (4), 443–450.

Roughgarden, Tim and Stephen H. Schneider (1999), 'Climate change policy: quantifying uncertainties for damages and optimal carbon taxes', *Energy Policy*, **27** (2), 415–429.

Scheffran, Jürgen, Michael Brzoska, Jasmin Kominek, P. Micheal Link and Janpeter Schilling (2012), 'Climate change and violent conflict', *Science*, **336** (6083), 869–871.

Schellnhuber, Hans-Joachim and Detlef F. Sprinz (1994), 'Umweltkrisen und Internationale Sicherheit' [Environmental Crises and International Security], in Karl Kaiser and Hanns W. Maull (eds), *Herausforderungen. Deutschlands neue Aussenpolitik*, II, Bonn, Germany: Forschungsinstitut der Deutschen Gesellschaft für Auswärtige Politik, pp. 239–260.

Schneider, Stephen H., Kristin Kuntz-Duriseti and Christian Azar (2000), 'Costing non-linearities, surprises, and irreversible events', *Pacific and Asian Journal of Energy*, **10** (1), 81–106.

Slettebak, Rune T. (2012), 'Don't blame the weather! Climate-related natural disasters and civil conflict', *Journal of Peace Research*, **49** (1), 163–176.

Stern Review (2006), Nicholas Stern et al., *The Economics of Climate Change*, London: HM Treasury and Cambridge, UK: Cambridge University Press.

Theisen, Ole Magnus (2012), 'Climate clashes? Weather variability, land pressure, and organized violence in Kenya, 1989–2004', *Journal of Peace Research*, **49** (1), 81–96.

Theisen, Ole Magnus, Nils Petter Gleditsch and Halvard Buhaug (2013), 'Is climate change a driver of armed conflict?', *Climatic Change*, **117** (3), 613–625.

Toset, Hans Petter Wollebæk, Nils Petter Gleditsch and Håvard Hegre (2000), 'Shared rivers and interstate conflict', *Political Geography*, **19** (8), 971–996.

Wolf, Aaron T. (1994), 'A hydropolitical history of the Nile, Jordan and Euphrates river basins', in Asit K. Biswas (ed.), *International Waters of the Middle East: From Euphrates-Tigris to Nile*, Oxford, UK: Oxford University Press, pp. 5–43.

Wolf, Aaron T. (1998), 'Conflict and cooperation along international waterways', *Water Policy*, **1** (2), 251–265.

Zhang, David D., Peter Brecke, Harry F. Lee, et al. (2007a), 'Global climate change, war, and population decline in recent human history', *PNAS*, **104** (49), 19214–19219.

Zhang, David D., Jane Zhang, Harry F. Lee and Yuan-qing He (2007b), 'Climate change and war frequency in Eastern China over the last millennium', *Human Ecology*, **35** (4), 403–414.

4. Space, time and scales of human security in climate change
Richard Matthew

For myself, I always write about Dublin, because if I can get to the heart of Dublin I can get to the heart of all the cities of the world. In the particular is contained the universal.

—James Joyce (in Ellman 1983: 505)

INTRODUCTION

Unfortunately, climate change researchers have typically sought to prove the particular through the universal rather than the other way around.

I use the term universal to refer to the science-based story of global climate change; in contrast, particular refers here to stories about climate change observed or experienced on a local or at least less than global scale. From the vantage of fields such as political philosophy and behavioral psychology, both of which I will introduce into this chapter later, there are many reasons why the universalist approach—which recounts the elegant, compelling, empirically-grounded story of global climate change caused by inefficient human practices—is not likely to catalyze an effective response.

This is not a critique of universal stories per se, some intimation that ultimately they are not valid, inspiring or useful. In fact, universal narratives are often very compelling, suggesting solidarity around important values like justice, human rights and peace, and a shared fate in relation to far reaching challenges such as pandemic disease, extreme poverty and climate change. At the same time, however, the complexity of things like justice and peace, and of challenges like extreme poverty, is inevitably grappled within very particular contexts, in response to very particular needs and observations, and results in very particular innovations or adjustments to local values, practices, and institutions.

In historical terms, universal stories typically emerge from particular ones. Christianity began, for example, as a very small cult with a highly particular account of justice, validating itself and trying to survive in the hostile, unequal and violent culture of the Roman Empire, and it took centuries of very local experiments before it could evolve and consolidate as a transformative universal story (Brown 2000; Matthew 2002). The Universal Declaration of Human Rights was constructed on the platform of centuries of particular events and experiments such as the Magna Carta and the rights of man discourse that emerged from the revolutions in France and the United States. The concept of universal human rights became very powerful after World War II and the Holocaust, and no doubt there are individuals and communities that have been able to leverage this story into behavioral change in particular contexts where a notion of rights is otherwise weak or absent. But this leverage comes from deploying a universal doctrine that emerged from, that was generalized from, some very high profile and successful sub-global episodes.

Does the reverse ever occur—that is, does a compelling universal narrative, like global climate change, which is not and could not be abstracted from particular accounts, ever trigger effective behavioral responses across many particular settings? There are moments when an issue that is widespread and urgent—like landmines or genocide—catalyzes a sense of injustice across many particular settings, and thus creates or orients a multi-level behavioral response around a powerful universal narrative. But these moments are rare and they do tap into and build upon particular accounts of justice. When something is worked through primarily at the global level, such as how to govern cyberspace or manage biodiversity loss, the outcome is likely to be abstract, fuzzy and non-binding. Insofar as climate change is concerned, I will argue that there are sound reasons to be skeptical of the likelihood that a global starting point will have a transformative impact on the ground, although ultimately I want to keep a sense of this possibility intact and vital.

A central argument in this chapter is that there are many reasons why the very compelling, and at times terrifying, universal narrative of climate change is failing to catalyze the serious and coordinated response its proponents believe is unambiguously essential for the wellbeing, and perhaps even survival, of humankind. There is clearly frustration among climate scientists who have identified patterns of global change, believe these to be tracking towards worst case scenarios, see technological and behavioral responses that could change this alarming trajectory, and cannot understand why so little is being accomplished, why markets, governments and citizens are so unresponsive to scientific information (e.g. Sachs 2005; Pachauri 2007; Gleick 2012; Gleick and Heberger 2012). Germanwatch, for example, which reports annually on what countries are doing in response to climate change, wrote in its 2012 report: "As in the years before, we still cannot reward any country with the rankings 1–3, as no country is doing enough to prevent dangerous climate change"

(2011: 4). In a similar vein, the journalist Fred Pearce writes that climate change "scares me, just as it scares many of the scientists I have talked to—sober scientists, with careers and reputations to defend, but also with hopes for their own futures and those of their children, and fears that we are the last generation to live with any kind of climatic stability" (2007: xxx).

This may help explain why an expectation has kindled among some climate scientists that as the accumulation of data, supercomputers and more sophisticated modeling technologies combine to enable more particular stories to be told and more particular futures to be predicted an effective global response will emerge. Typical of this sentiment:

> "It's a big deal," said climate scientist Linda Mearns of the National Center for Atmospheric Research in Boulder, Colo. "Yellowstone will help researchers calculate climate change on a regional, rather than continental, scale. With a better grasp of how warming may affect local water resources, endangered species and extreme winds, local and state governments will be able to plan more effectively." (Nash 2012)

How close scientists are to accurate local modeling is still debated. In their recent report, McElroy and Baker claim that "Regional prediction remains challenging and will require focused efforts to maintain and enhance Earth observations, especially of the oceans" (2012: 5; see also Lemos and Morehouse 2005; White et al. 2010).

In any case, I will argue that, sadly, these narratives could well fail in one important regard, revealing and consolidating fracture lines and incommensurable positions rather than nurturing coordinated and global responses. *In short, whether the story of climate change is told from the top down, or the bottom up, an effective and coordinated response will be extremely difficult to mobilize.*

I will make this basic argument using human security as a reference point, as a critical condition that climate change places at risk in communities around the planet. Human security is conventionally viewed in terms of freedom from fear of types of violence that could be neutralized, and freedom from want of types of goods and services that could be provided. Climate change impacts are often characterized as direct challenges to human security, as impacts that could generate both violence and scarcity.

The chapter begins with a discussion of universal versus particular narratives, discusses the narratives of climate change and human security, then explores the various scales at which climate change and human security can be linked, and concludes with an argument about why these compelling, empirically grounded, and broadly relevant narratives may not generate an effective global response.

THROUGH THICK AND THIN

In 1994 Michael Walzer wrote *Thick and Thin: Moral Argument at Home and Abroad*. He makes no mention of climate change, a topic already very much on the global agenda at that time, or of human security, which received a powerful articulation and endorsement that same year. He writes instead about truth, religion, genocide, self-determination, and justice, themes that unify his multi-book opus. Nonetheless, his basic arguments are useful to the discussion of climate change and human security.

Walzer's principal concern in this short study is with understanding the sources and dynamics of moral argument. He describes a common view about moral argument: "Men and women everywhere begin with some common idea or principle or set of ideas and principles, which they then work up in many different ways" (1994: 4). This is precisely the view that has informed the discourse of climate change. Climate science provides the basic observations about what is happening at the global level, and where this is tracking, and in response communities around the world are expected to develop and implement particular climate action plans.

The contemporary climate scientist thus seems a bit like Moses. Moses received a set of rules directly from God, and set out to introduce these into a particular community that viewed them as largely foreign and irrelevant. His task was to persuade them otherwise. Climate scientists have observed patterns that are invisible to the senses of any particular community. They want their understandings to be introduced into particular settings in meaningful ways that will catalyze action. Walzer notes that we tend to think that this is the way change takes place: the universal is actualized in, and thus transforms, particular settings. But he argues that this is not really what happens. At least insofar as moral argument is concerned, Walzer disagrees with the common view: "our intuition is wrong here. Morality is thick [i.e. particular] from the beginning, culturally integrated, fully resonant, and it reveals itself thinly only on special occasions, where moral language is turned to specific purposes" (p. 4).

Walzer's philosophical argument can be supported by construal level theory, developed in the field of social and behavioral psychology. The basic idea here is that distance matters. The events and trends described by climate science (or anything else) may be perceived by a particular community as near or far away spatially (Fujita et al. 2005) or near or far away temporally (Liberman and Trope 1998; Liberman et al. 2002; Trope and Liberman 2000, 2003), or as having a high or low probability of occurrence (Todorov et al. 2007; Wakslak et al. 2006), or as things likely to be directly experienced or likely to be experienced through a third party (Eyal et al. 2008). Those events and trends that are closer to ambient experience along these parameters are more likely to be viewed as concrete and actionable. In other words, a story that goes, climate change affects us all, we might be witnessing that begin to happen thousands of miles away, but it could happen here in a decade or maybe longer, this is really serious and the very viability of humankind might be in play—this is a story that is so fraught with conditionality that it may not catalyze much action. It is easy for someone to find another more concrete priority that requires immediate attention.

Later Walzer writes: "In moral discourse, thinness [i.e. universality] and intensity go together, whereas with thickness comes qualification, compromise, complexity, and disagreement" (p. 6). To matter at all, then, thin, intense, universal stories about a given value, X, have to find entry points into thick particular actualizations of that value, X. They are intense for a reason—they describe, in Walzer's case, moral affronts (genocide, torture, exploitation of children) that are truly egregious even if not experienced directly by a given community. They are very focused and they demand everyone's attention and participation.

Now, at first blush, this, too, might seem a reasonable alternative way to characterize climate discourse. Across the planet complex societies have matured, informed by sophisticated understandings of local climate. A case in point: the vast agricultural systems employing hundreds of millions of people in South Asia, and often integrated with a fishing season in the same sites, are shaped largely by local understandings of the monsoon. Complex local routines have developed governing access to the monsoon water, the land it saturates, and the natural capital that is extracted from this land. Climate science elaborates a universal story about global hydrological change, thin, intense, and clearly relevant to the particular understandings and practices that have grown up around the monsoon. The implication of the universal narrative is straightforward and unambiguous—local routines need to be reconsidered and amended in light of contemporary climate science (see, for example, Singh et al. 2011; Matthew 2012; National Research Council 2012).

But this is not quite what has happened, and part of the problem is that the universal story of global climate change, committed to the honesty of science, has been told with much "qualification, compromise, complexity, and disagreement" itself (Walzer 1994: 6). Thus there is a mismatch here that Walzer's argument makes very clear. If we characterize climate change science as an attempt to impose a thick universal story on humankind, then success will be very difficult—thick means complex and controversial. On the other hand, the possibility of a thin, intense, mobilizing universal narrative around climate change is being undermined because very quickly it is evident that the simple clarity that has animated some of the rhetoric of the past two decades is misleading—that is to say that, in fact, there is a lot of uncertainty and complexity surrounding the universal story. It does not have a thin formulation that matters to all the particular ones. So the universal climate change narrative cannot work as a thick narrative, or easily be deployed as a thin, intense, shared tool with which to assess and reconstruct thick, local practices.

In short, the combination of *thick* (embodying "qualification, compromise, complexity, and disagreement") and *universal* (which actually needs to be thin to be impactful) places climate change into the narrative form that is least likely to succeed. Can this be changed?

THE UNIVERSAL STORY OF CLIMATE CHANGE SCIENCE

The Swedish scientist Svante Arrhenius first suggested that carbon dioxide emissions could lead to global warming through a greenhouse gas effect in 1896. More than a century later, the Intergovernmental Panel on Climate Change (IPCC), a global network of climate scientists set up by the United Nations, affirmed that human activities, and especially those involved in generating CO_2 emissions, are almost certainly forcing global climate change.

The basic idea is that changes in the composition of the earth's atmosphere (that is, a net increase in CO_2 and other greenhouse gases) create conditions that retain more solar energy on the earth's surface and in the troposphere (the lowest level of the earth's atmosphere) than has been the case for at least several thousand years. Trapped solar energy warms the earth's surface, disrupting the planet's hydrological cycle. This process plays out as severe drought in some places, severe flooding in others, and severe storms in still others. Climate change science expresses the impacts of this change in the composition of the atmosphere in terms of a transformation of the statistical characteristics of weather aggregated over time and space.

Scientists are well aware that, understood in this way, climate change has been a regular feature of planet earth since it took shape from space debris some four and a half billion years ago. In fact, in geological time, much larger climate transformations took place in the pre-anthropocene. But while today's changes may be rather small compared to some of the periods of climate change that occurred in the past, such as the oxidation of the atmosphere, they are unique in that they are being stimulated by human activities, by everyday affairs such as consuming cheap and plentiful fossil fuels, and converting forest into farmland, firewood and framing for construction. These activities increase the concentration of carbon dioxide and other greenhouse gases in the atmosphere and reduce carbon storage on the earth's surface. The measurable outcome of this is that since the mid nineteenth century the earth's average surface temperature has increased by 1 degree centigrade (Solomon 2007).

This measurable outcome is grounded in considerable research and has fostered a very high level of scientific agreement (Weart 2008). This research begins with information from ice cores and tree rings, and with weather data that have been collected since the 1600s. The International Meteorological Organization, which was founded in 1873, prepared standards for collecting weather data and hence created the conditions for longitudinal analysis. Fifty years later, scientists like G.S. Callendar began accumulating time series data on temperature. This was supplemented with "The high-accuracy measurements of atmospheric CO_2 concentration, initiated by Charles David Keeling in 1958, [which] constitute the master time series documenting the changing composition of the atmosphere"(Le Treut et al. 2007: 98).

The universal narrative that has emerged from this research is today familiar worldwide. Contemporary global warming is being forced by human behavior. No matter what our response, this forcing will continue for decades and perhaps centuries because changes in the composition of the atmosphere and of land cover have triggered transformative processes that have not yet played themselves out and established new equilibria. However, action today could have an important mitigating effect. The best models suggest that the diacritical impact of climate change is an accentuation of familiar phenomena such as drought, flooding and extreme weather. In other words, arid areas will tend to receive less precipitation, whereas coastal areas and flood plains will tend to receive more, and storms will tend to worsen. How intense and how frequent may depend on how much more greenhouse gas we continue to add to the atmosphere. What we do really matters now.

Ecological systems and global biodiversity will be affected in multiple ways. However, these patterns will not be uniform across the planet but rather will display considerable variety. Some changes will not be foreseen at all. This is partly because climate change is affected by phenomena that have proven impossible to model, at least so far, such as changes in cloud cover, as well as by unpredictable feedback responses from both natural and human systems. It is partly because complex adaptive systems, such as forests, display non-linear behavior and generate new properties. In short, while much is known about anthropogenic climate change, and broad patterns of future climate change are being predicted, there also are important areas of uncertainty.

From this discussion we can identify two further features of the universal narrative that are problematic. First, the problem has been generated by the behavior of everyone across many generations. It is an unintended externality of all modern and contemporary human societies. So, why should any particular community transform unless all the others do as well? This is a classic problem of cooperation under anarchy. Ken Oye (1986) defined three variables that are key to cooperation—the number of actors, the likelihood of future interaction among them, and the payoff structure. In this case, there are a large number of actors, some are not likely to have much other or future interaction, and the payoff is unclear, especially short of a very high level of dedicated participation. In fact, there is a sense that the actions of a few would matter very little. So the grounds for cooperation are weak. Second, there is much uncertainty. This has led observers such as Bjorn Lomborg (2001) to question whether we should trust the prescriptions of scientists or instead focus on investing into areas that generally make us more resilient and powerful, like education, public health and infrastructure. A very powerful campaign has emerged challenging the recommendations of climate scientists.

These forms of complexity and uncertainty are only part of the problem in making the universal narrative an agent of transformation. The issue of natural variability is also problematic in this regard. Natural variability suggests that absent anthropogenic influence, the range of weather in a particular area can be very large, with some extreme forms of weather rare enough to be categorized as hundred or five hundred year events. What this means is that any severe storm or long drought or intense flood might actually be that hundred or five hundred year event that is expected within the scope of natural variability. So extreme weather events are simultaneously consistent with the expectations of climate change and conceivable as things that might have occurred anyway because they lie within the extremes of natural variability. This means that the patterns of climate change that are so clear to the scientific community are always experienced on the ground with conditionality—that is, this could be a natural event or it could be an event forced by climate change. Scientists might see the distinction as neutral insofar as response is concerned. Floods are devastating to humans, other species and property. We need to respond not to the origins of a given flood, but to the probability of more flooding. But conditionality can undermine a sense of urgency, providing a basis for demanding further investigation or adopting a wait and see posture. And given variability of impacts, there will often be a constituency content with the status quo and suspicious of what change might do to its position.

Complexity, uncertainty and natural variability, together with the fact that many particular understandings of climate are part of the narrative bedrock of societies around the world, help to explain why the universal narrative breaks down quickly and has failed so far to catalyze an effective global response (for an interesting perspective on this, see Hulme 2009).

THE UNIVERSAL STORY OF HUMAN SECURITY

Human security is structurally the converse of climate change. Here the global patterns are muted and uncertain, problematized by reasonable disagreements over what exactly defines an event as civil war or genocide or sexual assault or poverty, and in which direction the world is trending on any of these issues (e.g. Suhrke 1999; Tehranian 1999; Thomas and Wilkins 1999; Yuen 2001). On the ground, these experiences are far more vivid and explicit; global patterns and trends are largely exterior and even in some sense irrelevant to the direct, personal experience of violence or deprivation. A woman being raped in a conflict zone needs assistance and may not care whether there has been an X percent decline or increase in rape globally over the past decade. The humanitarian organizations that might help that woman may not care much for the global patterns either; they will gravitate towards the pockets of acute need, period. These are their reality.

Ironically, perhaps, the same set of international elites that has embraced climate change as a reality that must be addressed has been considerably more divided on whether human security is also a reality that requires immediate attention. This means that one of the routes through which climate change might be grounded locally is itself insecure. Part of the reason for this is that the concept quickly bifurcated after its introduction to the world community.

The concept of human security is generally attributed to Dr Mahbub ul Haq, an economist who moved from the University of Karachi, to the World Bank, back to Pakistan in a government position, and then to the United Nations Development Programme. Working with people like the fabled development economist Amartya Sen, ul Haq focused on the challenges of human development, and was the central figure shaping the human security focus of the 1994 UNDP *Human Development Report*. This report offered a definition that divided the concept of human security into two parts. "It means, first, safety from such chronic threats as hunger, disease and repression. And second, it means protection from sudden and hurtful disruptions in the patterns of daily life" (UNDP 1994: 23). The UNDP report, which became the constitutive moment for the concept, identified seven areas of global concern: economic, food, health, environmental, personal, community and political.

The UNDP report triggered two remarkably different foreign policy approaches. One, associated with Canada, linked human security to a particular concept of "freedom from fear," and focused on defining strategies for protecting people from egregious forms of violence such as landmines, genocide, slavery and civil war. Its tactics include humanitarian assistance, peacebuilding, and conflict prevention and mediation, and its signature campaigns have focused on improving human security through banning the use of landmines and child soldiers.

A second approach, elaborated by the Japanese government, has established a far more inclusive approach that emphasizes "freedom from want", and with it the more complex and elaborate vision of programming and investment that moves vulnerable people towards lives of security, sustainability, welfare and dignity. Strategies and projects informed by this perspective tend to focus on development assistance, and a signature campaign would be the Millennium Development Goals.

Perhaps because it has evolved along these two distinct trajectories, the concept of human security has been criticized as too fuzzy and all-inclusive to be analytically useful. Nonetheless, it has attracted considerable attention from scholars, policymakers and activists in the developing world and Europe (Nauman 1996; Suhrke 1999; Tehranian 1999; Thomas and Wilkins 1999; Yuen 2001). Reviewing the human security literature, Roland Paris concludes that while a high level of inclusiveness can "hobble the concept of human security as a useful tool of analysis," "[d]efinitional expansiveness and ambiguity are powerful attributes of human security ... human security could provide a handy label for a broad category of research ... that may also help to establish this brand of research as a central component of the security studies field" (2001: 102).

More pointedly, Tariq Banuri provided a concise argument in defense of human security shortly after the publication of the UNDP report:

> Security denotes conditions which make people feel secure against want, deprivation, and violence; or the absence of conditions that produce insecurity, namely the threat of deprivation or violence. This brings two additional elements to the conventional connotation (referred to here as political security), namely human security and environmental security. (1996: 163–164)

In this conception, structural insecurities and the multiple forms of violence associated with the marketization and financialization of the world economy, and enduring legacies of the patterns of fragmentation and exploitation crafted or intensified through colonial practices, together with newer modalities of violence and insecurity associated with global environmental change, combine to ensure that large portions of humankind—primarily in the south but not exclusively so—are rarely, if ever, free from danger and want.

Human security is a universal concern in the sense that, while some communities have a heightened vulnerability to threats to human security, anyone can become insecure in all or some of the senses noted above. So assessing climate change as a threat to human security spotlights vulnerable areas while retaining a thin resonance to everyone, a position well elaborated in the field of global environmental politics (e.g. Adger 1999; Floyd and Matthew 2012; Khagram et al. 2003; Lonergan 1999; Matthew et al. 2009; Naumann 1996).

In sharp contrast to climate change, human security looks much like old wine in a new bottle. It brings together threats with which we are all familiar, and suggests that, with the end of the Cold War and the growth of interconnectedness, there is an opportunity, and perhaps some moral and prudential reasons, to make more progress on reducing some of these. Hence countries can pick and choose from the human security agenda, and real progress does not typically require wide participation. Except, perhaps, on those threats to human security linked to climate change.

LINKING CLIMATE CHANGE AND HUMAN SECURITY

Climate change is a very long third act. We are decades into it and it is still not clear which kind of resolution lies ahead. But it is clearly affecting human security in significant ways in many parts of the planet today, and the trend is towards deeper and broader impacts in the years ahead (Matthew et al. 2009).

The linkage between climate change and human security often is crafted by layering different sources of information. The first layer focuses on the manifestations of climate change itself—where do and will the severe floods, droughts and storms occur? The second layer focuses on measures of a society's sensitivity to these type of events—usually economic, political and demographic variables are used to establish societal vulnerability. A third layer might highlight areas already prone to the sort of outcomes climate change impacts might be expected to generate, such as violent conflicts, population displacement, disease outbreaks, and development setbacks. The extent to which the places and stories that emerge from this layering are compelling depends on the quality of the information and of the analysis that draws connections across the layers. Just as climate science has matured over the years, so too has this type of analysis, especially where analysts have sought to link climate change to violent conflict, migration and health through mediating variables such as poverty and governance.

The most recent cascade of analysis was triggered by the overview of climate change provided in the 2007 reports of the Intergovernmental Panel on Climate Change (IPCC). In fact, these reports themselves offered the world some preliminary analysis of this linkage. For example, they underscored the high sensitivity to climate change impacts of parts of South Asia, the Middle East and sub-Saharan Africa (IPCC 2007). The basic idea is that some regions of the planet are more vulnerable to climate change than others because of both their geography (exposure) and their more fragile social institutions (sensitivity). Fragile communities and countries by definition are extremely constrained in their capacity to invest in prevention, adaptation and response. In these places hazards become disasters more quickly than elsewhere. Their inhabitants will therefore experience more droughts, floods, storms and heat waves than elsewhere. Inadequate public health systems, very rapid population growth and high urbanization rates, chronic extreme poverty, recent violent conflict, and agricultural economies work together to increase the vulnerability of these areas. As climate hazards become climate disasters, these regions may suffer extensive population displacement, livelihood and development setbacks, public health challenges, food insecurity, governance failure and violent conflict.

South Asia is one of the regions that received special attention in the IPCC reports and describing this case may make the argument more concrete. The IPCC reports note that climate models converge in predicting that South Asia will experience significant warming over the next few decades. Dry areas will become drier; wet areas will become wetter; glacial lake outburst floods (GLOFs) will cause extensive damage in mountain countries such as Nepal and Bhutan; the monsoon will change in terms of timing, location and intensity; and severe weather events will increase. These changes are either unique to the region or more accentuated than in other regions of the planet. The mean temperature increase, for example, is predicted to be 3.3 degrees centigrade this century, which is higher than the predicted global mean. Adding to the equation of despair, the countries of South Asia are poor and many have recent histories of very violent conflict (Afghanistan, India, Nepal, Pakistan and Sri Lanka in the past decade). These countries may not have the capacity to mobilize effective responses to climate change impacts, opening the door to enormous human security problems.

This bleak prognosis had been reiterated and elaborated upon outside the IPCC framework. Famously, for example, Sir Nicholas Stern suggested that hundreds of millions of people could be displaced permanently by the rising sea level, massive flooding, and long droughts predicted by climate change science, leading to a range of security problems (Stern 2007). Halsnæs and Verhagen have argued that

> there are numerous linkages between climate change impacts and the MDG's starting with the influence from climate change on livelihood assets and economic growth, and continuing with a number of serious health impacts including heat-related mortality, vector-borne diseases, and water and nutrition. Specific gender and educational issues are also identified as areas that indirectly will be impacted. (2007: 665)

This perspective was expressed more forcefully in a widely read CNA report entitled *National Security and the Threat of Climate Change*, which depicted "climate change acts as a threat multiplier for instability in some of the most volatile regions of the world" that might add "to tensions even in stable regions of the world" (2007: 6–7). The following year, the German Advisory Council on Global Change published *World in Transition: Climate Change as a Security Risk*, arguing that "Climate change will overstretch many societies' adaptive capacities within the coming decades" (2008: 1). In the same vein, Dan Smith and Janna Vivekananda of International Alert have identified "46 countries—home to 2.7 billion people—in which the effects of climate change interacting with economic, social and political problems will create a high risk of violent conflict" (2007: 3).

These arguments quickly made their way into official settings and documents. For example, Germany stated that during its tenure on the United Nations Security Council (UNSC) it would work to have climate change acknowledged as a security issue "in the broadest sense" (UNGA 2009). The *2010 National Security Strategy of the United States* describes climate change as an important security issue. This position was echoed in the *2010 Quadrennial Defense Review Report*, produced by the Department of Defense to explain the current security situation to Congress (and hence the rationale for its budget appropriation). In an even more recent Harvard University report, funded by the Central Intelligence Agency and entitled *Climate Extremes: Recent Trends with Implications for National Security*, McElroy and Baker conducted a careful assessment of past research. They concluded that,

> Small positive changes in the global mean annual temperature are causing an increased prevalence of local extreme weather conditions. Over the next few years—driven by a combination of natural variability, a warmer climate from the effects of greenhouse gases, and a more vulnerable world in general—the risk of major societal disruption from weather and climate-related extreme events can be expected to increase. These stresses will affect water and food

availability, energy decisions, the design of critical infrastructure, use of the global commons such as the oceans and the Arctic region, and critical ecosystem resources. They will affect both poor and developed nations with large costs in terms of economic and human security. (2012: 4)

Vivekananda has attempted to clarify the process through which climate change impacts could lead to human security problems.

The impact of climate change will challenge and reduce the resilience of people and communities to varying degrees. In some situations, it will cause extreme disruption with which people simply cannot cope, as it overwhelms them and renders their homes and livelihoods unviable. If the governance structures that the community regards as safeguards of their human security are not up to the task, climate change will weaken confidence in the social order and its institutions and damage the glue that holds societies together. In some contexts, this can increase the risk of instability or violence. This is a particular problem in conflict-prone or conflict-affected contexts where governance structures and institutions are often weak, regardless of climate change. (2011: 8)

Not all scholars agree with the arguments summarized above. Contributors to a recent issue of the *Journal of Peace Research* presented a wide range of criticisms of this type of analysis. Rune Slettebak contends that

Despite climate change, economic and political variables remain the most important predictors of conflict. Rather than over-emphasizing conflict as a result of climate change, I would recommend keeping the focus on societal development, including building resilience against adverse effects of climate change. While this promises the possibility of alleviating the danger of climate change, it can also lead to strengthened societies in the face of natural disaster and civil war. (2012: 175)

However, in the same issue, Tor Benjaminsen et al. argue that

we have observed two different and contrasting scenarios. First, the Sahelian droughts of the 1970s and 1980s led rice farming to move down the riverbed and encroach on the dry season burgu pastures. In this sense, a drought may play a role in causing confrontations between farmers and pastoralists, increasing intercommunal tensions and, quite possibly, escalating latent conflicts to the use of violence. Conversely ... good rainfall years with generous flooding might also induce more conflicts as the zone of potential contestation is expanded to areas with less established norms of ownership and control. (2012: 109)

While it is clear that the literature embodies much disagreement, three general conclusions can be suggested. The first is that while climate change is only visible to scientists at the global level, that is, as a pattern defined by decades of global measurements, its impacts are, at least to date, severe only in very local contexts, even though they may not be easily identifiable as such. Second, while climate change could affect human security anywhere, there is a strong bias towards areas in which human security threats are already significant—especially South Asia, the Middle East, and sub-Saharan Africa. Third and finally, climate change impacts on human security are significant in these fragile areas today and they are tracking towards much greater significance in the very same places in the years ahead.

It might seem that the universal narrative of climate science has penetrated local contexts effectively, imparting a sense of considerable urgency and creating conditions conducive to a coordinated response. Perhaps because climate change is a real material form of change, it is certain to become relevant in particular settings. In this sense it is different from human rights or the end of slavery. Perhaps, then, getting the universal story right is less important than I am suggesting. I hope this is true, and that an effective response emerges. In the concluding section, however, I will argue that this may not happen at all.

ONE WORLD, TWO SOLITUDES

In spite of much self-serving and congratulatory rhetoric, over the past 20 years, the major platform for addressing climate change at the global level, the United Nations Framework Convention on Climate Change (UNFCCC), has had powerful, data rich inputs that it has worked into remarkably weak policy outputs. Its signature output, the Kyoto Protocol, was negotiated in 1997, came into effect in 2005 and expired in 2012. Kyoto had the modest objective of reducing key greenhouse gas emissions by 5 percent below 1990 levels; by the time it expired (it has been extended for another eight years), these key emissions had increased by 58 percent (Paris 2012). These numbers do not surprise many observers: India and China were not asked to be a part of the Kyoto Protocol, and the United States and Australia would not join it. The signatories themselves fell short of their own target: Canada, for example, has had to use 2005 as its baseline instead of 1990 in order to show any progress.

Kyoto is only the tip of the melting UNFCCC iceberg. After two decades of meetings in exotic locales, almost all free of pronounced climate change impacts, no rigorous methodology was developed for measuring, reporting and verifying decreases in emissions—a step so fundamental that little can be achieved without it. Equally fundamental to an effective global response is participation on a global scale—but the UNFCCC has failed in this regard, as well, content instead with crafting the so-called

"Durban platform" at COP 17, tasking itself with bringing the developing world into the UNFCCC process by 2015. Further complicating matters, a core constitutive principle of the UNFCCC, the concept of "common but differentiated responsibilities and respective capacities" (CBDR+RC), has proven very difficult to operationalize as a basis for any action. And, of course, as in all international efforts, there is considerable floundering around the question of financing. At this time, developed countries have pledged $100 billion to support mitigation and adaptation efforts beginning in 2020, leaving activity in the next several years, which may be a critical period of massive growth in urban space, land modification and energy demand, without a funding mechanism.

Why has the UNFCCC process been fundamentally unproductive, given that it attracts thousands of skilled and influential people who generally believe that climate change is a serious and perhaps even existential threat to humankind? Part of the reason has been suggested above. The universal narrative is plagued by uncertainty, complexity, and especially spatial and temporal distance, features that do not tend to catalyze strong behavioral responses. This is why scientists have often assumed that developing compelling local narratives is the key to overcoming this inertia. Climate change plays out in local contexts in very different ways, with very different impacts on human security, a crucial fact that the universal story does not express very effectively. But while the playing out of climate change in local settings may trigger action, the differences are likely to work against much coordination. One inconvenient truth is that some places may actually experience considerable benefits from global warming—the Arctic nations like Russia and Canada, for example.

I have discussed in some detail the idea that narrative of climate change is at once both thick and universal, a combination that is ill-suited to mobilizing change. Two related problems are what I will call "climate colonialism" and "climate relativism." By climate colonialism I mean to suggest that the universal narrative is divisive in two ways. First, it reinforces a North-South divide by associating climate risk with all the weaknesses that have been invested into the concept of state fragility. States are fragile because of both social and geographical factors. In other words, places that rank high on fragile state indices, like Haiti and Sudan, are in fact twice burdened. The unfortunate arguments of an Ellsworth Huntington (1915), which join race and climate, float awkwardly in the background of this conception. At the same time, second, the narrative contains an equally potent critique of Western civilization—its wildly inefficient production processes, its poor waste management practices, its one-dimensional material culture, its ethical poverty and its willingness to displace many of the costs of its lifestyle across space or project them into the future have created an existential risk for which it steadfastly refuses to assume enough responsibility. The opportunities for blamecasting are endless. The South must recognize the weakness of its own geography, the North the violence of its own history.

By climate relativism I mean the uneven allocation of climate change impacts. While the developed countries and elites around the world (who so far are also the most responsible for emissions and land cover change) have been affected by storms, droughts and floods, the scale of their suffering is quite different from that of the world's least advantaged peoples in places such as Bangladesh, Nepal and Pakistan, or Ethiopia, Kenya and the Sudan. Harald Welzer has studied "the impact of climate change on global inequalities and living conditions." He concludes that "whether wars in the twenty-first century are directly or indirectly due to climate change, violence has a great future ahead of it" (2012: 6). As climate change challenges the capacity of some groups to survive, let alone flourish, Welzer believes that no approach to addressing the problem will be left off the table—including violence. "It is a modernist superstition that allows us to keep shrinking from the idea that, when people see others as a problem, they also think that killing them is a possible solution" (p. 23). The hope of scientists and climate change negotiators that clarity about climate change impacts will lead to cooperation is, for Welzer, hopelessly naïve. He argues that the academic world underestimates the likelihood of violence because "violence is only to a very limited degree, if at all, part of the experiential world of the academics who concern themselves with it. As a result, little explicit research has been devoted to this central area of human action, and even that is overloaded with moralism and fantasy" (p. 89).

The thick content of the universal climate change narrative is poorly structured to mobilize a coordinated global response. Further complicating matters, the thick content includes strands of what I call climate colonialism and climate relativism, strands that can easily be integrated into particular accounts of climate change and that, too, mitigate against an effective and coordinated global response. Who should pay for the unfortunate geography of the South, or the destructive history of the North? Why should climate change entrepreneurs pay to mitigate the effects of that process for the sake of climate victims? This language is very stark, but these two solitudes are already visible and they may become more so in the years ahead, separate geographies connected by a handful of real and virtual bridges but largely experiencing the impacts of climate change on their own.

Hence, the rich parts of the world, the early perpetrators of climate change, are where climate impacts are and may continue to be more varied and less intense (in part because of the way they respond to these stresses, building resilience into their zones of weakness and harvesting benefits quickly when they appear). They may find that cooperation and innovation are easiest when limited to their more or less shared condition.

As for the desperately poor, whose contributions to climate change are growing (which is rapidly dismantling their moral case as innocent victims), these places may find higher walls barring them from the more stable and prosperous parts of the world, and higher levels of crisis, violence and despair growing on their side of these walls. (For an informative related view, see Collier 2000, 2008; Diamond 1994; Homer-Dixon 1999; Kaplan 1994; and Sachs 2005.)

At the end of *Climate Wars* Welzer argues for developing "a critique of any limitation of survival conditions for others" (2012: 163). This would be a worthy goal for the UNFCCC, to establish global moral principles to guide thinking and particular responses to climate change. Absent this and, over the next few decades, the two solitudes I have described might develop along largely independent trajectories, solidifying a world of climate change entrepreneurs and climate change victims, a division that may ultimately increase the risk posed by climate change to all of humanity.

REFERENCES

Adger, W.N. (1999), 'Social vulnerability to climate change and extremes in coastal Vietnam', *World Development* (February), **27** (2), 249–269.

Banuri, T. (1996), 'Human security', in Naqvi Nauman (ed.), *Rethinking Security, Rethinking Development*, Islamabad: Sustainable Development Policy Institute, 163–164.

Benjaminsen, T.A., Alinon, K., Buhaug, H. and Buseth, J.T. (2012), 'Does climate change drive land-use conflicts in the Sahel?', *Journal of Peace Research*, **49**, 97–111. doi:10.1177/0022343311427343

Brown, P. (2000), *Augustine of Hippo: A Biography*, Berkeley and Los Angeles: University of California Press.

CNA (2007), *National Security and the Threat of Climate Change*, http://securityandclimate.cna.org/ (accessed 6 March 2013).

Collier, P. (2000), *Economic Causes of Civil Conflict and Their Implications for Policy*, Washington DC, USA: World Bank.

Collier, Paul (2008), *The Bottom Billion: Why the Poorest Countries Are Failing and What Can Be Done About It*, Oxford, New York: Oxford University Press.

Diamond, J. (1994), 'Ecological collapse of past civilizations', *Proceedings of the American Philosophical Society*, **138**, 363–370.

Ellman, R. (1983), *James Joyce*, Oxford: Oxford University Press.

Eyal, T., Liberman, L., and Trope, Y. (2008), 'Judging near and distant virtue and vice', *Journal of Experimental Social Psychology*, **44**, 1204–1209.

Floyd, R. and R. Matthew (eds) (2012), *Environmental Security: Frameworks for Analysis*, Oxford, UK: Routledge.

Fujita, K., Henderson, M., Eng, J., Trope, Y., and Liberman, N. (2005), 'Spatial distance and mental construal of social events', *Psychological Science*, **17**, 278–282.

German Advisory Council on Global Change (2008), *World in Transition: Climate Change as a Security Risk*, London: Earthscan.

Germanwatch (2011) *The Climate Change Performance Index: Results 2012*, http://germanwatch.org/klima/ccpi.pdf (accessed 30 April 2012).

Gleick, P.H. (2012), 'Climate change, exponential curves, water resources, and unprecedented threats to humanity', *Climatic Change*, **100**, 125–129.

Gleick, P.H. and Heberger, M. (2012), 'The coming mega drought', *Scientific American*, **306**, 1–14.

Halsnæs, K. and Verhagen, J. (2007), 'Development based climate change adaptation and mitigation—conceptual issues and lessons learned in studies in developing countries', *Mitigation and Adaptation Strategies for Global Change*, **12** (5), 665–684.

Homer-Dixon, T. (1999), *Environment, Scarcity and Violence*, Princeton: Princeton University Press.

Hulme, M. (2009), *Why We Disagree about Climate Change: Understanding, Controversy, Inaction and Opportunity*, Cambridge, UK: Cambridge University Press.

Huntington, Ellsworth (1915), *Civilization and Climate*, New Haven, USA: Yale University Press.

Intergovernmental Panel on Climate Change (IPCC) (2007), *Working Group II Report: Climate Change Impacts, Adaptation, and Vulnerability*, Cambridge, UK: Cambridge University Press.

Kaplan, Robert (1994), 'The coming anarchy: how scarcity, crime, overpopulation, tribalism, and disease are rapidly destroying the social fabric of our planet', *The Atlantic Monthly*.

Khagram, S., W. C. Clark and D. F. Raad (2003), 'From the environment and human security to sustainable security and development', *Journal of Human Development*, **4** (2) (July), 289–313.

Le Treut, H., R. Somerville, U. Cubasch, Y. Ding, C. Mauritzen, A. Mokssit, T. Peterson and M. Prather (2007), 'Historical overview of climate change', in Solomon, S., D. Qin, M. Manning, Z. Chen, M. Marquis, K.B. Averyt, M. Tignor and H.L. Miller (eds), *Climate Change 2007: The Physical Science Basis. Contribution of Working Group I to the Fourth Assessment Report of the Intergovernmental Panel on Climate Change*, Cambridge, UK: Cambridge University Press.

Lemos, M.C. and Morehouse, B. (2005), 'The coproduction of science and policy in integrated climate assessments', *Global Environmental Change and Human Policy Dimensions*, **15** (1), 57–68.

Liberman, N. and Trope, Y. (1998), 'The role of feasibility and desirability considerations in near and distant future decisions: a test of temporal construal theory', *Journal of Personality and Social Psychology*, **75**, 5–18.

Liberman, N., Sagristano, M. and Trope, Y. (2002), 'The effect of temporal distance on level of construal', *Journal of Experimental Psychology*, **38**, 523–535.

Lomborg, B. (2001), *The Skeptical Environmentalist: Measuring the Real State of the World*, Cambridge, UK: Cambridge University Press.

Lonergan, S. (1999), *Global Environmental Change and Human Security Science Plan*, IHDP Report 11, Bonn: IHDP.

Matthew, R. (2002), *Dichotomy of Power: Nation versus State in World Politics*, New York, USA: Lexington Press.

Matthew, R. (2012), 'Environmental change, human security and regional governance: the case of the Hindu Kush-Himalaya Region', *Global Environmental Politics*, **12** (3), 100–118.

Matthew, R., Barnett, J., McDonald, B. and O'Brien, K. (eds) (2009), *Global Environmental Change and Human Security*, Cambridge, USA: MIT Press.

McElroy, M. and D.J. Baker (2012), 'Climate extremes: recent trends with implications for national security', http://environment.harvard.edu/sites/default/files/climate_extremes_report_2012-12-04.pdf (accessed 15 February 2013).

Nash, S. (2012), 'Supercomputer will help researchers map climate change down to the local level', *The Washington Post*, May 28, http://articles.washingtonpost.com/2012-05-28/national/35457102_1_climate-projections-climate-change-yellowstone (accessed 10 January 2013).

National Research Council (2012), *Himalayan Glaciers: Climate Change, Water Resources, and Water Security*, Washington DC, USA: The National Academies Press.

Nauman, Naqvi (ed.) (1996), *Rethinking Security, Rethinking Development*, Islamabad, Pakistan: Sustainable Development Policy Institute.

Oye, K. (1986), *Cooperation Under Anarchy*, Princeton, USA: Princeton University Press.

Pachauri, R.K. (2007), Nobel Lecture. Oslo, December 10. Accessed at http://www.nobel prize.org/nobel_prizes/peace/laureates/2007/ipcc-lecture_en.html

Paris, Roland (2001), 'Human security: paradigm shift or hot air?', *International Security*, **26**, 87–102.

Paris, M. (2012) 'Kyoto climate treaty sputters to a sorry end', CBC News, http://www.cbc.ca/news/politics/story/2012/12/20/pol-kyoto-protocol-part-one-ends.html (accessed 20 January 2013).

Pearce, F. (2007), *With Speed and Violence: Why Scientists Fear Tipping Points in Climate Change*, Boston: Beacon Press.

Sachs, J. (2005), 'Climate change and war', http://www.tompaine.com/print/climate_change_and_war.php (accessed 6 March 2013).

Singh, S.P., Bassignana-Khadka, I., Karky, B.S., and Sharma, E. (2011), *Climate Change in the Hindu Kush-Himalayas: The State of Current Knowledge*, International Center for Integrated Mountain Development, http://www.icimod.org/publications/index.php/search/publication/773 (accessed 6 March 2013).

Slettebak, R.T. (2012), 'Don't blame the weather! Climate-related natural disasters and civil conflict', *Journal of Peace Research*, **49**, 163–176.

Smith, D. and Vivekananda, J. (2007), *A Climate of Conflict: The Links between Climate Change, Peace and War*, London, UK: International Alert. http://www.international-alert.org/pdf/A_Climate_Of_Conflict.pdf (accessed 6 March 2013).

Solomon, S. et al. (eds) (2007), *The Physical Science Basis: Contribution of Working Group I to the Fourth Assessment Report of the Intergovernmental Panel on Climate Change*, Cambridge, UK: Cambridge University Press.

Stern, N. (2007), *The Economics of Climate Change*, Cambridge, UK: Cambridge University Press.

Suhrke, A. (1999), 'Human security and the interests of states', *Security Dialogue*, **30**, 265–276.

Tehranian, Majid (ed.) (1999), *Worlds Apart: Human Security and Global Governance*, London, UK: I.B. Tauris.

Thomas, Caroline and Peter Wilkins (eds) (1999), *Globalization, Human Security and the African Experience*, Boulder, USA: Lynne Reinner.

Todorov, A., Goren, A. and Trope, Y. (2007), 'Probability as a psychological distance: construal and preferences', *Journal of Experimental Social Psychology*, **43**, 473–482.

Trope, Y. and Liberman, N. (2000), 'Time-dependent changes in preferences', *Journal of Personal and Social Psychology*, 79, 876–889.

Trope, Y. and Liberman, N. (2003), 'Temporal construal', *Psychology Review*, 110, 403–421.

UNDP (1994), *Human Development Report 1994*, Oxford, UK: Oxford University Press.

United Nations General Assembly (UNGA) (2009), 'Climate change and its possible security implications: report of the Secretary-General', 11 September, A/64/350.

Vivekananda, J. (2011), 'Practice Note: Conflict-sensitive responses to climate change in South Asia', International Alert, http://www.international-alert.org/sites/default/files/publications/201110IfPEWResponsesClimChangeSAsia.pdf (accessed 6 March 2013).

Wakslak, C.J., Trope, Y., Liberman, N. and Alony, R. (2006), 'Seeing the forest when entry is unlikely: probability and the mental representation of events', *Journal of Experimental Psychology: General*, **135**, 641–653.

Walzer, M. (1994), *Thick and Thin: Moral Argument at Home and Abroad*, Notre Dame: University of Notre Dame Press, 2006.

Weart, S. (2008), *The Discovery of Global Warming*, Revised and Expanded Edition, Cambridge, USA: Harvard University Press.

Welzer, Harald (2012), *Climate Wars: Why People Will Be Killed in the 21st Century*, Cambridge, UK: Polity Press.

White, I., R. Kingston and A. Barber (2010), 'Participatory geographic information sysems and public engagement within flood risk management', *Journal of Flood Risk Management*, **3** (4), 337–346.

Yuen, F. K. (2001), 'Human security: a shotgun approach to alleviating human misery?', *Global Governance*, **7**, 231–236.

PART II

THE DETERMINANTS OF HUMAN SECURITY IN THE CLIMATE CHANGE CONTEXT

PART II

5. The environmental determinants of human security in the context of climate change
David Simon

INTRODUCTION

> Human security is achieved when and where individuals and communities have the options necessary to end, mitigate, or adapt to threats to their human, environmental, and social rights; have the capacity and freedom to exercise these options; and actively participate in attaining these options. (Lonergan et al. 1999: 18)

> human security is to safeguard the 'vital core' of all human lives from critical pervasive threats, without impeding long term human fulfilment. (Bohle 2007: 14)

Despite their apparently different emphases, these definitions of human security integrate concerns with both the ability to meet short term needs and rights, and longer term capacities and fulfilment. Put differently, human security is about realizing – and having the capacity to realize – one's potential as an individual and as part of a wider community. This requires the exercise of human agency within broadly enabling societal *and environmental* contexts for the control of vulnerability and promotion of sustainability (of which the resilience of appropriate elements is one dimension). In keeping with the objectives of this handbook, this chapter focuses on environmental dimensions as one particularly important aspect of those broader human security contexts in relation to the growing challenges of climate change.

For this purpose, the broader term environmental change (EC) is actually more appropriate than simply climate change (CC). This is because it includes not only the specifically climatic variables of changing ambient temperatures and precipitation patterns but also changes to a wider range of important environmental conditions, such as sea-level rise, induced by increasing greenhouse gas emissions, as well as the bidirectional interaction (i.e. feedback loops) between human agency and the environment (often called human–environment relations) that are so central to understanding the phenomenon and addressing its implications. Because of the global-scale circulation of the atmosphere and oceans, such changes affect the whole world, with impacts often being felt very far from the source(s) of such change (the phenomenon known in climate science as teleconnections or telecoupling), the term global environmental change (GEC) is often used, also in part to distinguish it from ECs associated with 'normal' variations in prevailing conditions. For simplicity, EC is used here, with any particular aspects distinguished explicitly.

Environmental factors constitute important direct and indirect influences on human security in many contexts. While often perceived as being externalities or acts of God (or the will of the gods in polytheistic world views) to signify forces or processes beyond human control, the nature of their influence or impact on humans is frequently neither entirely random nor inexplicable. Although some people are inevitably just 'in the wrong place at the wrong time' when sudden extreme events like a tsunami or unanticipated earthquake strike, more generally patterns of exposure, vulnerability and resilience to environmental hazards reflect the way in which human societies are organized. In other words, where particular groups of people live and work in relation to topography, zones of actual or perceived amenity (however defined and valorized) and disamenity or hazard, how they travel between these areas, the kinds of work they perform and the resources they command in order to achieve these outcomes reflect social relations.

Historically there have been numerous systems and principles for organizing social and productive relations but almost everywhere today, capitalist relations of production and social reproduction have become a substantial, if not the dominant or sole, basis. Hence social groups differentiated increasingly in terms of the incomes and other assets or forms of capital that they control (whether clearly delineated into distinct classes or not), occupy different economic, political and environmental niches in society and within urban areas. In general, wealthy groups occupy relatively large areas of high amenity and low risk comprising low density, well constructed and durable housing on large plots. Other groups form a gradient (topographically and economically), with the poorest occupying the smallest, most flimsily built dwellings at high densities and on the lowest value land, which is often exposed to the greatest hazards and risks of flooding, landslides, erosion, subsidence or major industrial accidents, for example. Hence the least well resourced and endowed groups are exposed to the greatest risk from daily background health hazards of exposure to relatively or absolutely unhealthy and hazardous environments, and diverse sources of extreme events (both 'natural' and anthropogenic). EC, now recognized as increasingly anthropogenic in origin and severity, is very likely to exacerbate these patterns – and experiences – of unequal exposure and hazard, as several of the contributions to Pelling et al. (2012) demonstrate in particular contexts.

In high-income, late- or post-industrial societies it is easier than in poor countries to forget or underestimate the potential impact of environmental influences on human security because of the extent and general reliability of environmental management through both 'hard' and 'soft' interventions. The former comprise engineering and infrastructural approaches involving the construction or modification of physical structures and facilities such as seawalls, groyns, reservoirs, flood diversion channels and sophisticated waste treatment plants capable of handling and perhaps even recycling hazardous substances. By contrast, the latter ('soft') category refers to both institutional and community-based governance and management capacity and the availability of financial instruments like insurance that mediate risk and vulnerability while boosting resilience and recoverability from extreme events and other shocks.

Since the insurance and reinsurance industries operate on the basis of sophisticated actuarial risk assessments, phenomena like environmental change that alter 'normal' risk profiles substantially have the potential to undermine them. Hence it is no coincidence that these industries grasped the potential impacts of environmental change ahead of most others, and firms like Munich Re and Swiss Re have been at the forefront of research seeking to understand the most likely parameters of change and their implications for current and anticipated risk exposure profiles embodied in their (re)insurance portfolios (Canada Institute of the Woodrow et al. 2009; Munich Re 2012; Swiss Re 2012).

Of course, while insurance cover in high-income countries is widespread (though certainly not universal, especially among the poor), in low and lower middle-income countries, it is principally the large-scale commercial sector and elite and middle classes who are able to afford insurance premium payments; apart from large farms, plantations, tourist facilities and mines, insurance cover is also therefore heavily concentrated in urban areas. Hence, in all situations those people excluded from insurance markets – who are overwhelmingly the least well resource-endowed (including in terms of political connections to 'get things done') and hence most vulnerable to shocks – suffer a double disadvantage and are at greatest risk of ruin in the event of extreme events or other hazards and disasters. They may become destitute, especially as each successive shock is likely to erode further whatever resilience and recoverability they might (still) have, thereby increasing their vulnerability and leaving them dependent on any available government- or donor-funded emergency assistance and recovery programmes (Wisner et al. 2004).

In urban contexts, the principal EC challenges relate to securing the integrity of shelter, infrastructure, productive capacity and economic assets, recreational facilities, livelihoods, waste treatment facilities (especially for hazardous substances) and governance institutions and systems under GEC-induced duress. In low and lower middle-income countries, the livelihoods of the substantial proportion of absolutely and relatively poor urban households may encompass more diverse activities than in rural areas on account of their multi-activity survival strategies. That said, one of the most important trends over recent decades has been for many rural-dwelling households to engage in multi-local and multi-activity survival strategies, exploiting whatever opportunities arise. Often that involves different household members circulating between two or more of their rural homestead and associated land, rural wage-earning opportunities (e.g. on large farms or plantations), rural service centres, intermediate towns and large cities for different periods (commuting, seasonal work) and purposes (accessing health or educational services, temporary or longer term wage employment) in order to spread and minimize risk and maximize subsistence output, wage incomes and other needs.

Just as increasing recognition of the nature and importance of highly dynamic peri-urban zones or interfaces has blurred traditional urban-rural distinctions, so livelihood diversification over space and time has made the classification of households into one of these categories difficult. Reasons for such livelihood trends are complex and contingent. Originally identified as a consequence of increasing rural land pressure (including land alienation) and impoverishment, and the consequences of structural adjustment programmes in the 1980s and early 1990s (e.g. Holm 1995; Mbonile 1995), such complex mobility patterns are now being driven in part by environmental change in some areas.

This phenomenon has emerged first in the more marginal contexts of human habitation, such as arid and semi-arid zones (e.g. the Sahel and Horn of Africa), where climate variability, increasing temperatures and falling precipitation levels appear already to be undermining the viability of traditional rural-based livelihoods, and in flood-prone, low-lying areas such as the Ganges and Brahmaputra deltas vulnerable to the combined effects of Himalayan snow melt swelling the rivers and changed ocean currents and rainfall patterns in the Bay of Bengal (e.g. CARE Bangladesh 2003; Guèye et al. 2007; IPCC 2007a; McGranahan et al. 2008; ODI 2009; Hartmann et al. 2010; Simon 2012). In the following sections, the respective environmental challenges to human security in the context of climate/environmental change will be examined in relation to different settings at different scales and over different periods of time.

ENVIRONMENTAL CHANGE AND HUMAN SECURITY: THE LINKS

Scale Effects

The human security implications of environmental change are not scale-neutral and therefore demonstrate discontinuities and disjunctures when examined at different scales, which are, of course, intimately interconnected in the contemporary world, whether considering food security as the ultimate precondition for human security or environmental challenges bound up with urbanism (Lobell et al. 2008; Ingram et al. 2010; Marcotullio and McGranahan 2007). International and national discourses and policies tend to emphasize relatively abstract concepts of global or national interest, which generally translate into top-down sectoral interventions for 'the common good'. Concern with aggregate outcomes inevitably masks variations among particular groups or smaller spatial units. For regionally or locally defined institutions or social groups, such variations can make profound differences and represent a central concern, even though there are also almost invariably variations at still smaller scales. Even at the lowest level, that of individuals and households, power relations deriving from personal characteristics (see below) and other differences can translate into differential vulnerabilities and sources of resilience and recoverability in relation to the environment and other resources, but these are often 'invisible' and overlooked in top-down interventions. Conversely, scaling up does not easily account for the potential of collectivities to attain security through group action in situations where the whole is more than the sum of its constituent parts.

Categories of Environmental Risk and Human Insecurity

The range of environmental risks that might or do trigger human insecurity can be classified in different ways, none of them entirely unambiguous and discrete because of overlaps or discontinuities. This applies to distinctions such as between coastal and inland, highland and lowland, or urban and rural locations, for although there are some differences, other risks, e.g. landslides, floods and droughts, occur in both urban and rural areas. Such simple dichotomies are also increasingly seen as problematic in both theoretical and policy terms. For instance, the flourishing of research and policy attention to peri-urban areas over the last 15–20 years has demonstrated the far greater appropriateness of a conceptual continuum between the unambiguously rural and clearly urban for understanding changes and functional relationships (Simon et al. 2004; Simon 2008).

A great deal of the experience in this regard generated under the umbrella of disaster risk reduction (DRR) over the last 25 years or so, including during the UN Decade of Natural Disaster Reduction in the 1990s, has relevance here (e.g. Wisner et al. 2004; Pelling and Wisner 2009). While climate/environmental change does overlap with 'natural disasters', it is also partly distinctive in nature because it comprises two distinct elements. The first is the increasing severity and – as seems increasingly likely according to IPCC reports (cf. IPCC 2007a, 2007b, 2012) – possibly also frequency of extreme events, which does relate very closely to disasters, the literature on which is highly pertinent. However, the second dimension of environmental change is distinct, namely the slow-onset but semi-permanent or permanent changes to prevailing environmental parameters and conditions, such as rising sea level, and increased mean atmospheric and ocean temperatures.

Conceptually, therefore, these two broad categories or dimensions of EC might provide an appropriate framework by means of which to characterize environmental drivers of insecurity. However, since they overlap and, where they do, individual impacts are felt cumulatively in combination (e.g. damage from a storm surge that is more severe as a result of an elevated sea level as well as a spring high tide), in practice this becomes unhelpful.

A more useful approach would therefore be to modify the widely used definition of disaster risk (DR) as being a product of hazard x vulnerability, or in expanded form $DR = H \times [(V/C) - M]$, where H is the hazard, V stands for vulnerability, C is the capacity to protect oneself and M represents risk mitigation through collective action at other scales (Wisner et al. 2004; Wisner et al. 2012: 23–4). Another version of this approach includes exposure as an additional term on the left of the equation. The Disaster Risk Index is an attempt to scale up exposure and vulnerability to the global level (Peduzzi et al. 2009).

Such equations are, of course, at best crude approximations of complex realities in order to highlight the principal variables and their interrelationships. Conceptually it might seem attractive to attempt to parameterize the respective variables but it is doubtful whether any universal parameters could be identified empirically. Moreover, actual outcomes in a particular situation also depend on locally contingent power relations and other 'invisible' factors that would be very difficult to include except as notional (but probably unquantifiable) additional parameters.

In that spirit, my point in introducing the equations is not to argue for greater quantification or the research to underpin it but merely to underscore the differences between DR and EC by pointing out that, in the context of EC, modification of the DR equation(s) becomes necessary because of the differential duration of the various environmental changes and hence their respective impacts. Hence taking into account the intensity (severity) of the hazard, the level of vulnerability (itself a function of the parameters discussed above), both the degree and duration of exposure, and mitigation capacity, yields the following equation:

$$CR = H \times [I \times E]/D [(V/C) - M]$$

where CR represents climate risk, H is the hazard, I the intensity of that hazard, D is its duration, E is extent of asset/livelihood exposure, V is vulnerability, C is the capacity to protect oneself, and M is risk mitigation through collective action at other scales.

At the group level (however the group is defined), it also becomes important to know the number of people exposed and their respective durations, as well as the extent to which their assets/activities are exposed in order to understand how severely their current asset base and livelihood activities are affected.

The Extent of Exposure to Environmental Risk and Uncertainty

Geographical location, in the terms just outlined and as will be elaborated below, is one important variable affecting likely or possible degrees of exposure. In other words, place matters in terms of potential or actual environmental risk and insecurity. However, in practice, the degree of exposure for any individual, household or group at a moment in time depends on the interaction of a precise combination of factors or variables as they play out relationally over space.

Some of the other key variables are personal characteristics such as age, gender, position in household and community structures, extent of formal education and life and other skills, employment status, health status, nutritional status, access to and control over financial and other non-material resources, and political affiliation. Within culturally and socially diverse situations, these variables are likely to be cross-cut by social cleavages based on ethnic, caste or other ascribed occupational categories, class, religious and cultural differences. These can perhaps best be encapsulated as people's degree of empowerment, what Friedmann (1992) terms their access to the bases for accumulating social power.

As a broad generalization, the more that individuals, households and communities (however defined) depend on, and are exposed to, the environment, natural resources and ecosystem services, the greater their likely vulnerability to EC and hence the likely risks to their human security. That said, it is important to recognize that human security issues in relation to

EC have been considered almost exclusively in global, regional, national and sectoral terms. Somewhat surprisingly, there have to date been only initial attempts to 'urbanize' this agenda (e.g. Dalby 2009; Simon and Leck 2010, 2013) despite the fact that many poor, disempowered urban dwellers, in particular, also face substantial human insecurity arising through particular sets of urban environmental risks, which are and will be exacerbated by EC. These range from increased exposure to rising temperatures as a result of the urban heat island effect to possible uncertainty over the security of urban food supplies; water shortages if precipitation declines as urban demand rises; possible failures of electricity supply at periods of peak demand (see below); the impact of localized violence as heightened frustrations with conditions erupt at key moments; or even broader breakdowns of law and order if more widespread and sustained uprisings occur; not forgetting everyday forms of urban human insecurity linked directly to the lived environment. These will now be examined in greater detail.

IMPLICATIONS OF EC FOR HUMAN SECURITY IN THE BUILT ENVIRONMENT

The range of possible and likely urban impacts of EC across the great diversity of urban places worldwide is clearly very large, depending on numerous interacting variables under the precise EC scenario. Even within the same agro-climatic zone, a 'mature', stable town or city with good quality infrastructure and institutional capacity will have rather different prospects for anticipating and mitigating the impacts of particular ECs and adapting to the changing conditions than a rapidly growing one with inadequate shelter and infrastructure and low institutional capacity. An additional important factor is the extent to which a given urban area is locally, regionally and/or globally embedded and dependent in terms of its economic structure and processes and environmental resource demands and waste disposal. This is because the balance among different urban environmental burdens tends to shift in scale as mean incomes rise and economic production becomes more technologically sophisticated, from predominantly local to more regional and global (McGranahan 2007).

Outcomes will be contextually contingent: for instance, urban human security in a highly locally autonomous and self-reliant city will be high if regional and international production and distribution systems become vulnerable to EC; however, if the particular locality in which such a city is located is badly affected by a string of extreme events associated with EC then that same self-reliance will rapidly become a source of great vulnerability through the disruption of local food, water and energy supplies, plus likely damage to the city itself. What follows is therefore an outline of the principal known categories of urban EC impacts likely to affect parts or all of many urban areas in different regions. This does not mean that individual cities and towns experience EC uniformly. On the contrary, substantial intra-urban variations exist in terms of exposure, vulnerability and coping capacity and adaptability, which reflect the processes that produce and reproduce the built environment and which categories of people inhabit the different areas.

Intensified Heat Island Effects

Prolonged exposure to higher prevailing temperatures or short periods of unusually intense heat (heatwaves) pose a risk of heat stress and associated morbidity, and in extreme cases even death. As with all extreme conditions, it is young children, people with chronic health conditions (especially conditions affecting the airway and circulatory system) and the elderly who are most vulnerable. However, elevated risk occurs where such biophysical vulnerabilities cross-cut other personal or community characteristics listed above, such as poverty (and hence reduced affordability of electricity and electric appliances, for instance) and living alone. Hence, for example, the European summer heatwave of 2003 resulted in abnormally high levels of mortality among the elderly living alone or left alone while other members of the family were on their annual vacations during August (IPCC 2007b: 108). Such effects are also spatially mediated, with disproportionately large concentrations of poorer, elderly people living alone in inner city apartment blocks, where the heat island effects are often particularly intense and other amenities deficient. In Toronto, maps reveal very visually the substantial overlap between the geographies of elderly low-income people living alone and of peak summer mean temperatures (Gower 2011); such cartographic work is invaluable in assessing risk and hence prioritizing emergency responses during extreme weather, such as the Heat Alert and Response Systems (HARS) being piloted in several Canadian cities (Berry 2011) or more sustained interventions to mitigate the impacts and promote adaptation.

Such increasing mean annual and extreme summer temperatures not only influence morbidity and mortality patterns – which in temperate zones now seem to be becoming bimodal, with summer peaks as well as the traditional winter peaks resulting from elevated cold-weather death rates among vulnerable groups – but are changing the nature and timing of energy consumption. In many parts of Europe, North America and Japan, for instance, increasingly hot summers over the last decade have led to far more widespread use of electric fans and air conditioning, with the result that peak electricity demand in the summer months has matched or exceeded that during the cold winter months, when heating drives up energy consumption.

EC is likely to have similar impacts in the many tropical and subtropical areas also experiencing changing and more extreme temperatures. A dramatic illustration of what could happen is the massive power failure experienced so dramatically by over 600 million Indians in 20 of the country's 28 states for 48 hours on 30–31 July 2012 when parts of the national electricity grid could not cope with demand during hot conditions (BBC 2012). Electricity supplies have been struggling to keep pace with the combined impact of rapid industrial growth and rising incomes for a proportion of the growing population, more of whom now live in western-style apartments and houses, using air conditioners and other appliances.[1] If such trends are sustained, as seems very likely in view of IPCC and other climate change projections through this century, they will have profound implications for national and regional energy policy and planning. They will also increase the urgency of retrofitting existing buildings to improve energy efficiency and insulation against the predicted changes locally, and of ensuring that all new construction utilizes low-carbon technologies and products.

Urban Food Security

Food supplies might become less reliable and secure in areas where rural agricultural production (and such urban cultivation as may exist) is adversely affected by EC or in cities supplied from such affected areas (thus putting nutritional status and health at possible risk). In this respect, modelling the effects is relatively straightforward since the consequences of food insecurity as a result of shortages and famines are well understood from long experience with DRR and the deployment of famine early warning systems. These generally use increasing market prices and declining physical availability as indicators of impending problems, triggering the mobilization by relief and aid agencies of emergency supplies for positioning as close as possible to affected areas so that valuable time and hence lives are not lost when distribution of these supplies becomes necessary. However, again, the extent and intensity of shortages, and who is affected by them, usually reflects the interplay between changing availability, prevailing social structure in terms of affordability and other means of exercising entitlement through barter, use of social capital, including reciprocity relations, and the like.

Declines in access to food affect nutritional status unless appropriate substitutes (such as cassava, a tropical carbohydrate crop frequently of last resort that grows well almost anywhere, including in urban areas) are available, quite soon then also translating into reduced resistance to disease, reduced productivity or even ability to work, and hence falling money incomes in what could become a downward spiral if conditions persist and in the absence of relief interventions.

Conventional drought or famine conditions rarely last more than a few seasons but under conditions of EC, 'normality' itself will change and substantial adaptations may be needed to cope. Moreover, urban food shortages are known to be considerably more problematic politically than in rural areas because of the dense concentrations of people, many with higher literacy and skill levels, and the ability to protest much more vociferously directly to the seat of power at the parliament or presidential palace in so-called 'bread riots'.

Urban Unrest

Such riots are a well-documented phenomenon, often triggered by sudden food price rises resulting from shortages or, more commonly, from large reductions in, or the abolition of, price controls and subsidies. This occurred in many countries during the early years of structural adjust-ment programmes (SAPs) in the 1980s, designed to reduce government expenditure during the debt crisis, and more recently as part of market lib-eralization measures (e.g. Harris and Fabricius 1996; Riddell 1997; Simon 1995, 1997). Such considerations may help to raise the political priority accorded to addressing EC and its urban implications.

Other forms of urban unrest are most commonly triggered when under-lying grievances over inequality, discrimination, poverty, unemployment or vulnerability to daily micro-scale environmental hazards (as explained above) or extreme events are sparked by a particularly unpopular act such as an incident of police violence and brutality, especially against a member of an ethnic or religious group that feels discriminated against. Claims of 'otherness' – of being foreign, migrants/'strangers' or simply 'not from here' – are often associated with such perceptions and attitudes by dominant groups. Such sentiments may then be exploited for political ends. Periodic intercommunal violence in many northern Nigerian cities over recent years is a case in point, as are the riots that sometimes occur in US, British or French inner cities or *banlieues* (e.g. Dikeç 2007), gener-ally in the heat of summer when frustrations boil over. Even if defused or quelled rapidly, such episodes invariably involve the looting and/or arson of shops, offices and perhaps homes above them, and vehicles. While large firms and public institutions are generally insured, small shopkeepers and family businesses, especially in poor countries, may lack such cover or be underinsured and lose their livelihoods as a result.

Hostile Urban Environments

Even under less extreme circumstances, daily life for the urban poor and some minorities is often insecure in alienating and possibly hostile environments of soulless high rise housing estates, substandard formal or informal housing, poor if any social amenities and facilities, and considerable unemployment. Gang and/or drug cultures are often well established, accompanied by high levels of intimidation and violence or the threat of it. Many adults and the elderly live in fear, even when indoors. However, it is often teenagers and youths who are most at risk as a result of rival gang violence. In such cases, it is not dependence on environmental resources, as in rural areas, that creates the human insecurity but rather the combination of hostile physical environmental attributes coupled with inadequate access to other environmental resources for recreation, aesthetic enhancement or even basic utilization in a livelihood strategy.

Long experience demonstrates that official upgrading and related programmes which might improve residents' human security rarely achieve their objectives without full participation and co-management by residents. Moreover, attempting to move beyond amelioration of the symptoms of the inequalities that produce such conditions and acute human insecurity, as well as reliance on, or reinforcement of, clientelist political relations requires resident-led initiatives to redefine the terms of engagement among residents and between them and the local state, using methodologies such as pioneered by Shack/Slum Dwellers International (Mitlin 2012; see also Dalby 2009). Of course, wealthier and more empowered people are able to relocate out of such contexts or to secure their homes, livelihoods and themselves by various means.

BEYOND THE URBAN FRINGE: RESOURCE DEPENDENCE AND HUMAN (IN)SECURITY

In terms of the extent and intensity of exposure approach suggested above, it is most useful to consider the range of situations as forming a continuum in terms of vulnerability to the impacts of EC, with many modes of production and forms of social economy conventionally thought of as distinct, such as hunting-gathering and pastoralism, in fact merging into one another, with elements of two or more not uncommonly deployed as parts of complex multi-activity livelihood strategies, the precise combinations of which are dynamic according to prevailing environmental conditions and other circumstances.

Highly Vulnerable Groups and Livelihoods

The most vulnerable people and communities to EC are arguably small-scale hunter-gatherer clans such as the Australian aboriginals, Inuit of the Arctic, Kalahari Bushmen and baYaka/Bambenga 'pygmies' of the African equatorial rainforests, and numerous groups in Amazonia because of their total reliance on the natural environment and its resources. Any changes to prevailing conditions that further desiccate the Kalahari sand-veld or 'red centre' of Australia, for instance, or to loss of key elements of forest biodiversity on which they rely, could displace such groups or bring an end to their social economies and survival prospects. The rapid shrinking of the Arctic icecap over the last decade may also herald broader environmental changes that threaten the lifeworld of the Inuit.

In fact, very few of these 'First Peoples' still retain entirely traditional lifestyles, most having been westernized and sedentarized to some extent, either voluntarily or involuntarily. The latter category includes both coercive resettlement in the name of conservation and/or rural development, such as forced relocation out of their habitual home ranges in Botswana's Central Kalahari Game Reserve (e.g. Solway 1998; Hitchcock 2002) and the more numerous examples where the division of land into individual fenced farms has deprived communities of access to seasonally available resources, including water, and their transhumance routes following wildlife and water, or forest logging and water contamination through often illicit mining activities destroys their resource base. The result has often been impoverished and alienated lives as farm herders for others (e.g. Sylvain 2001), domestic servants and other menial jobs where available, accompanied by varying degrees of alcohol abuse and commercial sex work on the fringes of rural settlements or small towns. Extensive artisanal gold mining is now taking place deep in some of the Cameroonian and Congolese rainforest ranges of various 'pygmy' groups, involving them in some capacities but with still unknown future consequences for their economies, societies and autonomy, especially in relation to the integrity of the forest resource base if EC brings other changes. Many North American Inuit now live traditionally for only part of the year, keeping an urban home for winter, using snowmobiles and other equipment, and attending formal schools and urban health facilities at least some of the time. Unless assisted appropriately by the relevant authorities when lifestyles become unsustainable through environmental and other changes, this is how many of those still living on and from the land are likely to end up.

Overlapping with hunter-gathering lifestyles and their traditional ranges are those associated with nomadic or semi-nomadic pastoralism and dependent on their ability to drive their herds seasonally, most of which have also been severely affected by sedentarization (by design or default), individualization of land holdings and associated fencing, and livestock disease control measures. Indeed, it should be noted that hunting and gathering and herding were and are more commonly combined, either simultaneously or sequentially by particular groups as components of multi-faceted livelihood strategies in areas subject to variable prevailing conditions, than is generally appreciated (Wilmsen 1989; Gordon 1992; Smith et al. 2000).

Predicted ECs in the Sahel, East Africa and the Horn of Africa, for instance, suggest generally increasing temperatures and precipitation but which will largely cancel each other out. The net effect will probably exacerbate recent trends, increasing pressure on remaining water and grazing in most areas, reducing livestock carrying capacities and almost certainly intensifying existing conflicts over access to resources with private landowners, sedentary communities and within national parks. While the Maasai of southern Kenya and northern Tanzania are the most iconic example, the problems are widespread. Intensifying cattle-raiding conflicts between neighbouring (semi-)pastoralist groups in the northern Rift Valley, especially the Turkana, Pokot, Marakwet, Samburu and Karamojong, in western Kenya and eastern Uganda are being attributed anecdotally locally in part to the combined pressures of population increase and EC. As stock die and pastoral livelihoods become increasingly precarious, more such people will be driven into semi-sedentarized lives in urban and peri-urban areas, as already witnessed in Yabello, Ethiopia (Aberra 2006), unless appropriate policies can facilitate adaptability within what is a highly appropriate livelihood and means of production under climatically variable conditions in areas of marginal agricultural land. Such measures will need to include governance reforms to enhance security of land rights and increase representation and involvement of pastoralists; improved access to markets and related measures; and promotion of education and skills (ODI 2009; Hartmann et al. 2010).

Warming in the Arctic and sub-Arctic tundra belt is already adversely affecting the Sami (Lapp) reindeer herders, whose ranges are now also increasingly threatened by the new oil and gas prospecting rush in the region following the retreat of the ice (Tozer 2011; UNESCO 2011). Like the Inuit, their lifestyles have changed substantially in recent decades, becoming seasonally sedentary and utilizing many forms of modern technology. Nevertheless, a dramatic example of how an extreme (in this case anthropogenic) event can undermine their livelihoods at a stroke was provided by the radioactive fallout in parts of Lappland from the Chernobyl nuclear accident in the Ukraine, which rendered herds in the areas affected unfit for consumption and having to be culled (Stephens 2010). More than 25 years later, some areas that suffered contamination remain quarantined against use for grazing or cultivation.

Sedentary Agriculture, Fisheries and Forestry

Sedentary agriculture and livestock raising on a subsistence, smallholder or larger-scale commercial basis varies in vulnerability according to the precise combination of farmer, land and environmental variables discussed earlier, mediated by the degree and nature of autonomy over decision-making and operations and also of insertion into commercial markets. A true subsistence farmer or peasant able to market a small surplus might retain full autonomy and thus the flexibility of responding to seasonal variations in prevailing conditions rather than being locked into a multi-year outgrower or co-operative sales agreement, and to practice multi- and intercropping to diversify diets, minimize risk of crop failure and to optimize food and cash crop availability at different stages of the year rather than being required to grow a specified crop. However, that autonomy could become a source of vulnerability if fluctuating conditions linked to EC undermine sustainability because of the absence of access to credit, advice and technical assistance or other forms of support. Conversely, the nature of outgrower or supply contracts varies greatly, with some being more constraining and exploitative than supportive. Productivity, intensity of cultivation, and environmental sustainability are often higher on smallholder commercial farms than either subsistence or large-scale commercial farms. More generally, however, modern capital-intensive farming often involves strong economies of scale in certain capital investments, such as green revolution technologies for production of wheat, maize, rice or soya, for instance, with the result that small farmers are relatively or absolutely disadvantaged. The current and future impacts of EC for human security of general farming households and communities are too diverse to analyse individually here.

Forestry and both capture fisheries and aquaculture, so important to local and regional economies where widely practised at subsistence and especially commercial scales, also have environmental sensitivities. Freshwater fisheries are likely to be particularly severely affected by EC in areas where river flows diminish (affecting water chemistry and overall quality) or floods become more frequent, in both cases probably affecting mean water temperatures, to which many fish and crustacean species that form the basis of capture and aquaculture industries are highly sensitive. Changes in seawater temperatures and salinity will be more modest overall, although possibly significant in particular inshore localities, such as river mouths, estuaries, lagoons and sheltered bays and fjords in which key spawning grounds, high species diversity and aquaculture pens and cages are concentrated. Rising sea levels and increasingly severe and perhaps frequent storminess will pose particular hazards for fishing installations and inshore fisheries (e.g. FAO 2008).

In terms of forestry, individual tree species vary in their temperature, rainfall and other tolerance ranges and the levels of mean temperature changes predicted under the median to upper forecast scenarios in the IPCC AR4 will certainly exceed some of those limits. Increased vulnerability to pest infestation and disease, defoliation of both deciduous and evergreen tree species, leading to a rapid increase in wildfire risk and ultimately death of the trees are the likely consequences. The risk to human security of larger and more frequent wildfires of the sort sometimes experienced during hot, dry episodes in major forestry areas in parts of California, mainland Greece, South Africa, Kenya and southern Australia, for instance, or from seasonal tropical forest clearances in Southeast Asia or Amazonia that get out of control is considerable. Lessons learned in tackling such conflagrations, which often cause deaths and destroy homes and infrastructure as they rage out of control, will be important in coping with future scenarios.

Widespread forest die-offs and wildfires would have adverse environmental and ecological consequences in terms of a likely increase in erosion, loss of species diversity in indigenous forests and woodlands, and of output from forestry. If hardier, more tolerant species are available, substitution will be necessary (perhaps leading to an increase in utilization of exotic species); otherwise a change in land use will have to be implemented. Forest loss will also affect human security indirectly, through the loss of its carbon sequestration function, which could be locally very important as well as regionally or globally significant. International initiatives like Reduced Emissions from Deforestation and Forest Degradation (REDD and REDD+) are designed to tackle forest loss in poor countries on account of their ecosystem service functions as 'green lungs' in terms of carbon sequestration, and in biodiversity and water conservation. Quite apart from some controversial provisions in the schemes in terms of their impact on local forest-using communities, the extent to which they are likely to be able to enhance actual forest resilience to EC is limited.

Specialized Export-Oriented Commercial Agriculture

Other forms of commercial agricultural production with particular vulnerability to EC are those, regardless of physical scale, that are highly specialized rather than diversified, in the sense of having all their proverbial eggs in one basket. Wealthier, more sophisticated farmers producing single high value crops (e.g. hops) can protect themselves against hazards by means of appropriate investments in physical infrastructure and ecosystem services (e.g. protective barriers, improved drainage or irrigation, water conservation techniques, green fences, shade-giving and fruit-producing trees), insurance, and by having savings and access to commercial credit when required.

The hops example is particularly germane since it has quite specific growing requirements and is therefore sensitive to extremes of weather and is likely to become increasingly difficult to grow under conditions of EC. Because production is often concentrated in particularly suitable niche areas (perhaps one or two valleys in a specific locality), it may well become necessary to shift into other crops if ECs exceed hops' quite narrow tolerance range. Given the specialized high trellises or netting on which it grows, such crop switching will represent a considerable loss to the farmers concerned unless they have reached the end of their useful life or planned payback period. This is one example of the circumstances under which EC-related insurance claims of a new kind, linked to longer term changes in conditions rather than individual extreme events, are likely to be lodged and, no doubt, tested in court. In that sense, hops serves as a metaphor for many other specialized forms of agricultural and horticultural production involving considerable investment in facilities like greenhouses, polytunnels, sprinklers or other irrigation systems, and micro-climatic control systems for regulating temperature and humidity, the viability of which is likely to be threatened by sustained changes to prevailing conditions at the margins of current tolerance ranges.

While some of these systems were developed to enhance the quality of the produce and extend the growing season for local and regional markets, especially in Mediterranean and temperate climatic zones, others were installed specifically for intercontinental markets as part of production diversification drives into non-traditional exports linked to debt relief and economic liberalization programmes instituted in Latin America, Africa and elsewhere in the 1980s and 1990s. Indeed, this is the origin of the capital-intensive and large-scale Colombian, Ecuadorian, Peruvian, Kenyan and Zambian export industries for cut flowers, mange tout, snow peas, green beans, strawberries and the like to North America and central/ northern Europe (although some is consumed in the producer countries).

Most of these are geographically concentrated and now constitute *the* or a principal economic activity in those localities, which are generally situated within easy reach of international airports through which the produce is exported along intercontinental cold chain supply routes from farm to table in 36–48 hours.

While these agri-commodities arguably lie at the luxury end of the Euro-American consumption spectrum and hence might be relatively easily sacrificed if supply were threatened by EC, the most likely outcome under prevailing relations of globalized capitalism and agribusiness is that production would be shifted to other areas that remain or become suitable for those crops. Some such moves to 'repatriate' production to consumer countries are already occurring – somewhat ironically – partially as a result of environmental campaigns in Europe and North America to reduce fossil fuel consumption and hence 'food miles', although the main driver is likely to be changing relative production costs, of which fuel is certainly a significant component. Whether such relocations of production do actually cut total environmental footprints and greenhouse gas emissions is a moot point, with some research suggesting that the need to heat Dutch or British greenhouses, for instance, creates higher total emissions per bouquet of flowers in the vase or kilogram of mange tout on the dinner table than production in tropical environments and then intercontinental airfreight (MacGregor and Vorley 2006; Barclay 2012).

The impact of any widespread such relocations on the economies and hence human security of those communities in which production currently occurs will be serious, especially in the short and medium terms, unless new local and regional markets can be found or production switched to other crops for which there is local demand and, in the medium to longer term, which can tolerate changing environmental conditions. There are, however, important possible alternative scenarios in relation to environmental security, namely that if such export-oriented production is displacing locally required food production, is utilizing limited water supplies and hence diverting water from local requirements or depressing the water table to the detriment of the environment, and/or is contaminating the environment with chemical residues from fertilizers or pesticides, then reducing the scale or intensity of production – whether necessitated by EC or other factors – and reorienting it to locally more highly valorized crops may yield net environmental benefits by enhancing local environmental and hence human security.

In relation to water abstraction adversely affecting local environments and people, the concept of virtual water becomes pertinent. This refers to the water used in producing a crop (or industrial commodity) that is then exported. The value of such outputs is not just their immediate consumption value as reflected in market prices at their destination but also the value of the water used in their production. Thus the producer country is, in effect, exporting the water used in production (hence virtual water) and the consuming country importing it (Allan 2002, 2011). In the context of EC, such considerations are becoming increasingly important as a major human security issue, especially in countries anticipated to experience hotter and drier conditions, threatening water and food supplies to growing and increasingly urban populations (Ingram et al. 2010). Indeed, such concerns over what might become the ultimate human security challenge are leading already water-scarce Middle Eastern countries but also others including China to acquire vast tracts of land in parts of Latin America, Africa and Southeast Asia in highly controversial so-called 'land grabbing' deals that often displace local communities, in order to produce food and/or biofuels (which often supplant food crops) for their own markets (e.g. Schiavone 2009; Houtart 2010; Rosillo-Calle and Johnson 2010; Pearce 2012).

While production for global markets may improve labour conditions because of international scrutiny and compliance requirements, this is no guarantee, as periodic exposées of exploitation in the garment industry demonstrate. Yet, if unemployment exists, if there are few or no alternatives, or if conditions are better than in other sectors, such jobs will be filled. Nevertheless, because the origins of these industries lie in economic conditionalities imposed by multilateral agencies and donor governments in exchange for continuing development assistance, and because Northern consumers are the main destination markets, such cases raise moral and ethical issues of international equity and justice bound up with human, economic and environmental security in producer areas (i.e. the 'moral responsibility for distant strangers' of postcolonial development discourse – e.g. Corbridge 1998). This demonstrates the complexity of the issues involved and the potentially serious unintended consequences of simplistic environmental campaigns, however well meant, and of EC in the longer term.

CONCLUSIONS

Environmental factors continue to impact upon human activities and security in diverse and important ways, even in high-income countries with sophisticated 'hard' and 'soft' means for reducing their likelihood and mitigating their effects. Not everywhere or everyone faces such hazards or experiences them in the same way since their patterns of exposure and vulnerability or resilience in the face of impacts reflect how societies and their built environments are organized – nowadays increasingly based in part or whole on capitalist principles.

The two distinct but overlapping elements of environmental change (EC) are being or will be experienced in ways that reflect existing social, institutional, economic and physical urban structures. The poorest, least well resourced segments of society (however defined) will face the most extreme and sustained threats to their human security despite having the smallest ecological footprints and contributing least to the growth in greenhouse gas emissions that underpin EC.

This chapter has put forward a simple but robust approach to understanding the nature of EC risk that distinguishes it from conventional disaster risk through a focus on the extent, severity and duration of exposure to extreme events and slow-onset changes in environmental conditions. Similarly, simplistic dichotomies of urban versus rural and (semi-) nomadic and nomadic versus sedentary forms of social organization and production are avoided by means of a dynamic and more relational understanding of human adaptability in the face of changing circumstances. In this light, the chapter has surveyed many of the key likely dimensions of human security vulnerability in different EC contexts, ranging from coastal and inland urban environments in different agro-climatic zones to the progressive erosion of the productive environmental base underpinning remaining hunter-gatherer and pastoralist communities, freshwater and marine fisheries, and the particular complexities of specialized, globally integrated, capital-intensive, export-oriented commercial farming enterprises.

The particular environmental contexts and productive relations in which they are embedded exert strong influences over the extent to which resilience and adaptability are likely to be achievable, and over the likely consequences of exceeding such limits. Many uncertainties inevitably remain over the extent of predicted EC and its impacts, while precise outcomes will be locally contingent. However, a nuanced understanding of the complexities of production systems, of how people currently live in different contexts and cope with variations in conditions and with extreme events, can provide the basis for broad projections of likely vulnerabilities and adaptabilities.

NOTE

1. As in many countries, India's drive for modernity has been supported by an extensive programme of large dam construction, parts of which, particularly the series of dams on the Narmada River, have long been highly controversial on account of the large-scale

displacement of local people, with loss of livelihoods and inadequate compensation, as well as substantial environmental impacts (Roy 1999; McCulley 2001). This illustrates one particular form of the teleconnections or telecoupling (functional integration and interdependence) linking distant places, both urban and rural, through increasingly globalized urbanization and economic development processes (Seto et al. 2012) of direct relevance to the subject of this chapter. In this case, the displacement and forced resettlement of dam-affected communities, which has left the great majority worse off than previously, represents an extreme threat to their human security, derived not directly from their heavy natural resource and environmental dependence but from being deprived of their access to, and use rights over, those resources without appropriate compensatory arrangements by the more powerful urban-based interests driving the dam construction. In terms of the livelihoods and capabilities approaches, they have suffered a catastrophic loss of environmental entitlement, natural capital and livelihood assets, thus threatening their human security.

REFERENCES

Aberra, E. (2006), 'Alternative strategies in alternative spaces: livelihoods of pastoralists in the peri-urban interface of Yabello, southern Ethiopia', in D. McGregor, D. Simon and D. Thompson (eds), *The Peri-Urban Interface: Approaches to Sustainable Natural and Human Resource Use*, London, UK: Earthscan, pp. 116–133.

Allan, J.A. (2002), 'Water resources in semi-arid regions: real deficits and economically invisible and politically silent solutions', in A. Turton and R. Henwood (eds), *Hydro-Politics in the Developing World, A Southern African Perspective*, Pretoria, South Africa: AWIRU at Pretoria University, pp. 23–36.

Allan, J.A. (2011), *Virtual Water: Tackling the Threat to Our Planet's Most Precious Resource*, London, UK: IB Tauris.

Barclay, C. (2012), 'Food miles', *Standard Note SN/SC/4984*, London: House of Commons Library, http://www.parliament.uk/briefing-papers/SN04984.pdf, accessed 7 March 2013.

BBC (2012), 'Power restored after huge Indian power cut', http://www.bbc.co.uk/news/world-asia-india-19071383, accessed 7 March 2013.

Berry, P. (2011), 'Box 7.7 pilot projects to protect Canadians from extreme heat events', in C. Rosenzweig, W.D. Solecki, S.A. Hammer and S. Mehrota (eds), *Climate Change and Cities: First Assessment Report of the Urban Climate Change Research Network*, Cambridge, UK: Cambridge University Press, p. 207.

Bohle, H. (2007), *Living with Vulnerability: Livelihoods and Human Security in Risky Environments*, Bonn, Germany: United Nations University Institute for Environment and Human Security, Intersections Publication Series, No 6/2007.

Canada Institute of the Woodrow, V. Haufler and M.L. Walser (2009), 'Insurance and reinsurance in a changing climate', in C.J. Cleveland (ed.), *Encyclopedia of Earth*, Washington, D.C., USA: Environmental Information Coalition, National Council for Science and the Environment. First published in the Encyclopedia of Earth, 24 July 2009; last revised date 30 July 2012, http://www.eoearth.org/article/Insurance_and_reinsurance_in_a_changing_climate, accessed 7 March 2013.

CARE Bangladesh (2003), 'Report of a community level vulnerability assessment conducted in Southwest Bangladesh', A report prepared by the Reducing Vulnerability to Climate Change (RVCC) Project, Dhaka, Bangladesh: CARE Bangladesh.

Corbridge, S. (1998), 'Development ethics: distance, difference, plausibility', *Ethics, Place and Environment*, **1** (1), 35–53.

Dalby, S. (2009), *Security and Environmental Change*, Cambridge, UK: Polity Press.

Dikeç, M. (2007), *Badlands of the Republic; Space, Politics and Urban Policy*, Oxford, UK: Wiley-Blackwell.

FAO (2008), *Climate Change for Fisheries and Aquaculture: Technical Background Document*

from the Expert Consultation held on 7 to 9 April 2008, FAO, Rome, Report HLC/08/
BAK/6, Rome: Food and Agriculture Organization of the United Nations, ftp://ftp.fao.
org/docrep/fao/meeting/013/ai787e.pdf, accessed 7 March 2013.

Friedmann, J. (1992), *Empowerment*, Oxford, UK: Blackwell.

Gordon, R.J. (1992), *The Bushman Myth; The Making of a Namibian Underclass*, Boulder,
USA: Westview Press.

Gower, S. (2011), 'Box 7.6 Toronto, Canada: maps help to target hot weather response where
it is needed most', in C. Rosenzweig, W.D. Solecki, S.A. Hammer and S. Mehrota (eds),
*Climate Change and Cities: First Assessment Report of the Urban Climate Change Research
Network*, Cambridge, UK: Cambridge University Press, pp. 205–206.

Guèye, C., A.S. Fall and S.M. Tall (2007), 'Climatic perturbation and urbanization in
Senegal', *Geographical Journal*, **173** (1), 88–92.

Harris, N. and I. Fabricius (1996) (eds), *Cities and Structural Adjustment*, London, UK:
UCL Press.

Hartmann, I., A.J. Sugulle and A.I. Awalle (2010), *The Impact of Climate Change on
Pastoralism in Salahley and Bali-gubadle Districts, Somaliland*, Nairobi and Bonn:
Heinrich Böll Stiftung, Candlelight for Health, Education and Environment, and
Jaamacadda Camaad Amoud University, http://www.ke.boell.org/web/index-394.html,
accessed 7 March 2013.

Hitchcock, R.J. (2002), '"We are the First People": land, natural resources and identity in the
Central Kalahari, Botswana', *Journal of Southern African Studies*, **28** (4), 797–824.

Holm, M. (1995), 'The impact of structural adjustment on intermediate towns and urban
migrants: an example from Tanzania', Chapter 6 in D. Simon, W. van Spengen,
A. Närman and C. Dixon (eds), *Structurally Adjusted Africa: Poverty, Debt and Basic
Needs*, London, UK: Pluto, pp. 91–106.

Houtart, F. (2010), *Agrofuels: Big Profits, Ruined Lives and Ecological Destruction*, London,
UK: Pluto.

Ingram, J., P. Ericksen and D. Liverman (eds) (2010), *Food Security and Global Environmental
Change*, London, UK: Earthscan.

Intergovernmental Panel on Climate Change (IPCC) (2007a), *Climate Change 2007: The
Physical Science Basis; Contribution of Working Group I to the Fourth Assessment Report
of the Intergovernmental Panel on Climate Change*, Cambridge, UK: Cambridge University
Press.

Intergovernmental Panel on Climate Change (IPCC) (2007b), *Climate Change 2007: Impacts,
Adaptation and Vulnerability; Working Group II Contribution to the Fourth Assessment
Report of the Intergovernmental Panel on Climate Change*, Cambridge, UK: Cambridge
University Press.

Intergovernmental Panel on Climate Change (IPCC) (2012), *Managing the Risks of Extreme
Events and Disasters to Advance Climate Change Adaptation*, Special Report of the
Intergovernmental Panel on Climate Change, Cambridge, UK: Cambridge University
Press.

Lobell, D., M. Burke, C. Tebaldi, M. Mastrandrea, W. Falcon and R. Naylor (2008),
'Prioritizing climate change adaptation needs for food security in 2030', *Science*, **319**,
607–610.

Lonergan, S.C., M. Brklacich, C. Cocklin, N. Petter Gleditsch, E. Gutierrez-Espeleta,
F. Langeweg, R. Matthew, S. Narain and N. Soroos (1999), *Global Environmental Change
and Human Security: GECHS Science Plan*, Bonn, Germany: International Human
Dimensions Programme on GEC (IHDP) Report No. 11.

MacGregor, J. and B. Vorley (2006), 'Fair miles? The concept of "food miles" through a sustainable development lens', *Sustainable Development Opinion*, London, UK: International
Institute for Environment and Development.

Marcotullio, P.J. and G. McGranahan (eds) (2007), *Scaling Urban Environmental Challenges;
From Local to Global and Back*, London, UK: Earthscan.

Mbonile, M. (1995), 'Structural adjustment and rural development in Tanzania: the case
of Makete District', Chapter 8 in D. Simon, W. van Spengen, A. Närman and C. Dixon

(eds), *Structurally Adjusted Africa: Poverty, Debt and Basic Needs*, London, UK: Pluto, pp. 136–158.

McCulley, P. (2001), *Silenced Rivers; The Ecology and Politics of Large Dams*, 2nd edition, London, UK: Zed.

McGranahan, G. (2007), 'Urban transitions and spatial displacements of environmental burdens', in P. Marcotullio and G. McGranahan (eds), *Scaling Urban Environmental Challenges: From Local to Global and Back*, London, UK: Earthscan, 18–44.

McGranahan, G., D. Balk and B. Anderson (2008), 'Risks of climate change for urban settlements in low elevation coastal zones', in G. Martine, G. McGranahan, M. Montgomery and R. Fernández-Castilla (eds), *The New Global Frontier: Urbanization, Poverty and Environment in the 21st century*, London, UK: Earthscan, 165–181.

Mitlin, D. (2012), 'Lessons from the urban poor: collective action and the rethinking of development', in M. Pelling, D. Manuel-Navarrete and M. Redclift (eds), *Climate Change and the Crisis of Capitalism: A Chance to Reclaim Self, Society and Nature*, London and New York: Routledge, pp. 84–98.

Munich Re (2012), *Climate Change*, http://www.munichre.com/en/group/focus/climate_change/default.aspx, accessed 7 March 2013.

ODI (2009), 'Pastoralism and climate change: enabling adaptive capacity', Humanitarian Policy Group Synthesis Paper, April, London: Overseas Development Institute, http://www.odi.org.uk/resources/details.asp?id=3304&title=pastoralism-climate-change-adaptation-horn-africa, accessed 7 March 2013.

Pearce, F. (2012), *The Land Grabbers: The New Fight Over Who Owns the Earth*, London, UK: Bantam.

Peduzzi, P., H. Dao, C. Herold and F. Mouton (2009), 'Assessing global exposure and vulnerability towards natural hazards: the Disaster Risk Index', *Natural Hazards and Earth System Sciences*, **9**, 1149–1159.

Pelling, M. and B. Wisner (2009), *Disaster Risk Reduction: Cases from Urban Africa*, London UK: Earthscan.

Pelling, M., D. Manuel-Navarrete and M. Redclift (eds) (2012), *Climate Change and the Crisis of Capitalism: A Chance to Reclaim Self, Society and Nature*, London and New York: Routledge.

Riddell, J.B. (1997), 'Structural adjustment programmes and the city in tropical Africa', *Urban Studies*, **34** (8), 1297–1307.

Rosillo-Calle, F. and F.X. Johnson (2010), *Food Versus Fuel: An Informed Introduction to Biofuels*, London and New York: Zed Books.

Roy, A. (1999), *The Cost of Living*, London, UK: Flamingo.

Schiavone, C. (2009), 'The global struggle for food sovereignty: from Nyéléni to New York', *Journal of Peasant Studies*, **36** (3), 682–689.

Seto, K., A. Reenberg, C.G. Boone, M. Fragkias, D. Haase, T. Langanke, P. Marcotullio, D.K. Munroe, B. Olah and D. Simon (2012), 'Urban land teleconnections and sustainability', *Proceedings of the National Academy of Sciences*, **109** (20), 7687–7692.

Simon, D. (1995), 'Debt, democracy & development: Sub-Saharan Africa in the 1990s', in D. Simon, W. van Spengen, C. Dixon and A. Närman (eds), *Structurally Adjusted Africa: Poverty, Debt & Basic Needs*, London, UK: Pluto, pp. 17–44.

Simon, D. (1997), 'Urbanisation, globalisation and economic crisis in Africa', in C. Rakodi (ed.), *The Urban Challenge in Africa: Growth and Management of its Large Cities*, Tokyo, Japan: United Nations University Press, pp. 74–108.

Simon, D. (2008), 'Urban environments: issues on the peri-urban fringe', *Annual Review of Environment and Resources*, **33**, 167–185.

Simon, D. (2012), 'Hazards, risks and global climate change', in B. Wisner, I. Kelman and J.-G. Gaillard (eds), *The Routledge Handbook of Hazards and Disaster Risk Reduction*, London and New York: Routledge, pp. 207–219.

Simon, D. and H. Leck (2010), 'Urbanizing the global environmental change and human security agendas', *Climate and Development*, **2** (3), 263–275.

Simon, D. and H. Leck (2013), 'Cities, human security and global environmental change', in L. Sygna, K. O'Brien and J. Wolf (eds), *A Changing Environment for Human Security: Transformative Approaches to Research, Policy and Action*, London and New York: Earthscan from Routledge (pp. 170–180).

Simon, D., D. McGregor and K. Nsiah-Gyabaah (2004), 'The changing urban-rural interface of African cities: definitional issues and an application to Kumasi, Ghana', *Environment and Urbanization*, **16** (2), 235–247.

Smith, A., C. Malherbe, M. Guenther and P. Berens (2000), *The Bushmen of Southern Africa: A Foraging Society in Transition*, Cape Town and Athens, OH: David Philip and Ohio University Press.

Solway, J. (1998), 'Taking stock in the Kalahari: accumulation and resistance on the Southern African periphery', *Journal of Southern African Studies*, **24** (2), 425–441.

Stephens, S. (2010), 'Chernobyl fallout: a hard rain for the Sami', *Cultural Survival* website, 19 February, http://www.culturalsurvival.org/ourpublications/csq/article/chernobyl-fall out-a-hard-rain-sami, accessed 7 March 2013.

Swiss Re (2012), 'Road to Rio+20: building a sustainable world', http://www.swissre.com/ rethinking/climate/, accessed 7 March 2013.

Sylvain, R. (2001), 'Bushmen, boers and baasskap: patriarchy and paternalism on Afrikaner farms in the Omaheke Region, Namibia', *Journal of Southern African Studies*, **27** (4), 717–737.

Tozer, J. (2011), 'Arctic special: Sami reindeer herders struggle against Arctic oil and gas expansion', http://www.theecologist.org/investigations/climate_change/1097154/sami_rein deer_herders_struggle_against_arctic_oil_and_gas_expansion.html, accessed 7 March 2013.

UNESCO (2011) 'On the frontlines of climate change: Sami reindeer herders', UNESCO Media Services 19 December, http://www.unesco.org/new/en/media-services/single-view/ news/on_the_frontlines_of_climate_change_sami_reindeer_herders/, accessed 6 August 2012.

Wilmsen, E. (1989), *Land Filled with Flies*, Chicago, USA: Chicago University Press.

Wisner, B., P. Blaikie, T. Cannon and I. Davis (2004), *At Risk, Natural Hazards, People's Vulnerability and Disasters* (Second Edition), London and New York: Routledge.

Wisner, B., J.-G. Gaillard and I. Kelman (2012), 'Framing disaster: theories and stories seeking to understand hazards, vulnerability and risk', in B. Wisner, I. Kelman and J.-G. Gaillard (eds), *The Routledge Handbook of Hazards and Disaster Risk Reduction*, London and New York: Routledge, pp. 18–33.

6. The social dimensions of human security under a changing climate
Jürgen Scheffran and Elise Remling

1. INTRODUCTION

Anthropogenic climate change is a complex and cross-cutting issue potentially affecting all aspects of human life. Climate-related phenomena such as extreme weather events and natural disasters or gradual change of environmental conditions, can create multiple stresses on people in many regions of the world. This will most likely add to prevailing problems such as population growth or competition on natural resources which are vital for human wellbeing and security. As a result, global warming could undermine human development and the stability of social systems that support human needs, including water, food, health and energy services, agriculture, land use and urban infrastructure. In spite of being a global phenomenon, the consequences of climate change on people's lives are specific for each region. The impacts on human security depend on the vulnerability of individuals, communities and countries, their adaptive capacities and actual responses which are shaped by the specific economic, social and political contexts of each region.

This chapter focuses on the social dimensions, conditions and determinants of human security and related theoretical concepts. Although the degree of future climate change is uncertain, there is consensus that the effects on social systems will be significant and diverse. This chapter lays a focus on the impacts of environmental change on livelihoods of people and how they adapt to the challenges in order to protect human security. The main objectives are (1) to identify and examine the impacts of climate change on human security and societal stability; (2) to understand the availability and lack of livelihood assets and capabilities of people to react and adapt to the changes; and (3) to examine the role policies and institutions can play in adapting and building resilience. The analysis expands on the concept of social vulnerability, as well as on the capability and livelihood approaches, putting particular emphasis on access to resources, local strategies and institutions. By doing so, we wish to understand what capabilities people are lacking to be able to react and adapt to the changes, and finally, examine what kind of policy response would be necessary to ensure adaptation.

2.   THE FRAMEWORK OF HUMAN SECURITY

2.1   Defining Human Security

With the end of the Cold War and increasing globalization, the meaning of security has significantly changed, comprising many actors and factors which are shaping the security discourse in a complex way (Scheffran 2008a, 2011). Comprehensive security concepts consider economic, political, social, technical and ecological dimensions (Brauch 2009). While negative meanings of security build on the ability to protect against danger, threat and doubt, positive security concepts aim for developing opportunities to preserve and expand core values. Combining both aspects, security can be seen as a difference between gains and losses, between chance and risk, developing core values and avoiding harmful interference at the same time. To operationalize the concept of human security, it is important to determine the subject whose security is of concern, the values that are affected; the causes of risk; the vulnerability to losses; and the capability to protect against them.

Human security shifts the focus from nation states to the peoples of the world; from national sovereignty to human wellbeing and survival; from identifying the "other" as a threat to "us" as the main cause and the victims. The human security discourse within academia (Ulbert and Werthes 2008) was initiated by the 1994 Human Development Report (UNDP 1994: 24) which broadly defined human security as "freedom from fear and freedom from want". The UN Commission on Human Security (CHS 2003: 4) defined the goal of human security as "to protect the vital core of all human lives in ways that enhance human freedoms and human fulfilment. [. . .] It means using processes that build on people's strengths and aspirations. It means creating political, social, environmental, economic, military and cultural systems that together give people the building blocks of survival, livelihood and dignity". In his report on human security (A/64/701) of 8 March 2010 the Secretary-General stated that "broadly defined, human security encompasses freedom from fear, freedom from want and freedom to live in dignity" (UNSG 2010: 2).

Combining positive and negative security aspects as mentioned above, one approach to operationalize human security is based on "shielding people from acute threats and empowering people to take charge of their own lives" (CHS 2003: iv). In this definition the task of "shielding" aims at protecting people and their core values against risks, and the task of "empowering" seeks to enhance people's coping capacities in fulfilling basic human needs. Essential is to protect the vital core of human lives (including fundamental needs and rights) and maintain the freedom and capacity of people to live with dignity and make informed choices to pursue their interests (Barnett 2010; Gasper 2007). According to the Human Security Network, "[h]uman security and human development are thus two sides of the same coin, mutually reinforcing and leading to a conducive environment for each other" (HSN 1999:1).

While narrow concepts of human security focus on direct threats to people's lives, broader concepts also consider potential risks and fears to the core of people's values from a variety of sources. These include illness, poverty, personal safety, economic crises, social and environmental problems that are more incremental and indirect (King and Murray 2002; Owen 2004; Human Security Report Project 2010). Accordingly, O'Brien and Leichenko (2007: 3) relate human security "not only to security from physical violence, but also to food security, livelihood security, environmental security, health security and energy security". Thus, human security can be affected through multiple pathways and many dimensions of values and losses, each having their likelihoods and acceptabilities from the perspective of the affected entities. Summarizing the above discussion, human security rests on three pillars (Brauch and Scheffran 2012):

1. "freedom from fear" addressing the conflict, violence and humanitarian law agenda;
2. "freedom from want" in the context of the human development agenda; and
3. "freedom to live in dignity" referring to human rights, rule of law and good governance.

Due to its diversity, the concept of human security combines the environmental, peace and development communities and can serve as a "conceptual bridge between the [. . .] fields of humanitarian relief, development assistance, human rights advocacy, and conflict resolution" (Owen 2004: 377). However, due to the wide range of issues combined in human security, it has, by some, been criticized as a vague concept (Paris 2001). Within international relations, the human security concept has remained controversial (see Security Dialogue 2004). This disagreement has undermined communication with policy-makers and their efforts to move from declaratory statements to concrete policy initiatives and actions (Brauch and Scheffran 2012). To overcome the dispute between the proponents of a narrow and a wide human security concept, Owen (2004: 381) suggests: a threshold-based approach "that limits threats by their severity rather than their cause."

It is being increasingly recognized that new challenges are being posed on human security by environmental changes (Barnett 2001; Brauch

2005), which has led to a proposal for a fourth pillar "freedom from hazard impacts" (Fuentes Julio and Brauch 2009: 997). Barnett and Adger (2007) discuss how climate change may affect and even undermine human security, and how human insecurity may increase the risk of violent conflict as well as call on the role of states in human security and peace building. Another emerging issue is to understand human security in the social context of human action and interaction, going beyond the individual level towards collective levels. Future research challenges are to understand the linkages between climate change, vulnerability, adaptation and human security as well as the social dimensions of human security which will be discussed in the following sections.

2.2   Linkages between Climate Vulnerability, Adaptation and Human Security

Climate change affects various dimensions of human security in multiple ways. Some of the climate-induced stresses may directly threaten human health and life, such as floods, storms, droughts and heat waves, others gradually undermine the wellbeing of people over an extended period, including food and water scarcity, diseases, weakened economic systems and degraded ecological systems. Once critical thresholds are exceeded, the risks of climate change may turn into existential threats to human security, to what extent will depend on the vulnerability of those affected.

Vulnerability
Vulnerability is a broad term, which has been adopted by different disciplines and various scholarly communities (for an overview see Adger et al. 2009; Mearns and Norton 2009; Scheffran 2011). In an early attempt, Blaikie et al. (1994: 275) define vulnerability as the "characteristics of a person or group in terms of their capacity to anticipate, cope with, resist, and recover from the impact of a natural hazard". Thus, vulnerability depends on the societal unit, the type of event and the actions taken against the hazard. Gallopín (2006: 294) states that "[d]epending on the research area, it has been applied exclusively to the societal subsystem, to the ecological, natural, or biophysical subsystem, or to the coupled SES [socio-ecological systems], variously referred [to] also as target system, unit exposed, or system of reference". While traditionally natural hazards have been studied by natural scientists, recent research has increasingly included perspectives from social sciences, which examine the socioeconomic side of vulnerability (Bohle and Glade 2008; Füssel 2007, 2009). Similar to the human security concept, the theoretical diversity in defining vulnerability leads to a lack of cohesion in a common conceptual

framework. As Füssel (2007: 156) argues, "'vulnerability' can only be used meaningfully with reference to a particular vulnerable situation". Thus, vulnerability is closely linked to a place as well as to the natural and social environment in which a specific system exists.

Adaptation and adaptive capacity

Most commonly referred to is IPCC (2007: 27) which defines vulnerability as a function of "the character, magnitude, and rate of climate change and variation to which a system is exposed, its sensitivity, and its adaptive capacity." Thus, the vulnerability concept involves the capacity to adapt and respond to "actual or expected climatic stimuli or their effects, which moderates harm or exploits beneficial opportunities" (ibid.). Matching the positive and negative aspects of human security, successful adaptation allows reduction of losses or establishment of positive values. Adaptive capacity (or adaptability) relates to the system's potential or capacity to react to the impacts or transformations related to climate change, moderate potential damages, take advantage of opportunities, or cope with the consequences. More generally, it describes "the ability of a system to evolve in order to accommodate environmental hazards or policy change and to expand the range of variability with which it can cope" (Adger 2006: 270). For real world systems, adaptive capacity is influenced by a number of factors, such as poverty, state support, economic opportunities, the effectiveness of decision-making, social cohesion and other societal conditions. This points to the fact that adaptation takes place within an economic, social and political context that can either facilitate or constrain adaptation (Adger and Vincent 2005).

Once vulnerability exceeds critical thresholds, human security is threatened due to inacceptable or existential risks. The decisive question is: at which point will the costs of adaptation exceed the benefits or the potential loss avoided? Human systems are equipped with a coping range, a capacity the system has established to accommodate variations – in our case of climate conditions – which relates to specific stressors (Carter et al. 2007). The core of the coping range contains beneficial outcomes while outside of the range outcomes become intolerable (Figure 6.1). If despite changing conditions the ability to cope is held constant (Option 1), vulnerability will increase to extreme levels.[1] Adaptations can ameliorate the adverse effects by widening the coping range (Option 2). Simplified, the aim of adapting is to ensure that an adequate coping range is established, keeping a damage threshold to which a system can accommodate, adapt to, and recover from some deviations from "normal" conditions. It varies among systems and regions, and does not necessarily remain static (Smit and Pilifosova 2001; Smit and Wandel 2006). Within the given range of adaptive capacity,

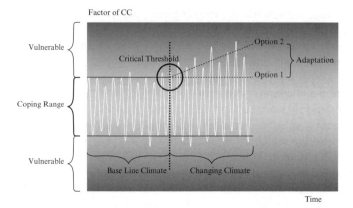

Source: Remlin 2011

Figure 6.1   Idealized version of a coping range describing the relationship between a factor of climate change and threshold exceedance, and how adaptation can establish a new coping range, reducing vulnerability to climate change

human beings have some freedom to choose possible responses to environmental stress. This means that an event causing increased vulnerability can be ameliorated by responses to protect against it or avoid it in the first place.

The above discussion indicates that vulnerability is a dynamic concept which changes over time according to the stimulus and its impact, the sensitivity and response of the respective system (Smit and Wandel 2006). Adger (2006: 274) appropriately describes vulnerability as "a dynamic phenomenon often in a continuous state of flux, both the biophysical and social processes that shape local conditions and the ability to cope are themselves dynamic".

The form of social interaction can significantly determine the efficiency of adaptation, and either restrain or expand the adaptive capacity for each individual. Existential threats to individuals or groups may increase temptations to switch to non-legal and violent acts, but alternatively could also force people to work together through collective action and thus to improve the chances for survival. Some responses could help to minimize the risks, others may cause more problems. Such inappropriate responses to climate change – sometimes called maladaptations – (e.g. through uncertainties about the degree of future climate change or inadequate consideration of local circumstances), instead of reducing vulnerability may inadvertently lead to increased vulnerability in the long term (Jones et al. 2010; Agrawal 2008). For instance, migration is a possible adaptive response not only to poverty and social deprivation but also to environmental hardships.

If impacts of climate change provoke responses that affect large parts of society, the consequences may also become an issue of national, international, or even global security, thereby contributing to the securitization of the climate discourse, for example by triggering speech acts and justifying extraordinary measures in response to threats (according to securitization theory developed by the Copenhagen school, see Buzan et al. 1998). Some impacts may cause governments and the military to take action, for example for disaster management, in response to massive refugee flows, or in conflicts induced by environmental stress. Vulnerability and lack of adaptation to environmental change may contribute to human insecurity which in turn could create risks to national security.

The following section introduces a number of concepts that help to understand the social context of human security under climate change.

3.   SOCIETAL CONDITIONS AND CONTEXTS OF HUMAN SECURITY

3.1   Human Needs, Capabilities and Sustainable Livelihoods

Human wants and needs

The impact of climate change on human values fundamentally affects vulnerability and human security. Values represent the preferences of human beings concerning certain pathways or outcomes of actions, and tend to influence attitudes and behaviour. While rational actors select actions that improve or even optimize their values, there are circumstances which restrain the freedom and rationality of choice, including dependency on established action paths or the influence of the social environment that shapes individual actions. Among human values it is important to distinguish human *wants* from human *needs*: While wants can be infinite and insatiable, needs are few, finite, and classifiable; a deficiency in fundamental needs could have severe consequences such as dysfunction or death. A taxonomy of fundamental human needs includes subsistence, protection, affection, understanding, participation, leisure, creation, identity and freedom (Max-Neef et al. 1991).

Distinguishing between human wants and needs is important when facing climate change. While restraints on wants are experienced as a loss by individuals and lead to dissatisfaction, impinging on fundamental needs threatens the survival of people and thus provokes more drastic responses that can directly affect human security, including displacement or violence (Scheffran et al. 2012b). Climate change is expected to affect systems and processes that provide for human needs, including water, food and energy supplies, agriculture and land use, health, and urban life. When human health and wellbeing is massively affected, this may also undermine the functioning or stability of social systems, for example through weakened economies, infrastructures or institutions.

Capabilities
The consequences of climate change for human security depend on the individual and collective responses to value losses, which require a certain capability to act. Capability is generally understood as "the capacity or ability to do something, achieve specific effects or declared goals and objectives" (Wikipedia 2012). If actions are directed at producing something of value, capability is often associated with the "capital" used to create goods and services (Bohle 2009). In economic theory, the most common dimensions of capital are natural capital (usable resources of an ecosystem), physical capital (assets made by humans for production), and financial capital (monetary wealth). Increasingly, human production factors are included, such as human capital (workers' skills and abilities), social capital (collective action in social networks), political capital (instruments and institutions in political decision-making), and cultural capital (knowledge, skills, education status, and personal advantages) (Scheffran et al. 2012b).

   In the context of human development and welfare economics, the capability approach has been developed in the 1980s by Amartya Sen (1985) and has served as a basis for the human development index (HDI). In distinction to other economic approaches that focus on utility, income, commodities or access to resources, true development requires every person's access to "freedoms which people have to achieve prioritized outcomes" (Gasper 2006: 3). Sen (1999) distinguishes between functionings on the one hand, the things a person might value to do or be in his life, such as being healthy or actively practising a religion, and capabilities on the other, which are the "alternative combination[s] of functionings that are feasible for [the person] to achieve" (Sen 1999: 75). The set of capabilities available to a person are the real opportunities, or attainable alternative lives this person has (Gasper 2006). In other words, capability is the freedom of every person to find out which culture and form of life he or she would like to pursue. These include the abilities to live a long life, engage in economic transactions, or participate in political activities. Accordingly, having the capabilities at one's disposal, which are necessary to lead a fully functioning life, is essential (ibid.).[2] Which functionings a person realizes, then, depends upon her own will, but only when the relevant capabilities to function are present, can human dignity be assured.

Sen (1999) mentions five "instrumental freedoms": political freedom, economic facilities, social opportunities, transparency guarantees, and protective security, as well as a variety of economic and social rights. Specifying Sen's rather broad approach, Nussbaum (2003) provides a clear list of ten central capabilities that are "abilities or opportunities to act and choose" (ibid.: 25) and "fundamental entitlements inherent in the very idea of minimum social justice, or a life worthy of human dignity" (Nussbaum 2011: 24ff). These capabilities should undergo a permanent validation process whereas the implementation is left to each community (e.g. a nation) itself to be able to take into account local and cultural differences. This implies that political participation is central, both as an entitlement in itself, and also as a tool for a transfer from theory to practice.

Sustainable livelihoods

The approach framework gained popularity among development and poverty researchers since the late 1990s (Bohle 2009). Similar to human security, livelihood represents a bottom-up perspective on development which is people-centred and promotes participatory principles By so doing, it proposes action at community level to address poverty and prevent future vulnerability, rather than concentrating on the government-level (Bohle 2009; Jones et al. 2010). Expanding the individual perspective of the capability approach towards the societal environment, livelihoods provide the basis for the set of capabilities available, while at the same time capabilities enable livelihoods to be gained. Livelihood approaches deal with the opportunities people have and what they do with their resources (e.g. money, labour, land, crops, livestock, knowledge, and social relationships).

According to the terminology proposed by Chambers and Conway (1991: 5), "[a] livelihood in its simplest sense is a means of gaining a living". In more detail, they suggest that, "[a] livelihood comprises the capabilities, assets [. . .] and activities required for a means of living: a livelihood is sustainable which can cope with and recover from stresses and shocks, maintain or enhance its capabilities and assets" (ibid.: 6) while not undermining the natural resource base for future generations. Sustainability is "a function of how assets and capabilities are utilized, maintained and enhanced so as to preserve livelihoods" (Bohle 2009: 525). This also implies that sustainable livelihoods are more capable of accommodating shocks and consequently, adapt to climate change.

The promotion of "sustainable livelihood security is very closely connected with the concept of human security, putting people at the centre and taking equity, human rights, capabilities and sustainability as its normative basis" (Bohle 2009: 528; see also Bohle and O'Brien 2007). Livelihood strategies consist of a range and combination of activities and choices people make to meet their livelihood goals. Classical examples are agriculture, pastoralism or wage labour (Jones et al. 2010); however, in practice they often consist of "repertoires" (Chambers and Conway 1991: 7) or "mixes" of activities. By their very nature, livelihood strategies are dynamic and changeable over time, adjustable with new knowledge and information, and constrained by resource access. People's livelihood choices are made under the considerable influence of the governance environment (policies, institutions and processes), as well as the asset base people have. These assets are resources that people can use in order to achieve their livelihood goals, extending beyond material categories, and in the absence of them, can hamper livelihoods. According to Bohle (2009: 527) "[o]ne of the major strengths of the livelihood approach [is] that it views the poor and vulnerable not as passive victims, but highlights the active or even proactive role to secure their living in contexts of uncertainty, risk, stress and shocks" and as such marks a shift towards actor-oriented perspectives, usually at the unit of households and communities. Even when people have few assets, they possess agency and can take an active role in responding to and enforcing change.

The capability and livelihoods approaches are useful for understanding vulnerability and human security under a changing climate as they emphasize the importance of individual and social capacities for responding to threats and managing risks, as well as the external threats to livelihood security, such as climate change. A loss in livelihood can become a threat to human security, as it limits people's opportunities to act. If livelihoods are secure and sustainable, people are less vulnerable (Ribot 2010; Chapter 7 this volume). As individuals and communities are facing both rapid change and increasing uncertainty, livelihoods and the wellbeing of people are severely impacted, which, as discussed above can also have significant implications for human security. Bohle (2009: 525) highlights the fact that "state security will be precarious (and empty) unless based on and consistent with the security of individuals", thus providing a link between livelihood security and the stability of a state. The enhancement of adaptive capacity involves similar requirements to promotion of sustainable development and equity (Smit and Pilifosova 2001). Joint approaches in the context of a livelihood perspective could bridge the disconnection between community needs and the policy process. An integrated framework calls for maintaining or enhancing capabilities and assets in order to safeguard livelihood security and enhance the capacity to cope with climate-related shocks.

3.2 Social Capital, Networks and Resilience

Social capital and networks

As we have seen in previous sections, the impacts of climate change on human beings have to be considered within people's broader social environment. This also includes the social interactions and networks in which humans participate as well as institutions and governance structures. Societies require structures to realize the benefits of cooperation, maintain and enforce accepted rules and establish effective and predictable social interactions. These *social contracts* are based on the consent of a large majority of citizens to rules, regulations and institutions that maintain social order and serve as the glue which keeps society together. The stability of societies refers "to the durability of political institutions and to robust social structures", while destabilization "can be understood as a process that (gradually) causes an originally stable political and social situation to break down" (WBGU 2008: 236). Societal structures that lose credibility and support from its citizens, become weak, less cohesive and sensitive to societal and political instability.

Social linkages and networks are vital parts of the social capital of vulnerable communities. Going back to the work of Porter (1998) and Granovetter (1973), social capital and social networks have evolved as diverse and complex concepts in the social science literature (Sanginga et al. 2007). Social capital stands for features of social organizations (such as social networks, interactions, norms, trust, and reciprocity) that facilitate coordination and cooperation and that enable people to act collectively for mutual benefits (Putnam 1993). Cohesive social capital based on collaboration, transparency and community participation contributes to human development, effective collective action and management of common property. When communities are affected by hardships, mechanisms for building social capital may encourage not only self-help and self-reliance, but also community empowerment and participation as well as respect for local values, customs and traditions (Sanginga et al. 2007). They can also help to strengthen political and administrative structures and become an important precondition for sustainable development and conflict resolution.

An important ingredient of social capital are social networks that comprise actors (e.g. individuals, organizations) as network nodes, and linkages that represent the interconnections and ties between them (Wasserman and Faust 1995). Social network analysis is a technique operating across many levels, from families to nations. Analysing the topology of social networks, it is important to understand a wide range of phenomena, i) including the power of influential actors with many network connections; ii) the vulnerability and robustness of the network to hazards and attacks that eliminate nodes and connections; and iii) the spread and diffusion of attributes, such as innovations, practices, diseases and conflicts. The structure of a social network also determines its efficiency in addressing the challenges of climate change, in particular the individual and collective responses to hazards and the distribution of values, resources and capabilities among network agents. The challenge is to design social networks to allow affected communities to better survive and live under climate risk.

Social resilience
Related is the concept of social resilience that seeks to protect social capital and strengthen the capability of social networks in their creative and collective efforts to handle the problems associated with climate change (e.g. Adger 2003). First developed for ecological systems (Holling 1973) the resilience concept is increasingly applied to socio-economic systems, in particular with regard to climate change (for an overview of the resilience concept, see Folke 2006; also the review in Campana 2010). Resilience points to the ability of a system to absorb disturbances and still retain its original structure and function. Having emerged from a more ecological and natural science-driven perspective, resilience is closely related to vulnerability and adaptive capacity. Accordingly, social resilience is "the ability of a community to withstand external shocks and stresses without significant upheaval" (Adger et al. 2002: 358), which implies that communities "may be able to absorb these shocks, and even respond positively to them" (ibid.). In a resilient social environment, actors are able to cope with and withstand the disturbances caused by environmental change in a dynamic and flexible way that preserves, rebuilds, or transforms their livelihood. If stabilizing mechanisms fail, societies become prone to internal failure, with negative implications for all individuals. Resilient communities are not only able to absorb and survive external shocks and surprises, but as active agents can design their environment, anticipate and resist future shocks and stresses, and recreate and rebuild themselves to preserve their identities, which are subject to internal motivations and capabilities.

Resilience research studies disturbances which a system experiences and its response to the impacts related to its adaptive capacity. Since resilience and adaptation are often used in similar contexts, there has been confusion within the scientific community regarding the similarities and differences between the two concepts (Adger 2006; Gallopín 2006; Renaud et al. 2010). What matters both in adaptation and social resilience is the capacity of the community to cope with the magnitude of environmental change. To survive hardships and strengthen resilience many households diversify their livelihoods and pursue alternative action pathways (McDowell and de Haan 1997). Innovative strategies to respond to climate change involve technical innovations, investments in physical, human and social capital as well as institutional mechanisms to mitigate conflict and facilitate cooperation. The challenge is to develop adequate institutional frameworks that help to overcome the barriers and create favourable conditions for coordinated collective action, using synergistic mechanisms to merge capabilities, efforts and actions, and adjust values, goals and rules for multiple actors.

In the following we will draw on the concepts introduced in this section to describe how they enable us to better understand the impacts of climate change on human security.

4.   IMPACTS OF CLIMATE CHANGE ON HUMAN SECURITY AND SOCIETAL STABILITY

In many parts of the world human living conditions are harsh due to a number of stresses, such as poverty, hunger, corruption, war, lack of access to health services, education or participation in political decision-making. Together these combine to weaken the capability of people and communities. Climate change will particularly increase the burden on those people who are already under stress from other problems and are lacking essential capabilities and freedoms to take actions that reduce vulnerability and protect human security. Due to low social capital and lack of resilience they may be easily overwhelmed by the multidimensional impacts of climate change which further disrupt their capability and societal stability. These issues will be discussed in the following.

4.1   Deprivation of Capabilities for Human Security

4.1.1.   Poverty and marginalization

Poverty is a severe deprivation of human capabilities which can, for instance, result from unemployment and low income, government oppression and instability. Sen's capability approach "is not a theory that can *explain* poverty, inequality or wellbeing; instead, it provides concepts and a framework that can help to *conceptualize* and *evaluate* these phenomena" (Robeyns 2006: 353, italics in original). Poverty is a key social determinant for climate-induced impacts on human security: It is widely assumed that the most serious climate-induced risks and conflicts are expected in poor communities which are more vulnerable and dependent on climate-sensitive resources, have fewer assets to invest in adaptation and thus cannot take the responses needed. Thus, "[g]roups [which] are already marginalised bear a disproportionate burden of climate impacts, both in the developed countries and in the developing world" (Adger 2006: 273). A progression of climate change will increase the vulnerability of weak and fragile communities and states and further reduce their adaptive capacities. Some regions such as Bangladesh and the African Sahel are particularly vulnerable due to their geographic and socio-economic conditions and the lack of adaptation capabilities. But poverty also increases vulnerability in developed countries, as the case of Hurricane Katrina demonstrated in 2005 in the USA. On the other hand those with ample resources "will be more able to protect themselves against environmental degradation, relative to those living on the edge of subsistence who will be pushed further towards the limit of survival" (Nordås and Gleditsch 2005: 20). Climate change can thus reinforce obstacles to development and enhance poverty which will further increase climate vulnerability.

4.1.2.   Hunger, food insecurity and health problems
Lack of food is a major concern in several developing regions. More than 850 million people are undernourished (FAO 2011), and many agricultural areas are overexploited. Climate change will likely reduce crop productivity and food security in many hunger-prone areas, thus worsening malnutrition and health problems, with significant variations from region to region (IPCC 2007; WBGU 2008). Given the significant impact of climate variability on the natural environment, agricultural societies are more affected than industrial countries due to their heavy reliance on natural resources and ecosystem services. However, in recent food crises, climate change played only a marginal role. Increased food prices are much more affected by the oil price, food speculations, inadequate local markets and failed social security and distribution mechanisms. Lack of access to food may continue to trigger regional food crises and undermine economic performance and human security in weak and unstable states.

4.1.3.   Inequality and injustice
The potential risks, costs and benefits related to climate change will be unevenly distributed among the world's populations, which raises critical issues of social justice, between north and south, rich and poor, current and future generations. The resources, capabilities, benefits and risks for each social entity are shaped by allocation and distribution mechanisms, market processes, and power structures. Although greenhouse gas emissions are mainly emitted by the industrialized countries, the impacts of climate change are expected to be more severe in developing countries. Vulnerabilities and adaptive capacities vary greatly between nations and regions, for a number

of reasons (Mertz et al. 2009): (1) Large parts of developing countries lie in regions in which temperature rise adds to already high temperatures and rainfall losses; (2) Developing countries are often highly dependent on the agricultural sector as a source of income, livelihoods and employment; (3) As many people are impoverished, they are being hit harder by the impacts of climate change and are thus more vulnerable; (4) The adaptive capacity is weakened due to limited economic and technological capacities. Inequalities are particularly relevant for the three dimensions of vulnerability (exposure, sensitivity and adaptive capacity). Marginalized communities often live in regions more exposed to climate change (e.g. in flood and storm zones of rivers and coastal areas or in drought and wildfire regions), and some regions (such as Bangladesh, the Middle East and the African Sahel) are more exposed due to their geographic conditions. Communities which are highly dependent on agriculture and ecosystems will be more sensitive to hazards. Even within one community, the individual vulnerabilities can vary to a great extent: "The poor and wealthy, women and men, young and old, and people of different social identities or political stripes experience different risks while facing the same climate event" (Ribot 2010; see also Chapter 7 this volume). An illustrative example referring to flood in a community is presented in Gallopín (2006: 289):

> The most precarious homes are hit harder by a flood than the solid ones (sensitivity). Oftentimes, the poorest homes are located in the places most susceptible to flooding (exposure). The families with the greatest resources have a greater availability of means to repair water damage (response capacity). The magnitude of the final impact will also depend on the intensity, magnitude, and permanence of the flood (attributes of the perturbation).

By affecting the weak, poor and fragile communities to a greater extent, and further reducing their already limited capabilities, climate change increases social divisions and disparities in many societies. Wealthy communities will be better able to protect themselves against environmental degradation, within their coping range.

4.2 Complex Interactions of Societal Instability and Conflict at Different Scales

Environmental conditions provide constraints and opportunities for the development of social systems, which in turn exploit, pollute, and manage natural systems. As we have seen in the previous sections, a changing climate that significantly alters these conditions will cause stress to social systems and thus have an impact on human life and livelihoods. By causing large-scale transformations in the biophysical systems which consequently affect natural resources and ecosystem services across the planet, climate change not only affects human security directly but could also weaken societal stability which then indirectly undermines the conditions required for human security. Since ecosystem services are valuable for individuals and society at large (Leemans 2009; IPCC 2007), environmental changes caused by global warming may have larger societal effects; for example, by undermining livelihood assets and societal infrastructure, or by inducing destabilizing human responses and interaction patterns of social systems. At the global level of decision-making, the main actors are governments of nations or groupings among them. At the local level, individual citizens are key players who affect or are affected by climate change. The multi-level process between local and global decision-making is connected through several layers of aggregation, with each layer having its own characteristic decision procedures (Scheffran 2008b).

At micro-level, assets determine how people can react to change and adjust their livelihoods. By impacting ecosystems and the services they provide, climatic changes could put increasing pressure on the asset base of households, particularly of those that have a high dependency on natural resources. Climate change undermines what people do to make a living during normal times and thus presents a livelihood disturbance that can impact livelihood security when the coping range against climate change is exceeded. Individuals who experience personal losses (life, income, property, job, health, family and friends) and find their personal identity and human security at stake, may be more tempted to violate rules and select non-legal actions including violence, especially if these actions offer benefits at low risk.

At medium to macro level, global warming may lead to national and international security risks and social instability if it outpaces the adaptive capacity of states and societies. Key issues are the extent of climate change to which societies can adapt and how effective and creative they are in developing strategies to deal with altered environmental conditions. If such change exceeds the available capability to act, actors are no longer able to prevent or fully compensate for the risks of climate change. This demonstrates that for significant environmental change, actors with low capability are not able to adapt to the changing environmental conditions and associated value losses. Powerful actors may not be able to adapt either, if the magnitude or speed of environmental change exceeds their capability to adjust. The implications of environmental change become even more significant and long-lasting if they not only affect value but also reduce capability itself, which in turn then affects the ability to respond. A recent example is the case of the buses and hospitals in New Orleans that were flooded by Hurricane Katrina, which diminished the overall societal capability to adequately help affected people.

Increased environmental stress and resource scarcity could induce structural changes and make states susceptible to disruptive climate-related events that destroy or disable vital assets and societal infrastructures, such as systems for the supply of water, food, wood, human health, housing, energy, transportation and financial services. The associated impacts can undermine the functioning of communities, the effectiveness of institutions and the ability of governments to satisfy and protect the needs of citizens. Disruptive climate change could overwhelm the already limited capacity of governments to respond effectively to the challenges. This could lead to increasing frustration and tensions between different groups within countries and to political radicalization, which together with the other factors could destabilize countries. "Failed states" cannot guarantee the core functions of government, such as law and public order, welfare, participation and basic public services (e.g. infrastructure, health and education), and the monopoly on the use of force (Starr 2008). These core functions are pillars of security and stability, and if any of them is missing, state failure under climate change becomes more likely.

Whether a state fails is determined by long-term structural factors, including the endowment with natural resources, the economic infrastructure, power constellations, demographic development and ethnic diversity. States on the edge of failure are vulnerable to abrupt changes that can trigger spirals of corruption, crime and violence. The erosion and collapse of state structures leaves a power vacuum that is often filled by non-state actors such as private security companies, terrorist groups or warlords. A prominent example is Somalia, a country controlled by powerful clans. In sub-Saharan Africa many countries are considered at risk of state failure.

A number of authors have expressed concerns that increasing societal instability from climate change could undermine the security of nations or people and lead to violent conflicts (see Scheffran and Battaglini 2011). However, empirical studies testing the relationship between climatic variables (temperature, precipitation) and conflict-related variables (number of armed conflicts or casualties) did not support a clear causal link (for an overview see Gleditsch 2012; Scheffran et al. 2012a). While historical case studies find significant statistical correlations (e.g. during the Little Ice Age), more recent data provided mixed results. Since the end of the Cold War the number of armed conflicts has largely declined, despite a temperature increase. Conflicts are complex and depend on a wide range of factors and causal pathways connecting both phenomena. Several conflict constellations have been particularly mentioned (WBGU 2008), including degradation of freshwater resources, decline in food production, increase in storm and flood disasters, and environmentally induced migration. When climate change reduces vital resources and infrastructures essential for the provision of human security, societal instability and conflict may become more likely, which could then further threaten human security. Armed conflicts involve destruction of infrastructure and productive assets, population displacement, loss of rural livelihoods, and greater vulnerability to natural disasters. While developed countries may be better able to avoid this vicious cycle, with growing temperatures their own infrastructures and networks for energy and water supply, transportation and communication, production and trade could become disrupted. In addition, they may be drawn into violent conflicts and refugee movements elsewhere. The destabilization of society could ultimately also affect wealthy people who might have the individual capability to survive but feel threatened if everything around them falls apart.

   In the worst case, climate change could trigger a cycle of environmental degradation, economic decline, social unrest and political instability that could accumulate to become security threats and aggravate conflicts, leading to cascading events. For instance, an increase of forced migration by climate change and conflicts could spread migration hotspots around the world that become nuclei for social unrest (WBGU 2008). Tipping elements in the climate system could induce a sequence of instabilities that trigger other tipping elements in social systems, and seemingly 'minor' events could provoke major qualitative changes. A self-reinforcing "chain reaction" could increase the potential risk of social cascades that put the whole system at risk.

5.   SOCIETAL RESPONSES, STRATEGIES AND SYNERGIES OF HUMAN SECURITY

To reduce the risks of climate change to human security and livelihood, communities need the capacity to respond and take action that diminishes harm or compensates for it by establishing positive values. This is in accordance with the above mentioned IPCC definition of adaptation that "moderates harm or exploits beneficial opportunities" (IPCC 2007: 869). Whether societies are able to cope with the impacts of climate change will depend on the individual and collective responses within their coping range. Adaptive decisions are embedded in the given social, economic, institutional and political structures that are framing the lives of people. As discussed, relevant factors are the assets people own, but also the accountability and effectiveness of institutions as well as access to "education, health care, agricultural services, justice systems and conflict resolution mechanisms" (Jones et al. 2010: 3). Or as Adger and Vincent (2005: 401) put it, "[a]daptive capacity in effect gives a picture of the adaptation space within which adaptation decisions are feasible".

It is essential that people have the freedom to determine the adaptations necessary for functioning in their own communities. Farmers and pastoralists have developed various adaptation mechanisms to deal with a changing climate (Amekawa et al. 2010; Freier et al. 2012; Schilling et al. 2012). They may shift to other economic activities in response to adverse climatic conditions such as drought. Another form of adaptation is seasonal migration. Less desirable is permanent distress migration caused by hostile conditions in the home location (such as the loss of vital assets). What is possible and adequate depends on the resilience, innovation and capability for self-help, adaptive capacity and social organization of the community (Scheffran et al. 2012c).

Coalitions in social networks can combine the capabilities of multiple actors to jointly reduce the value losses of those affected. To protect the most vulnerable and weakest members of a community collective action is required as well as principles of solidarity and justice. Strengthening the adaptation process requires organizational, structural and institutional mechanisms, including rules and regulations.

Kelly and Adger (2000) consider the "entitlements" people have within the structural framework that determines their ability to respond and cope with climate stressors. The concept of entitlements is compatible with Nussbaum's (2003) extension of the capabilities approach which also considers community capability and functioning. Examining vulnerability to climate change through a capabilities framework broadens the perspective, draws the attention towards the core problems hindering the ability of people to respond and highlights the role of institutions, people's wellbeing, class, social status and gender as key variables. Regarding gender for example, Roy and Venema (2002) suggest that by transferring Sen's and Nussbaum's approach to vulnerability assessments, the situation of rural women can be improved and their vulnerability to the risks of climate change be reduced while at the same time enabling them to gain the capabilities necessary to act as their own agents of change.

As demonstrated in previous sections of this chapter, a group's vulnerability and adaptive capacity is not determined solely or primarily by climate, but rather by a range of social, economic and political factors inherent to the system. The social environment can either provide support and strengthen cooperation between actors to mitigate damage, or aggravate the stresses if the losses by one actor are exploited by others. For communities vulnerable to climate change and lacking core capabilities, various measures can be taken to strengthen and protect their capabilities and livelihoods.

5.1   Social and Economic Development

Recognizing that wealthy regions are less vulnerable and more robust against climatic changes, economic development is an important strategy to reduce security risks and stabilize societies. Developed societies have more advanced technological capabilities, financial assets and institutional mechanisms to respond to climate change and resolve related conflicts. It is plausible that growing wealth per capita increases adaptive capacities to counter climate exposure and sensitivity. For instance, being faced with frequent droughts, rich farmers will be better able to invest in irrigation technology or diversify crops than poor and less educated farmers. There are however economic limits that restrain the affordability of adaptation measures. The Stern Review concludes that the costs of climate change for developed countries "could reach several percent of GDP as higher temperatures lead to a sharp increase in extreme weather events and large-scale changes" (Stern 2007: 157). Furthermore, economic growth aggravates the problem as climate change is the result of a non-sustainable economic system that threatens the natural resource base, the wellbeing and capabilities of people as well as the economic growth that is causing it. This indicates a complex feedback loop: economic growth and development are needed for human security but (unless carbon-neutral) cause further climate change which undermines the conditions for development that could help people to diminish the risks to human security. The challenge is to design social and economic development in a sustainable way that avoids climate change and its severe consequences. Sustainable development "meets the needs of the present without compromising the ability of future generations to meet their own needs" (WCED 1987: 43). Thus, human needs and capabilities of present and future generations are to be satisfied within natural resource limits, making societies more resistant to climate change. To manage the transition to a low-carbon society, various measures could be applied throughout the life-cycle of goods and services (e.g. resource exploration, exploitation, transport, transformation, use and recycling). These require behavioural and societal changes, economic, financial and technical instruments, regulatory schemes and public-private collaboration in North and South (Santarius et al. 2012). Altogether they could strengthen adaptive capacity and improve the conditions for survival, strengthen human security and build problem-solving capabilities.

5.2   Governance and Institutional Mechanisms

Political institutions, including governments and governance mechanisms, are essential to provide a political framework for human security under climate change. In the most fragile spots, special attention from the international community is required to assist countries in reducing the social stresses that emerge when state institutions are overstretched and the delivery of basic services is inadequate. In the Secretary-General's Report human security is based

> on a fundamental understanding that Governments retain the primary role for ensuring the survival, livelihood and dignity of their citizens. [. . .] This helps Governments and the international community to better utilize their resources and to develop strategies that strengthen the protection and empowerment framework needed for the assurance of human security and the promotion of peace and stability at every level – local, national, regional and international. (UNSG 2010: 1)

Key is to engage and empower affected stakeholders into policy-making. Adger (2006: 276) highlights the fact that "[v]ulnerable people and places are often excluded from decision-making and from access to power and resources." Consequently, policy interventions to reduce vulnerability need to address the social and political context and redress marginalization as a cause of social vulnerability which challenges the "design of good governance to promote resilience to minimize exclusion" (ibid.). Selection of adaptation strategies must be based on the knowledge of local conditions framing the impact, including local perceptions providing important complementary knowledge to climate science. To balance different interests, conflict resolution is essential to improve the efficiency of institutions and collective actions. In recent decades international efforts to prevent and manage conflicts have been strengthened, which considerably reduced the number of armed conflicts and battle deaths. However, a remaining challenge is the lack of capacity of conflict management institutions to deal with subnational conflicts and multiple crises simultaneously. Effective institutional frameworks, governance mechanisms and democratization are often seen as an important precondition for peaceful management of conflict. While the number of democratic states has grown in waves over the past half century and armed conflicts have declined, low-level violence and the number of fragile states with weak institutions has grown in the last few decades (Stewart and Brown 2010; Marshall and Cole 2011). After the global financial crisis of 2008, the pressure on social security systems, humanitarian aid and development assistance has increased. Thus, there is a risk that institutions could be overwhelmed by climate change related crises (WBGU 2008).

To deal with the concern over climate injustice, those who contribute most to the problem should also contribute most to its solution while those who are more affected deserve more protection against the risks. This requires a fair transfer of resources, technologies, know-how and investments between industrialized and developing countries. Fair and efficient burden-sharing in a global framework of climate justice would balance responsibilities and impacts among countries and communities. A global diplomacy could help to contain climate-induced distributional conflicts and strengthen cooperation, as well as the development of compensation mechanisms for the victims of climate change. A useful framework for international governance on human security has been defined by the Millennium Development Goals (MDGs) that were agreed upon by UN member states and development aid agencies. It includes eight goals, ranging from fighting extreme poverty to reducing the spread of HIV/AIDS and providing universal primary education.

5.3   Legal Framework and Human Rights

Anthropogenic climate change has the potential to violate human rights in many ways, in particular of those who are most vulnerable, and often also live in regions with already extreme human rights violations. Thus, national and international legal frameworks that protect human rights against risks, threats and challenges are key elements of human security. Kaldor (2011) argued that a human security approach should be based on the rule of law and effective law enforcement, which implies for international organizations an expanded international presence; new human security forces; and a legal framework. Human rights are inherently related to the capability approach. Nussbaum (2011) sees a "conceptual connection" between entitlements (basic human rights) and the responsibility of a nation state. From her point of view, a central duty of states is to legally enforce and defend its citizens' central capabilities. A society is unjust unless the capabilities to function are present. A nominal right based in a constitution is worthless without exercising this right in the sense of a citizen's capability (Nussbaum 2011). However, in addition to the state's responsibility, there is also a "collective obligation" (ibid.: 26) meaning that the central capabilities must also be secured by the people themselves. The human security report of Secretary-General Ban Ki-moon suggested "a set of freedoms that are fundamental to human life, and as such it makes no distinction between civil, political, economic, social and cultural rights, thereby addressing security threats in a multidimensional and comprehensive manner" (UNSG 2010: 7). Some studies examine the legal issues around human rights regarding climate change (Posner

2007; Humphreys 2010) and discuss cases that have tested these rights in practice, in particular of indigenous peoples. For example, Caney (2008) argues that the right to life, to health, and to a minimum subsistence level of material wellbeing are threatened by climate change, in addition to the rights to development and to residence. Others question the human rights approach in climate policy (e.g. Depledge and Carlane 2007).

To conclude, adequate policies for human security under climate change require a long-term planning horizon and an integrative framework to better understand the actual and potential adaptation needs of affected countries and communities from local to global levels. Adaptation offers the chance to develop innovative strategies that could strengthen the adaptive capacity of communities to gain new capabilities and revive their livelihoods. This would involve not only technical innovations that facilitate sustainable management of natural resources under climate change, but also social and institutional innovations that organize and coordinate community actions to exploit the "beneficial opportunities" of adaptation which depend on the societal conditions in the affected regions.

ACKNOWLEDGEMENT

Research for this study was funded in part by the German Science Foundation (DFG) through the Cluster of Excellence CliSAP (EXC177).

NOTES

1. This may even hold true when, for example, a pastoralist community is highly adapted to its environmental surroundings, in other words has a wide coping range. Such a community might nevertheless have little internal capacity to adapt to new, changed environmental conditions, stresses and shocks (Robinson and Berkes 2011).
2. As one example, Sen argues that a person (a) who chooses to fast might have the same functioning concerning eating habits, but a person (b) who is forced to starve is lacking the capability-set of person (a) and does not have the freedom to choose. "Fasting is not the same thing as being forced to starve" (Sen 1999: 76), it is the choice of not eating while there is food available.

REFERENCES

Adger, W. N. (2003), 'Social capital, collective action, and adaptation to climate change', *Economic Geography*, **79** (4), 387–404.
Adger, W.N. (2006), 'Vulnerability', *Global Environmental Change*, 16, 268–281.
Adger, W.N. and K. Vincent (2005), 'Uncertainty in adaptive capacity', *C. R. Geoscience*, **337**, 399–410.

Adger, W.N, P.M. Kelly, A. Winkels, L.Q. Huy and C. Locke (2002), 'Migration, remittances, livelihood trajectories, and social resilience', *Journal of the Human Environment*, **31** (4), 358–366.

Adger, W.N., I. Lorenzoni and K. O'Brien (2009), 'Adaptation now', in N. Adger, I. Lorenzoni and K. O'Brien (eds), *Adapting to Climate Change. Thresholds, Values, Governance*. Cambridge, UK: Cambridge University Press, pp. 1–22.

Agrawal, A. (2008), *The Role of Local Institutions in Adaptation to Climate Change*, Paper prepared for Social Dimensions of Climate Change meeting, Social Development Department, World Bank, Washington, 5–6 March.

Amekawa, Y., H. Sseguya, S. Onzere, I. Carranza (2010), 'Delineating the multifunctional role of agroecological practices: toward sustainable livelihoods for smallholder farmers in developing countries', *Journal of Sustainable Agriculture*, **34** (2), 202–228.

Barnett, J. (2001), *The Meaning of Environmental Security: Ecological Politics and Policy in the New Security Era*, London: Zed Books.

Barnett, J. (2010), 'Adapting to climate change: three key challenges for research and policy – an editorial essay', *Wiley Interdisciplinary Reviews: Climate Change*, **1**, 314–317.

Barnett, J. and W.N. Adger (2007), 'Climate change, human security and violent conflict', *Political Geography*, **26**, 639–655.

Blaikie, P., T. Cannon, I. Davis and B. Wisner (1994), *At Risk: Natural Hazards, Peoples' Vulnerability and Disasters*, London: Routledge.

Bohle, H.-G. (2009), 'Sustainable livelihood security. Evolution and application', in H.G. Brauch et al. (eds), *Facing Global Environmental Change: Environmental, Human, Energy, Food, Health and Water Security Concepts*. Berlin, Germany: Springer-Verlag, pp. 522–528.

Bohle, H.-G. and T. Glade (2008), 'Vulnerabilitätskonzepte in Sozial- und Naturwissenschaften', in C. Felgrentreff and T. Glade (eds), *Naturrisiken und Sozialkatastrophen*. Berlin, Germany: Springer-Verlag, pp. 99–119.

Bohle, H.-G. and K. O'Brien (2007), 'The discourse on human security: implications and relevance for climate change research: a review article', *Climate Change and Human Security*, Special issue of *Die Erde*, **137** (3), 155–163.

Brauch, H.G. (2005), 'Environment and human security. Freedom from hazard impact, InterSecTions', 2/2005 (Bonn: UNU-EHS) accessed 20 September 2012; at: <http://www.ehs.unu.edu/file.php? id=64>.

Brauch, H.G. (2009), 'Human security concepts in policy and science', in H.G. Brauch et al. (eds), *Facing Global Environmental Change*. Berlin, Germany: Springer, pp. 965–989.

Brauch, H.G. and Scheffran, J. (2012), 'Introduction: climate change, human security, and violent conflict in the Anthropocene', in J. Scheffran, M. Brzoska, H.G. Brauch, P.M. Link and J. Schilling (eds), *Climate Change, Human Security and Violent Conflict*, Berlin: Springer Verlag, Hexagon Series, **8**, pp. 3–40.

Buzan, B., O. Wæver and J. de Wilde (1998), *Security. A New Framework for Analysis*, Boulder/London: Lynne Rienner.

Campana, S. (2010), 'Climate change and the Mediterranean: reframing the security threat posed by environmental migration', Master's Thesis, Universiteit Amsterdam, available at: clisec.zmaw.de.

Caney, S. (2008), 'Human rights, climate change, and discounting', *Environmental Politics*, **17** (4), 536–555.

Carter, T.R., R.N. Jones, X. Lu, S. Bhadwal, C. Conde, L.O. Mearns, B.C. O'Neill, M.D.A. Rounsevell and M.B. Zurek (2007), 'New assessment methods and the characterisation of future conditions', in IPCC (2007), pp. 133–171.

Chambers, R. and G. Conway (1991), *Sustainable Rural Livelihoods: Practical Concepts for the 21st Century*. Brighton, UK: Institute of Development Studies.

CHS (Commission on Human Security), (2003, 2005), *Human Security Now, Protecting and Empowering People*, New York: Commission on Human Security, accessed 20 September 2012; at: <http://www.unocha.org/humansecurity/chs/finalreport/English/FinalReport.pdf>.

Depledge, M. and C. Carlane (2007), 'Sick of the weather: climate change, human health and international law', *Environmental Law Review*, **9** (4), 231–240.

FAO (2011), *The State of Food Insecurity in the World 2011*. Rome, Italy: Food and Agriculture Organization of the United Nations.

Folke, C. (2006), 'Resilience: the emergence of a perspective for social-ecological systems analyses', *Global Environmental Change*, **16**, 253–267.

Freier, K.P., Brüggemann, R., Scheffran, J., Finckh, M. and Schneider, U.A. (2012), 'An eco-sociological model concerning future livelihood strategies – climate change and traditional transhumance in semi-arid Morocco', *Technological Forecasting and Social Change*, **79**, 371–382.

Fuentes Julio, C.F., and H.G. Brauch (2009), 'The human security network: a global north-south coalition', in H.G. Brauch et al. (eds), *Facing Global Environmental Change: Environmental, Human, Energy, Food, Health and Water Security Concepts*. Berlin, Heidelberg, New York: Springer-Verlag, pp. 991–1002.

Füssel, H.-M. (2007), 'Vulnerability: a generally applicable conceptual framework for climate change research', *Global Environmental Change*, **17**, 155–167.

Füssel, H.M. (2009), 'Development and climate change', Background note to the *World Development Report 2010. Review and Quantitative Analysis of Indices of Climate Change Exposure Adaptive Capacity, Sensitivity and Impacts*. Potsdam, Germany: Potsdam Institute for Climate Impact Research.

Gallopín, G.C. (2006), 'Linkages between vulnerability, resilience, and adaptive capacity', *Global Environmental Change*, **16**, 293–303.

Gasper, D. (2006), 'What is the capabilities approach? Its core, rationale, partners and dangers', The Hague: Institute of Social Studies.

Gasper, D. (2007), 'What is the capability approach? Its core, rationale, partners and dangers', *Journal of Socioeconomics*, **36** (3), 335–359.

Gleditsch, N.P. (2012), 'Whither the weather? Climate change and conflict', *Journal of Peace Research*, **49**, 3–9.

Granovetter, M.S. (1973), 'The strength of weak ties', *American Journal of Sociology*, **78** (6), 1360–1380.

Holling, C.S. (1973), 'Resilience and stability of ecological systems', *Annual Review of Ecology and Systematics*, **4**, 1–23.

HSN (1999), 'A perspective on human security', Chairman's Summary, 1st Ministerial Meeting of the Human Security Network, Lysøen, Norway, 20 May 1999.

Human Security Report Project (2010), *Human Security Report 2009/2010: The Causes of Peace and the Shrinking Costs of War*. Vancouver, Canada: Oxford University Press.

Humphreys, S. (ed.) (2010) *Climate Change and Human Rights*. Cambridge, UK: Cambridge University Press, p. 348.

IPCC (2007), *Climate Change 2007: Impacts, Adaptation and Vulnerability. Contribution of Working Group II to the Fourth Assessment Report of the Intergovernmental Panel on Climate Change*. Cambridge, UK: Cambridge University Press, pp. 869–883.

Jones, L., S. Jaspars, S. Pavanello, E. Ludi, R. Slater, A. Arnall, N. Grist and S. Mtisi (2010), 'Responding to a changing climate. Exploring how disaster risk reduction, social protection and livelihoods approaches promote features of adaptive capacity', Working Paper 319, London, UK: Overseas Development Institute.

Kaldor, M. (2011), 'Human security in complex operations', *PRISM*, **2** (2), 3–14.

Kelly, P.M. and W.N. Adger (2000), 'Theory and practice in assessing vulnerability to climate change and facilitating adaptation', *Climatic Change*, **47**, 325–352.

King, G. and C.J.L. Murray (2002), 'Rethinking human security', *Political Science Quarterly*, **116** (4), 585–610.

Leemans, R. (2009), 'The Millennium Ecosystem Assessment: securing interactions between ecosystems, ecosystem services and human well-being', in H.G. Brauch et al. (eds), *Facing Global Environmental Change: Environmental, Human, Energy, Food, Health and Water Security Concepts*. Berlin, Heidelberg, New York: Springer-Verlag, pp. 53–62.

Marshall, M.G., and B.R. Cole (2011), *Global Report 2011 – Conflict, Governance, and State Fragility*. Vienna, USA: Center for Systemic Peace.

Max-Neef, M., A.A. Elizalde, and M. Hopenhayn (1991), *Human Scale Development: Conception, Application and Further Reflections*. New York, USA: Apex.

Mearns, R., and A. Norton (eds) (2009), *Social Dimensions of Climate Change: Equity and Vulnerability in a Warming World*. Washington, DC, USA: World Bank.

McDowell, C. and A. de Haan (1997), *Migration and Sustainable Livelihoods: A Critical Review of the Literature*. Sussex, UK: IDS.

Mertz, O., K. Halsnaes, J.E. Olesen and K. Rasmussen (2009), 'Adaptation to climate change in developing countries', *Environmental Management*, **43**, 743–752.

Nordås R. and N.P. Gleditsch (2005), 'Climate conflict: common sense or nonsense?', Paper presented at the international workshop Human Security and Environmental Change, Oslo, Norway 21–23 June.

Nussbaum, M.C. (2003), 'Capabilities as fundamental entitlements: Sen and social justice', *Feminist Economics*, **9**, 33–59.

Nussbaum, M.C. (2011), 'Capabilities, entitlements, rights: supplementation and critique', *Journal of Human Development and Capabilities*, **12**, 23–37.

O'Brien, K.L. and R.M. Leichenko (2007), 'Human security, vulnerability, and sustainable adaptation', Background Paper commissioned for the *Human Development Report 2007/2008: Fighting Climate Change: Human Solidarity in a Divided World*. New York, USA: United Nations Development Programme.

Owen, T. (2004), 'Human security: conflict, critique and consensus: colloquium remarks and a proposal for a threshold-based definition', Special Section: What is Human Security?, *Security Dialogue*, **35** (3), 345–387.

Paris, R. (2001), 'Human security: paradigm shift or hot air', *International Security*, **26** (2), 87–102.

Porter, A. (1998), 'Social capital: its origins and applications in modern sociology', *Annual Review of Sociology*, **24**, 1–24.

Posner, E. (2007), 'Climate change and international human rights litigation: a critical appraisal', *University of Pennsylvania Law Review*, **155**, 1925–1945.

Putnam, R.D. (1993), *Making Democracy Work: Civic Traditions in Modern Italy US*, Princeton, USA: Princeton University Press.

Renaud, F.G., J. Birkmann, M. Damm and G.C. Gallopín (2010), 'Understanding multiple thresholds of coupled social-ecological systems exposed to natural hazards as external shocks', *Natural Hazards*, **55**, 749–763.

Robeyns, I. (2006), 'The capability approach in practice', *The Journal of Political Philosophy*, **14**, 351–376.

Robinson, L.W. and F. Berkes (2011), 'Multi-level participation for building adaptive capacity: formal agency-community interactions in northern Kenya', *Global Environmental Change*, **21**, 1185–1194.

Roy, M. and H.D. Venema (2002), 'Reducing risk and vulnerability to climate change in India: the capabilities approach', *Gender & Development*, **10**, 78–83.

Sanginga, C.P., R.N. Kamugisha and A.M. Martin (2007), 'The dynamics of social capital and conflict management in multiple resource regimes: a case study of the southwestern highlands of Uganda', *Ecology and Society*, **12** (1).

Santarius, T., J. Scheffran and A. Tricarico (2012), *North South Transitions to Green Economies – Making Export Support, Technology Transfer, and Foreign Direct Investments Work for Climate Protection*. Heinrich Böll Foundation, accessed 20 September 2012 at: http://www.santarius.de/wp-content/uploads/2012/08/North-South-Transitions-to-Green-Economies-2012.pdf.

Scheffran, J. (2008a), 'The complexity of security', *Complexity*, **14** (1), 13–21.

Scheffran, J. (2008b), *Preventing Dangerous Climate Change*, in V.I. Grover (ed.), *Global Warming and Climate Change*, Science Publishers, **2**, 449–482.

Scheffran, J. (2011), 'The security risks of climate change: vulnerabilities, threats, conflicts and strategies', in H. Brauch et al. (eds), *Coping with Global Environmental Change, Disasters and Security*. Berlin, Germany: Springer, pp. 735–756.

Scheffran, J. and A. Battaglini (2011), 'Climate and conflicts: the security risks of global warming', *Regional Environmental Change*, **11**, 27–39.

Scheffran, J., M. Brzoska, J. Kominek, M. Link, and J. Schilling (2012a), 'Climate change and violent conflict', *Science*, **336**, 869–871.

Scheffran, J., P.M. Link and J. Schilling (2012b), 'Theories and models of climate-security interaction: framework and application to a climate hot spot in North Africa', in J. Scheffran et al. (eds), *Climate Change, Human Security and Violent Conflict*, Berlin, Germany: Springer, pp. 91–132.

Scheffran, J., E. Marmer and P. Sow (2012c), 'Migration as a contribution to resilience and innovation in climate adaptation: social networks and co-development in Northwest Africa', *Applied Geography*, **33**, 119–127.

Schilling, J., K.P. Freier, E. Hertig and J. Scheffran (2012), 'Climate change, vulnerability and adaptation in North Africa with focus on Morocco', *Agriculture, Ecosystems and Environment*, **156**, 12–26.

Security Dialogue (2004), 'Special Section: What is Human Security?', **35** (3).

Sen, A. (1985), *Commodities and Capabilities*. Amsterdam, Netherlands: North-Holland.

Sen, A. (1999), *Development as Freedom*. Oxford, UK: Oxford University Press.

Smit, B. and O. Pilifosova (2001), 'Adaptation to climate change in the context of sustainable development and equity', in J.J. McCarthy, O.F. Canziani, N.A. Leary, D.J. Dokken and K.S. White (eds), *Climate Change 2001: Impacts, Adaptation, and Vulnerability – Contribution of Working Group II to the Third Assessment Report of the Intergovernmental Panel on Climate Change*. Cambridge, UK: Cambridge University Press, pp. 877–912.

Smit, B. and J. Wandel (2006), 'Adaptation, adaptive capacity and vulnerability', *Global Environmental Change*, **16**, 282–292.

Starr, H. (ed.) (2008), 'Failed States, Special Issue', *Conflict Management and Peace Science*, **25** (4).

Stern N. (2007), *The Economics of Climate Change. The Stern Review*, New York: Cambridge University Press.

Stewart, F. and G. Brown (2010), 'Fragile states: CRISE Overview 3', Centre for Research on Inequality, Human Security and Ethnicity (CRISE), Oxford.

Ulbert, C. and S. Werthes (eds) (2008a), 'Menschliche Sicherheit. Globale Herausforderungen und regionale Perspektiven', Baden-Baden: Nomos.

UNDP (1994), *Human Development Report 1994. New Dimensions of Human Security*, New York/Oxford/New Delhi: Oxford University Press, accessed 20 September 2012; at: <http://hdr.undp.org/reports/global/1994/en/pdf/hdr_1994_ch2.pdf>.

UNSG (2010), *Human Security*, Report of the UN Secretary General Ban Ki-moon, (A/64/701), 8 May. New York, USA: United Nations, accessed 20 September 2012 at: http://responsibilitytoprotect.org/human%20security%20report%20april%206%202010.pdf.

Wasserman, S. and K. Faust (1995), *Social Network Analysis: Methods and Applications*. Cambridge, UK: Cambridge University Press.

WBGU (2008), *World in Transition – Climate Change as a Security Risk*, Wissenschaftlicher Beirat der Bundesregierung Globale Umweltveränderungen (German Advisory Council on Global Change). London, UK: Earthscan.

WCED (1987), *Our Common Future*, Report of the World Commission on Environment and Development.

Wikipedia (2012), 'Capability management', accessed 20 September 2012 at: <http://en.wikipedia.org/wiki/Capability_management>.

7. Vulnerability does not just fall from the sky: toward multi-scale pro-poor climate policy

Jesse Ribot

AUTHOR'S PREFACE: CAUSE AND BLAME IN THE ANTHROPOCENE – VULNERABILITY STILL DOES NOT JUST FALL FROM THE SKY

> God is dead. God remains dead. And we have killed him. How shall we comfort ourselves, the murderers of all murderers? What was holiest and mightiest of all that the world has yet owned has bled to death under our knives: who will wipe this blood off us? What water is there for us to clean ourselves? What festivals of atonement, what sacred games shall we have to invent? Is not the greatness of this deed too great for us? Must we ourselves not become gods simply to appear worthy of it?
>
> – Nietzsche, *The Gay Science*, Section 125

> . . . "risk" has become *the* organizing concept that gives meaning and direction to environmental regulation.
>
> – Jasanoff 1999: 135

Sea level off of New York has risen 20 centimeters this century and when Sandy struck the US East Coast in October 2012 it wreaked expectable but unexpected havoc. Sandy showed that under a changing climate the rich could lose their houses and the poor their lives and livelihoods. Sandy brought global climate change home in a way that the remaining deniers just look stupid and the believers look complacent. It put the state in gear defending past actions – dune and seawall construction, tunnel flood gates – and proposing aid to facilitate recovery. Few are asking why God did this. President Obama did not proclaim "but for the grace of God, there we go", as he had for the 2010 Haiti quake (Wood, *New York Times*, 23 January 2010). Sandy was also not seen as natural. People are viewing Sandy as an anthropogenic superstorm (Kaplan, *New York Times*, 3 December 2012). People are asking "who did what when" and "why did this happen" People seek to understand cause of risk, its failure to be regulated, and then to attribute blame (Lipton and Moss, *New York Times*, 10 December 2012; Preston, Fink and Powell, *New York Times*, 3 December 2012). The politics of cause and blame are becoming central to the new normal of American climate politics.

Recent climate events are also reshaping geopolitics. Mohammed Chowdhury, who represents the group of poorest nations at the December 2012 Doha climate negotiations, saw how much assistance President Obama asked from the US Congress for Hurricane Sandy, and noted ". . .we won't get that scale and magnitude of support" (John Broder, *New York Times*, 9 December 2012). Links are now drawn globally between climate change and disaster. The behavior of emitters is seen to cause the pain of climate victims. Links are drawn through the morality of inequality of mitigation, adaptation and response. The causal chain is also being traced from disaster to the preparations of governments and their post-event responses, and to the excesses and moral vacancy of unprecedented inequalities of a free-market society. Pain and suffering are inspiring fuller analyses of causality, indicating of responsibility, and attribution of blame – supporting claims of liability from direct causality and claims of moral obligation by people to people. With anthropogenic climate change making climate social, the social nature of disaster is clear. Disaster is socially produced – on the ground and through the sky. We can only blame ourselves. We must now look to society – at all scales – for causes, responsibilities, blame, and solutions.

"Vulnerability does not just Fall from the Sky" (Ribot 2010 being reprinted here) explores causal structures of vulnerability and the relation of vulnerability to climate change. Many analyses of vulnerability shy away from historical political-economic analyses of causality. Instead, they focus on identifying *who* is vulnerable rather than *why*. This is no surprise. Causality is threatening. It implies responsibility, blame and liability. The discourses on climate change have shifted toward adaptation as a means of addressing climate-related vulnerabilities – but through a forward looking analysis of how to enable adjustment rather than a historical analysis of the structures that generate risk. Yet generative causes of vulnerability remain important for redressing vulnerability. Causes of insecurity are important for producing security. The chapter focuses on analysis of *vulnerability* and does not use the language of *human security*. Nevertheless they are closely related. Vulnerability and insecurity are socially generated and socially redressed. They do not just fall from the sky. We would do well to understand their causes in order to propose enduring transformative solutions. We would also do well to develop a new sociology of risk that helps us to understand what risk does to society, and how, when and why cause and blame are analyzed, presented, and acted upon.

The enlightenment replaced God with nature, priests with scientists, and theodicy with the study of risk. What remained constant was society's need to explain pain and suffering – to identify risk and to attribute blame. All human cultures are faced with explaining excessive suffering in the world; all people in all times struggle to reduce pain and to make sense of human experience (Wilkinson 2010). Weber saw such rationalization as the basis of cultural or social change – the need to reconcile belief and experience resulting in the transformation of culture. Douglas believed that explanations or risk also served to define and consolidate community by drawing lines between good and bad, us and them, and by providing a basis of organizing for self-protection.[1]

Rose, using Foucault's governmentality, sees risk management as a means of social control – through conduct of conduct. Governments produce risk subjects by conducting "at risk" populations and society to see themselves as the causal agents of risk, thus disciplining individual subjects to take on the project of adjustment – deflecting liabilities from the state, larger society and political economy that produced their condition (Rose 1999). We also often observe politicians addressing risk and security to hold onto power – as angry people vote and protest. Risk, cause and associated blame, following many social theorists, are central to political organization, social organization and social change (Bordieu 1977; Douglas 1985, 1992; Beck 1986; Rose 1999; Jasanoff 1999; Adam et al. 2000; Wilkinson 2010).

What does climate-related risk do in and to society? How does the analysis and interpretation of the causes of risk shape social life? How does risk analysis produce solutions and change social organization? Sen's (1981) causal "entitlements" model explains vulnerability to drought-related famine as a result of the failure of legal market-based means of attaining food. He shows that during droughts well-functioning markets allocate food away from the hungry – despite the presence of sufficient food to nourish everyone. Cause and blame are focused on the legal-economic means of households to use their assets to obtain sufficient food. It indicates a need for capabilities support, market regulation, and social protections to enable household asset formation and prevent entitlement collapse. To explain assets and entitlements within that system, Watts and Bohle (1993) examine the generative political-economy and broad social inequalities through which assets, laws and resulting entitlements are formed. Their analysis indicates the role of political mobilization or "empowerment"– and in my own framing, this includes political representation – in shaping the very protections required to enhance household security. These explanations place cause and blame for climate-related risks within society, attributing social responsibility and requiring social response.

By tracing causality to what Fraser (2008: 28) calls the "generative framework", these analyses point to the potential for transformative intervention – the kind that can restructure the processes producing vulnerability. Such transformative solutions require changes in the power relations that shape the political economy that shape entitlements.[2] They present deep challenges to the status quo. While understanding causality is a necessary element of response, explanation quickly generates conflict – of theory, method, historiography, interpretation; but more fundamentally, the conflicts are over implication and interest. Causality is a contentious category of mind. Causes indicate blame and liability, linking damages to social organization and human agency. The tracing of causality from any instance of crisis is a threat to those who might have played a role – of ignorance, of negligence, of intent, of greed or avarice – in the production of pain. It is a threat to those who benefit, passively or actively, from unacceptable but everyday relations of production, exchange, and consumption.

Those exposed to blame often avoid or deny generative analyses. It is no wonder that even the well-meaning choose causes of least contention. Rather than looking back at place-based histories and causes of vulnerability, those wishing to reduce vulnerability in the face of climate change prefer to look forward toward "adaptation" – a naturalized neutral space of imaginary futures devoid of social cause or blame (Ribot 2011).[3] Their forward gaze silently blames the hazard while accepting current configurations on the ground, contributing to a "death of the social" – a failure to acknowledge generative processes of differentiation that produce unequal protection and opportunity (Rose 1999). It seems that any explanation that does not blame God, nature, or the victim (e.g. the blaming of risk subjects *á la* Foucault and Rose) – or distant drivers of climate change – is suspect and avoided by policy makers, implementers and many activists alike. The continued shunting of blame back to climate does the double work of occluding local causality while continuing to displace blame onto the hazard – as act of God, nature, or today anthropogenic climate change.[4]

Of course, under an anthropogenic climate, blaming the hazard does indicate social liability by tracing cause to greenhouse gas emitters and the political-economic system that enables them to emit. So does this mean that risk is located in hazards or that vulnerability now falls from the sky? Who is responsible for climate events – and associated hunger, famine, dislocation and economic losses? Do we need to rethink causal structure to include causality articulated through the sky? Like with Sen's (1981) and Watts and Bohle's (1989) explanation of famine, the conditions on the ground that translate climate events into disasters remain social and remain in place. Hence, climate events still cannot be blamed. Hazards – no matter how generated – do not explain people's assets or entitlements. They do not explain why following Hurricane Sandy, predominantly white, gated communities expect government funds to rebuild their plush homes while Blacks and Hispanics in the projects can only demand support sufficient to reproduce their initial level of poverty (Berger 2012). Nor can analysis of the hazard explain why the same magnitude drought or storm is fatal in one place and time while being a mere nuisance in another.

Nevertheless, something happens to causality with social nature – when climate hazards become anthropogenic. Why can't we now attribute causality and blame to these events in the sky? What does anthropogenic climate change do to cause and blame? Social liability, it is true, can now travel through the sky. But, it is still not a product of the sky. Risk of damage in the face of a biophysical event is still caused by the place-based social and political-economic histories that put people on the threshold of disaster. Climate events – anthropogenic or natural – still find vulnerable people in place. They don't put them there. Because the biophysical events are anthropogenic, however, the causal explanation of risk must now account for the human intentionality and interest behind the stressor. This becomes even more acute with the advent of geo-engineering (Klein 2012). Even if disasters were never acts of God or nature, climate events, which could have been seen as external to the social world, are now cultured. Climate events have become traceable to acts of social systems and agents (Jones and Edwards 2009; Arthur 2012). So once again, liability resides on the ground, not in the sky. The causal structure of vulnerability remains within society. The sky is merely a medium. We don't blame a car for running someone over.[5] We cannot wash our hands. A new politics of cause and blame characterize our current era. Welcome to the anthropocene.[6]

Cause, and therefore blame, in the anthropocene are now bifurcated. Of course, it is not as if society could ever – with or without anthropogenic climate change – have washed its hands of the production of vulnerability, which is a result of the differentiation of a world system of political-economic and social relations among countries and among people within countries. The vulnerability on the ground is (and always has been) as much a product of far-away forces as the changes we now see in the skies. Risk articulates through climate events due to protected actions of real people in real places who, without direct liability through the rules, structures and subjectivities of differentiation, shape patterns of inclusion and exclusion that externalize the cost of their desires and their profit on others far away. The structure of vulnerability is still purely social. The causes of vulnerability – above all the differentiated causes in a given place – can still be traced from that place through the social relations of production, exchange, domination, subordination, governance and subjectivity. They still have to be analyzed and understood starting from the instance of crisis in a real place and real time. But, acknowledged anthropogenesis provides a new pathway for attributing social causality, and therefore, blame and liability – and claims for redress and compensation (Jones and Edwards 2009; Hyvarinen 2012).[7] Like vulnerability, the causes of new climate stress also do not fall from the sky.

While anthropogenesis remains on the ground, it profoundly changes the meaning of climate events. Humans are now demonstrably responsible – not only for the vulnerability on the ground, but for the stressors that arc across the sky. Blaming the sky – and its Godliness or its nature – can no longer absorb, divert or occlude liability. Indeed, it adds a new dimension to a connectivity of the globe that has long been apparent to historians and to social and political-economic theorists (Wolf 1981). Social causes of place-based vulnerability and of stressors in the sky – the two strands of cause and blame – are interlinked. Inequality in access to the production of climate-changing greenhouse gasses is partly responsible for the poverty and marginality that places some people in secure standing and others at risk. Those who can consume well beyond subsistence are less vulnerable than those who cannot (see Watts 1983; Agarwal and Narain 1991). Unfettered access to resources and goods – protected through a differentiated global political economy with rules and social relations that protect some actors and subordinate others – enables the excess consumption that is changing the climate and increasing the stresses on those at risk. Social stratification and inequalities that are behind vulnerability on the ground are contributing to stress articulated through a changing climate system.[8]

Historical causal analysis is not the only path toward productive strategies for reducing vulnerability and increasing human security, but it is important. This preface helps frame analysis of the origins of vulnerability and insecurity in the era of anthropogenic nature as a way of updating the article. As early as 1994, the UNDP proposed Human Security as a people-centered concept in which "security . . . means safety from the constant threat of hunger, disease, crime and repression" as well as "protection from sudden and hurtful disruptions in the pattern of our daily lives" (UNDP 1994). In developing this concept further, Ogata and Sen (2003: 2) called for "security centered on people – not states." This redefinition of security brings the focus to the individual, household and community levels, precisely where analyses of vulnerability must always begin. So, to reduce vulnerability and build security requires a deep understanding of experienced crisis and its origins. To achieve human security, to protect "the vital core of all human lives in ways that enhance human freedoms and fulfillment" (Ogata and Sen 2003), requires an understanding of how these lives arrive at thresholds of disaster. Climate-related security needs to be centered on people – not in the sky. Rather than looking to causal analysis, it is easier to naturalize blame, shift it to God, or, in the anthropocene, to displace it through the clouds onto people far off. This last is a productive social analysis of causality. The causes of place-based vulnerability, however, which have local and distant roots, need also to be revealed and treated.

Rationalization of pain – understanding its causes and attributing blame – shapes relations in and among societies. In the anthropocene we face new plausible pathways of damage and blame with the potential to reconfigure geopolitics. Al. Qaida can blame the West for some of their suffering based on their experience of global political-economic marginalization filtered through their own theodical lens.

Now societies around the world have grounds on which to blame the industrial world for storms and droughts. What filters will soften or harden their understanding? What will they do in response? The new channels of global-scale blame are emerging in the anthropocene. Long-distance cause and effect, of course, is nothing new. It has long been attributed through analysis and imagination. A Haitian taxi driver in Newark told me that the 2010 Haiti quake was caused by problems with a secret tunnel being dug from Miami to Port-au-Prince. He understood there was a causal relation between actions in the US and Haiti's pain. He was not wrong. The history that made Haiti vulnerable was connected to French slavery and subsequent indemnities collected by Citibank, trade blockades, the US occupation, externally backed dictators, and other forces from overseas that left Haiti at risk. He understood and his imagination filled in. In the anthropocene, a new channel of blame, backed by the quite credible priests of science, has been added to the repertoire of global division. God is dead, nature is cultured, we need even less imagination to see how the burdens of human agency can turn back upon us.

Acknowledgements

Many thanks to Tim Forsythe, Helen Epstein, Zsuzsa Gille, and Malini Ranganathan for insightful comments on this preface.

Notes

1. For Mary Douglas, following Durkheim and Weber, the risks that people fear are not the sociological focus. She explores only the social *functions* of risk. She argues that when social bonds that hold people together are weak, people commonly become obsessed with disaster (perhaps implying that weak bonds are a condition of our own modernity). This

focus on disaster has a positive function for society. By finding a common threat, communities come together, and organize around common social objectives. They protect their group from harm. Simultaneously, as part of group survival, comes the search for explanation and blame – which helps further consolidate the group by defining insiders and outsiders. It identifies those who belong and creates boundaries that define the community. It also defines the "other." Like Weber, Douglas sees discourses of "risk" as having taken the place of discourses of theodicy. The language of "risk" supplanted language of "sin," both being languages of blame. Where people used to talk about being "sinned against," they now speak of being "at risk": sin appeals to the authority of priest; risk appeals to the authority of science and modern rationality (Douglas 1985). On the production of "at risk" subjects also see Rose 1999.

2. Focused on injustice, Fraser (2008: 28) outlines two approaches to remedy – affirmative and transformative. She argues that "by affirmative remedies for injustice I mean remedies aimed at correcting inequitable outcomes of social arrangements without disturbing the underlying framework that generates them." I would place many approaches to adaptation in this camp. She continues "By transformative remedies, in contrast, I mean remedies aimed at correcting inequitable outcomes precisely by restructuring the underlying generative framework."

3. People's possibilities of action are always being structured by others – their conduct is being conducted. Conducting conduct, I would argue, is the function of concepts like "adaptive capacity." Viewing risk as a product of limits that structure options points the individual and group to aim to restructure their circumstance by changing the political economy and power relations in order to expand their possibilities of action. Viewing risk as determined by adaptive capacity points the individual to adapt within their internal or inherent capacity, making *invisible* the production of their possibilities of action – naturalizing them and creating a subjectivity of action that is the responsibility of the individual.

4. Here anthropogenic climate change is at risk of becoming a means of directing attention to distant causes at the expense of continued analysis of the production of local vulnerabilities. So, it too can serve, like God and Nature to deflect responsibility.

5. One reviewer felt this statement was too close to the US gun lobby's statement that "guns don't kill people, people kill people." Indeed, they are right. We don't blame a gun for killing someone – we blame legislators who make them freely available, among many other social factors.

6. I take this phrase from the video "Welcome to the Anthroposcene": http://sociology. leeds.ac.uk/sites/environment/2012/03/27/welcome-to-the-anthroposcene/.

7. Blaming humans for biogeophysical events is by no means new. It defines communities, defines divides, places blame and locates the origins of pain for some people for some reason. It is worth taking seriously. It is as reasonable as the notion of a natural disaster or an act of God. All of these explanations do "work" – they produce meaning and mobilize action, avoid or attribute blame, etc. They are about people trying to make sense of extraordinary suffering, as in the case of the Haitian taxi driver at the end of this preface. Of course fantasies, nature and god (evoked by the taxi driver, insurance companies, Pat Robertson, and Barak Obama) are all means to locate blame and shift it away from the history of subordination that put Haiti at risk (Ribot 2010). Further, magic in many cultures is a means of attributing human causality to events that may or may not be of human origin.

8. The production of that suffering resides in unequal access to pleasure and plenty – social stratification. It is commonly but spuriously said that the Chinese character for crisis is composed of danger plus opportunity. The popular interpretation of this misconception is that change is both painful and promising – and crisis should be a welcomed opportunity for personal growth (yet another logic that blames the victim). What these wishful thinkers fail to recognize is that danger and opportunity may be related, but more likely across separate segments of society. Some people are faced with danger while others reap related opportunity. Danger for some is opportunity for others – society is stratified.

References

Adam, Barbara, Ulrich Beck and Joost Van Loon (2000), *The Risk Society and Beyond: Critical Issues for Social Theory*. London, UK: Sage.

Agarwal, Anil and Sunita Narain (1991), *Global Warming in an Unequal World*. New Delhi, India: Centre for Science and Environment.

Arthur, Charles (2012), 'Revealed: how the smoke stacks of America have brought the world's worst drought to Africa', Editorial in *The Independent*, 12 June. http://www.freerepublic.com/focus/news/699049/posts.

Beck, Ulrich (1986), *Risk Society: Towards a New Modernity*. Los Angeles, USA: Sage.

Berger, Joseph (2012), 'Enclaves, long gated, seek to let in storm aid', *New York Times*, 26 November.

Bordieu, Pierre (1977), *Outline of a Theory of Practice*. Cambridge, UK: Cambridge University Press.

Douglas, Mary (1985), *Risk Acceptability According to the Social Sciences*. New York, USA: Russell Sage Foundation.

Douglas, Mary (1992), *Risk and Blame: Essays in Cultural Theory*. London, UK: Routledge.

Fraser, Nancy (2008), 'From redistribution to recognition? Dilemmas of justice in a "post-socialist" age', in K. Olsen (ed.), *Adding Insult to Injury Nancy Fraser Debates Her Critics*. London, UK: Verso.

Hyvarinen, Joy (2012), 'Loss and damage caused by climate change: legal strategies for vulnerable countries', October 2012 report of the Foundation for International Environmental Law and Development (FIELD), London, UK. http://www.field.org.uk/files/field_loss__damage_legal_strategies_oct_12.pdf.

Jasanoff, Sheila A. (1999), 'The songlines of risk', *Environmental Values*, **8**, 135–152.

Jones, Tim and Sarah Edwards (2009), 'The climate debt crisis: why paying our dues is essential for tackling climate change', Report of the World Development Movement and Jubilee Debt Campaign. http://wdm.org.uk/sites/default/files/climatedebtcrisis06112009_0.pdf.

Kaplan, Thomas (2012), 'Most New Yorkers think climate change caused hurricane, poll finds', *New York Times*, 3 December.

Klein, Naomi (2012), 'Geoengineering: testing the waters', *New York Times*, 27 October, p. SR4.

Lipton, Eric and Michael Moss (2012), 'Housing agency's flaws revealed by storm', *New York Times*, 10 December.

Nietzsche, Friedrich (1974), *The Gay Science*. New York, USA: Vintage Books.

Ogata, Sadako and Amartya Sen (2003), *Human Security Now*, Report of the UN Commission on Human Security. http://ochaonline.un.org/humansecurity/CHS/finalreport/index.html.

Preston, Jennifer, Sheri Fink and Michael Powell (2012), 'Behind a call that kept nursing home patients in storm's path', *New York Times*, 3 December, p.1.

Ribot, Jesse (2011) 'Vulnerability before adaptation: toward transformative climate action', *Global Environmental Change*, **21**, 4.

Rose, Nicholas (1999), *Powers of Freedom: Reframing Political Thought*. Cambridge, UK: Cambridge University Press.

Sen, Amartya (1981), *Poverty and Famines: An Essay on Entitlement and Deprivation*. Oxford, UK: Oxford University Press.

UNDP (1994), *Human Development Report*. New York, USA: Oxford University Press.

Watts, Michael J. (1983), *Silent Violence*. Berkeley, USA: University of California Press.

Watts, Michael J. and Hans Bohle (1993), 'The space of vulnerability: the causal structure of hunger and famine', *Progress in Human Geography*, **17** (1), 43–68.

Wilkinson, Iain (2010), *Risk, Vulnerability and Everyday Life*. London, UK: Routledge.

Wolf, Eric (1981), *Europe and the People without History*. Los Angeles, USA: University of California Press.

Wood, James (2010), 'Between God and a hard place', *New York Times*, 23 January.

ARTICLE: VULNERABILITY DOES NOT JUST FALL FROM THE SKY: TOWARD MULTI-SCALE PRO-POOR CLIMATE POLICY*

A society is ultimately judged by how it treats its weakest and most vulnerable members.

– Hubert Humphrey

If a free society cannot help the many who are poor, it cannot save the few who are rich.

– John F. Kennedy

Introduction

If some combination of narcissistic morality and raw self interest does not help reduce vulnerability, then perhaps some good analysis and political engagement may.

Analysis of vulnerabilities can help answer where and how society can best invest in vulnerability reduction. Analysis may not motivate all decision makers to make those investments, but can give development professionals, activists, and affected populations fodder to promote or demand the rights and protections that can make everyone better off. Climate variations and changes present hazards to individuals and to society as a whole. The damages associated with storms, droughts, and slow climate changes are shaped by the social, political, and economic vulnerabilities of people and societies on the ground. Impacts associated with climate can be reduced through measures falling anywhere on a spectrum from climate change mitigation to reduction of the vulnerabilities of individuals and groups (McGray et al. 2008: 35). This chapter calls for evaluation of the relatively neglected social and political-economic drivers of vulnerability at one end of this spectrum. The objective is to enable consideration of a full range of vulnerability-reducing policy responses. The article is concerned with the reduction of the everyday vulnerabilities of poor and marginal groups exposed to climate trends and events.

The world's poor are disproportionately vulnerable to loss of livelihood and assets, dislocation, hunger, and famine in the face of climate

* I want to thank Robin Mearns for permission to reprint this article, which first appeared as: Ribot, Jesse C. (2010) "Vulnerability does not just fall from the sky: toward multi-scale pro-poor climate policy", in Robin Mearns and Andrew Norton, *Social Dimensions of Climate Change: Equity and Vulnerability in a Warming World*, Washington DC: The World Bank. Https://openknowledge.worldbank.org/handle/10986/2689 License: CC BY 3.0 Unported. This reprint contains modifications, particularly concerning the notion of so-called "social constructionist" approaches to vulnerability.

variability and change (Cannon et al. n.d.: 5; Anderson et al. 2010; Heltberg et al. 2010). Living with multiple risks, poor and marginalized groups must manage the costs and benefits of overlapping natural, social, political and economic hazards (Moser and Satterthwaite 2010). Their risk-minimizing strategies can diminish their income even before shocks arrive, while shocks can reinforce poverty by interrupting education, stunting children's physical development, destroying assets, forcing sale of productive capital, and deepening social differentiation from poor households' slower recovery (Heltberg et al. 2010). The poor may also experience threats and opportunities from development or climate action itself, such as efforts to reduce greenhouse-gas emissions in sectors such as household energy, land, and forest management (Turner et al. 2003: 8076; O'Brien et al. 2007: 84; ICHRP 2008: 1–2; White et al. 2010).[1]

The good news is that policy can drastically reduce climate-related vulnerability. While the best global data indicate human suffering and economic loss are worsening in the face of natural hazards,[2] the number of people per total population affected is declining (Kasperson et al. 2005: 151–2). This reduction in vulnerability is most pronounced in high-income countries, where higher levels of wellbeing along with better infrastructure, policy, and planning are successfully mediating the relation between climate trends or events and outcomes. Effective climate action can further widen this gap between climate stressors and the risk of hardship.

In 1970, when Cyclone Bhola hit Bangladesh with six-meter tidal surges, some 500,000 people perished (Frank and Husain 1971). In 1991 the similar Cyclone Gorky, struck Bangladesh with 140,000 deaths. Yet, in 2007 when Cyclone Sidr, which was stronger than either Bhola or Gorky, hit Bangladesh with ten-meter tidal surges, fatalities dropped to 3,406. Although population density increased in this area during this time, the death toll was dramatically reduced (Government of Bangladesh 2008). The reduced damage was due to Bangladesh's shift from a focus on disaster relief and recovery to hazard identification, community preparedness, and integrated response efforts (CEDMHA 2007). Most important were sophisticated early warning and evacuation systems (Ministry of Food and Disaster Management of Bangladesh 2008; Bern et al. 1993; Batha 2008), which made Sidr 150 times less fatal than Bhola.[3] This is an example of effective climate action.

While there are notable policy successes, vulnerability of poor, marginalized, and under-represented people remains widespread. In cases like Bangladesh, women, the poor, and other marginalized groups are disproportionately and unacceptably vulnerable (Mushtaque et al. 1993). When facing droughts in Northeast Argentina, industry-dependent tobacco growers are more vulnerable than independent agroecological farmers, whose farms are more bio-diverse, more technologically equipped, less exposed to external markets, and have greater political negotiating power (Kasperson et al. 2005: 158–9). In Kenya, privatization of pasturelands has improved security of some while making the poorer and landless much more vulnerable (Smucker and Wisner 2008). In Northeast Brazil the poor remain vulnerable due to dependence on rain-fed agriculture combined with little access to climate neutral employment (Duarte et al. 2007: 25). Poorer people excluded from access to services, social networks, and land experience intensified climate-related vulnerabilities and losses due to unequal social relations of power and representation. These kinds of problems are also a target for climate action.

The vast differences in damages associated with similar climate stressors in the same place at different times, from place to place or among different social strata, reflect the complex and non-linear relation between climate and outcomes. The damages associated with climate events result more from conditions on the ground than from climate variability or change. Climate events or trends are transformed into differentiated outcomes via social structure. The poor and wealthy, women and men, young and old, and people of different social identities or political stripes experience different risks while facing the same climate event (Wisner 1976; Sen 1981; Watts 1987; Swift 1989; Hart 1992; Agarwal 1993; Blaikie et al. 1994: 9; Demetriades and Esplen 2010; Moser and Satterthwaite 2010). These different outcomes are due to place-based social and political-economic circumstance. The inability to sustain stresses does not come from the sky. It is produced by on-the-ground social inequality, unequal access to resources, poverty, poor infrastructure, lack of representation, and inadequate systems of social security, early warning, and planning. These factors translate climate vagaries into suffering and loss.

Poverty is the most salient of the conditions that shape climate-related vulnerability (Prowse 2003: 3; Cannon et al. n.d.: 5; Anderson et al. 2010; Heltberg et al. 2010). The poor are least able to buffer themselves against and rebound from stress. They often live in unsafe flood- and drought-prone urban or rural environments, lack insurance to help them recover from losses, and have little influence to demand that their governments provide protective infrastructure, temporary relief, or reconstruction support (ICHRP 2008: 8). Indeed, their everyday conditions are unacceptable even in the absence of climate stress. Climate stresses push these populations over an all-too-low threshold into an insecurity and poverty that violates their basic human rights (ICHRP 2008: 6; Moser and Norton 2001).

Since the adaptation side of climate action aims to reduce human vulnerability, it cannot be limited to treating incremental effects from climate change so as to maintain or bring people back to their pre-change deprived state (also see Heltberg et al. 2010).[4] As Blaikie et al. (1994: 3) point out, "despite the lethal reputation of earthquakes, epidemics, and famines, many more of the world's population have their lives shortened by unnoticed events, illnesses, and hunger that pass for normal existence in many parts of the world. . ." (also see Kasperson et al. 2005: 150; Bohle 2001). It is this "normal" state that effective climate action must aim to eradicate if climate variation and change are to be downgraded from deadly threats to mere nuisances.

Following a brief review of vulnerability theory, this chapter frames an approach for analyzing the diverse causal structures of vulnerability and identifying policy responses that might reduce vulnerability of poor and marginal populations. The chapter argues that understanding the multi-scale causal structure of specific vulnerabilities – such as risk of dislocation or economic loss – and the practices that people use to manage these vulnerabilities can point to solutions and potential policy responses. Analysis of the causes of vulnerability can be used to identify the multiple scales at which solutions must be developed and can identify the institutions at each scale responsible for producing and capable of reducing climate-related risks.

The literature pays insufficient attention to the social causal factors that shape the needs for and potential elements of vulnerability-reduction interventions, policies and programs.[5] This chapter outlines a policy-research agenda on causal structures of multiple vulnerabilities in different environmental and political-economic contexts so that causal variables can be aggregated to help develop higher-scale vulnerability-reduction policies and strategies. The focus on causality builds on insights from successes of existing project approaches, such as social funds, social safety nets, or community-driven development (Heltberg et al. 2010), and successful adaptation support based on coping and risk-pooling practices (Agrawal 2010; Anderson et al. 2010). A focus on causal structure adds systematic attention to root causes at multiple scales. It identifies the proximate responses to risk, ordinarily conducted via projects and people's own coping arrangements, while also attending to the more distant social, political, and economic root causes of vulnerability.

Vulnerability analysis and policy development are only first steps in a multi-step iterative governance process. The chapter concludes with a discussion of governance, arguing that to tilt decision making in favor of the poor will require systematic representation of poor and marginal voices in climate decision-making processes.

Linking Climate and Society: Theories of Vulnerability

> It is widely noted that vulnerability of environmental change does not exist in isolation from the wider political economy of resources use. Vulnerability is driven by inadvertent or deliberate human action that reinforces self-interest and the distribution of power in addition to interacting with physical and ecological systems (Adger 2006: 270).

Vulnerability analysis is often polarized into what are called risk-hazard and social constructivist frameworks (Füssel and Klein 2006: 305; also see Adger 2006; O'Brien et al. 2007: 76). Risk-hazard is characterized as the *positivist* (or realist) school while the entitlements and livelihoods approaches are lumped together as *constructivist*. I, however, will call this latter category entitlements or livelihoods approaches – since neither are founded on social constructivist perspectives.

The "social constructivist" label here is a misnomer. For the positivists, "risk . . . is a tangible by-product of actually occurring natural and social processes. It can be mapped and measured by knowledgable experts, and within limits, controlled" (Jasanoff 1999: 137). In social constructivist views, "risks do not directly reflect natural reality but are refracted in every society through lenses shaped by history, politics and culture" (Jasanoff 1999: 139). It falsely contrasts a positivist or "realist" view, which these authors attribute to natural sciences, with a social constructivist view which these authors attribute to the social sciences.

It is evident to any social scientist that both the risk-hazards and the entitlements and livelihoods approaches can be positivist as decribed above by Jasanoff. Both analyses can also be subject to or can integrate a social constructs view, which would certainly shed light on our understanding of risk and its assessment. If one distinguishes between constructivism as ontology, referring to the nature of things, and constructivism as a methodological stance, a constructivist analysis does not have to suggest that conditions and causes of vulnerability are not "real" (Leach 2008: 7). Indeed, there is no reason why a methodological constructivist approach cannot respect the phenomenology of vulnerability. It would also be perfectly positivist to assert that the socially constructed meanings that emerge from differently positioned actors shape causality (see Rebotier 2012).[6] In short, we need to discard this false dichotomy introduced, it would seem, to discredit social analysis.[7]

One concrete distinction between the two schools is that the risk-hazards models tend to evaluate the multiple outcomes (or "impacts") of a single climate event (Figure 7.1), while the entitlements and livelihoods approaches, characterize the multiple causes of single outcomes (Figure 7.2) (Ribot 1995; Adger 2006) – both of which can be done in a

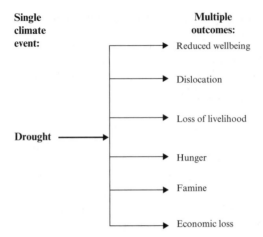

Figure 7.1   Impact analysis

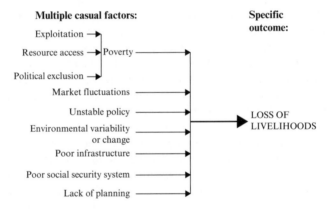

Figure 7.2   Vulnerability analysis

positivist manner or applying constructivist lenses. The risk-hazards traces a linear causal relation back to the environmental hazard itself while the entitlements and livelihoods approaches tend to trace cause to multiple social and political-economic factors. The entitlements livelihoods approach locates causality in agency and hence tends to see natural phenomena as playing a role but not as having "caused" the risk of damage in the face of an event. A third category, integrative frameworks, have grown mostly from the entitlements and livelihoods approaches, yet treat environment as a causal factor.

The two archetypal approaches ask different questions. The risk-hazard approach, which defines vulnerability as a "dose-response relation between an exogenous hazard to a system and its adverse effects" (Füssel and Klein 2006: 305) is concerned with predicting the aftermath or "impact" of a given climate event or stress, and estimating the increment of damage caused by an intensification from "normal" climatic conditions to the conditions expected under climate change scenarios. They view people as vulnerable *to* hazards – locating risk in the hazard itself. This approach is usually portrayed as inadequately incorporating social dimensions of risk (Adger 2006: 270; also see Cannon 2000).

The entitlements and livelihoods schools are concerned with what causes vulnerability. They consider people to be vulnerable *to* undesirable outcomes – loss of a valued asset. They are also concerned with the likely aftermath of a climate event or trend. They view climate events and trends as external phenomena and view the risk of disaster and suffering as social, therefore they place the burden of explanation of vulnerability within the social system. They locate risk within society. The entitlements and livelihoods approaches are described as depicting "vulnerability as lack of entitlements" or a lack of sufficient means to protect or sustain oneself in the face of climate events where risk is shaped by society's provision of food, productive assets, and social protection arrangements (Adger 2006: 270). The entitlements approach is often depicted as ignoring biophysical factors.

Integrative frameworks link these two views. These frameworks tend to borrow from entitlements and livelihoods models, rather than being purely risk-hazard based. Integrative frameworks view vulnerability as depending on both biophysical and human factors. One views vulnerability as having "an external dimension, which is represented by the 'exposure' of a system to climate variations, as well as an internal dimension, which comprises its 'sensitivity' and its 'adaptive capacity' to these stressors" (Füssel and Klein 2006: 306). The IPCC views internal and external aspects as separate dimensions of vulnerability. These notions of external and internal aspects of vulnerability, however, are entirely contingent on how one draws the boundaries of the system under analysis.

Turner et al. (2003; also see Blaikie 1985 and Watts and Bohle 1993) have adopted an approach that avoids this boundary problem by tracing the causes of vulnerability from specific instances of risk – explaining why a given individual, household, group, nation, or region is at risk of a particular set of damages (see Figure 7.2). By tracing causality out from each unit at risk, their model views the entire system as one integrated whole. Analyses of vulnerability must then account for all factors –

biophysical and social – contributing to the stresses affect the unit of concern (Kasperson et al. 2005: 159–161). This causality-based integrative approach to vulnerability informs the available integrative analytic approaches described in the next section. It allows a multi-scale multifactor analysis of vulnerability.

Vulnerability Analysis

Two objectives of any vulnerability analysis for climate action are to identify who is vulnerable and how to assist them. Analysts need to ask: *Where* should we spend public funds earmarked for climate adaptation, and *in what kinds of projects* should we invest in these places? The first question, how to target expenditures, requires identifying which regions (where), social groups (who) and things of value (what) are vulnerable. The question of what we need to invest in requires an understanding of the characteristics of their vulnerability and reasons (why) these places, people, and things are at risk, so we can assess the full range of means for reducing that vulnerability. *Where, who* and *what* are very different questions than *why*. Knowing *where, who* and *what* tells us how to target expenditures. Knowing *why* tells us what to modify or improve in these targeted places and communities. *Why* also indicates the complexity and cost of short- and long-term solutions to vulnerabilities associated with climate variability and change.

While risk-hazard style impact assessments can indicate that a place might be affected by a predicted climate change under given static on-the-ground circumstances (a given level of exposure and ability to respond), it rarely tells us *why* the places and people or ecosystems are sensitive or lack resilience. Knowing likely "impacts" can help us target funding to particular places or to particular social groups or ecological systems. It cannot, however, tell us how to spend that money once we get there. Analysis of causes can help direct funds into vulnerability reducing projects and policies. Climate action should be guided by both types of analysis. Much attention has been given to impact assessment, indicators, and mapping for targeting.[8] This section trains our attention on the elements of an analysis of causal structures of vulnerability.

The causal structure of vulnerability

The two most common approaches to analyzing causes of vulnerability use the concepts of entitlements or livelihoods.[9] These approaches analyze the sensitivity and resilience of individual, household, or livelihood systems, and in some instances, the linked human-biophysical system. They tend to bring attention to the most-vulnerable populations – the poor, women, and other marginalized groups. These approaches provide a starting point for analyzing the causes of climate-related vulnerability.

Entitlements and livelihoods approaches – putting vulnerabilities in place
Sen (1981, 1984; also see Drèze and Sen 1989) laid the groundwork for analyzing causes of vulnerability to hunger and famine. Sen's analysis begins at the household level with what he calls "entitlements." Entitlements are the total set of rights and opportunities with which a household can command – or through which they are "entitled" to obtain – different bundles of commodities. For example, a household's food entitlement consists of the food that the household can command or obtain through production, exchange, or extra-legal legitimate conventions, such as reciprocal relations or kinship obligations (Drèze and Sen 1989). A household may have an endowment or set of assets including: investments in productive assets, stores of food or cash, and claims they can make on other households, patrons, chiefs, government, or on the international community (Swift 1989: 11; cf. Drèze and Sen 1989; Bebbington 1999). Assets buffer people against food shortage. They may be stocks of food or things people can use to make or obtain food.[10] In turn, assets depend on the ability of the household to produce a surplus that it can store, invest in productive capacity and markets, and use in the maintenance of social relations (cf Scott 1976; Berry 1993; Ribot and Peluso 2003).

Vulnerability in an entitlements framework is the risk that the household's alternative commodity bundles will fail to buffer them against hunger, famine, dislocation, or other losses. It is a relative measure of the household's proneness to crisis (Downing 1991; also see Downing 1992; Watts and Bohle 1993: 46; and Chambers 1989: 1). By identifying the components (that is, production, investments, stores, and claims) that enable households to maintain food consumption, this framework allows us to analyze the causes of food crises.[11] Understanding causes of hunger can shed light on policies to reduce vulnerability (Blaikie 1985; Turner et al. 2003). By analyzing chains of factors that produce household crises, a whole range of causes are revealed. This social model of how climate events might translate into food crisis replaces eco-centric models of natural hazards and environmental change (Watts 1983). By showing a range of causes, environmental stresses are located among other material and social conditions that shape household wellbeing. Hunger, for example, may occur during a drought because of privatization policies that limit pastoral mobility, making pastoralists dependent on precarious rain-fed agriculture (Smucker and Wisner 2008).

By locating environment (including climate) within a social framework, the environment may appear to become marginalized – set as one among many factors affecting and affected by production, reproduction, and development (also see Brooks 2003: 8). But, this does not diminish the importance of environmental variability and change. Indeed, it strengthens environmental arguments by making it clear how important – in degree and manner – the quality of natural resources is to social wellbeing. These household-based social models also illustrate how important it is that assets match or can cope with or adjust to (as in buffer against) these environmental variations and changes so that land-based production activities are not undermined by and do not undermine the natural resources they depend on.[12] Leach et al. (1999) later called these environmental inputs to household sustenance "environmental entitlements" (also see Leach et al. 1997; and Leach and Mearns 1991).

"Environmental entitlements refer to alternative sets of utilities derived from environmental goods and services over which social actors have legitimate effective command and which are instrumental in achieving wellbeing" (Leach et al. 1999: 233). In this definition these authors make four innovations. First, they expand Sen's concept of entitlements from an individual or household basis up to the scale of any social actors – individuals or groups. This enables analysis to be scaled to any relevant social unit (or exposure unit in the case of climate related analyses) – such as individuals, households, women, ethnic groups, organizations, communities, nations, or regions. Second, they introduce the notion of a sub-component entitlement, a set of utilities that a particular resource or sector contributes to wellbeing – e.g. environment.[13]

Leach et al.'s (1999: 233) third innovation also draws on Sen to show that "environmental entitlements enhance people's capabilities, which is what people can do or be with their entitlements." Lastly, they expand the idea of rights such that things may be "claimed" rather than just legally "owned." In this framing, claims may be contested – something Sen fails to capture. For example, when hunters near Mkambati Nature Reserve in South Africa are banned from the reserve by state law, they continue hunting based on customary rights which they view as legitimate. They claim their rights, contesting the state's claim (Leach et al. 1997: 9). Hence endowments such as natural resources that are not classically owned within a household can still be accessed through social relations that may introduce cooperation, competition, or conflict mediated by systems of legitimization other than state law. With this insight, they introduce the notion that rights, which Sen takes as singular and static, may also be plural (*á la* von Benda-Beckman 1981; Griffiths 1986) and are based on multiple, potentially conflicting, social and political-economic relations of access (*á la* Blaikie 1985; Ribot and Peluso 2003).

Watts and Bohle (1993) also place Drèze and Sen's (1989) analysis of household entitlements in a multi-scale political economy. They argue that vulnerability is configured by the mutually constituted triad of entitlements, empowerment, and political economy. Here, empowerment is the ability to shape the higher-scale political economy that in turn shapes entitlements. For example, democracy or human rights frameworks can empower people to make claims for government accountability in providing basic necessities and social securities (Moser and Norton 2001: xi). Drèze and Sen (1989: 263) have observed the role of certain types of political enfranchisement in reducing vulnerability, specifically the role of media in creating crises of legitimacy in democracies. Watts and Bohle go far beyond media-based politics to show that empowerment through enfranchisement puts a check on the inequities produced by ongoing political-economic processes. While not outlined in their model, their approach indicates that direct representation, protests and resistance, social movement, union, and civil society pressures can all shape policy and political processes or the broader political economy that shapes household entitlements (Ribot 1995). Moser and Norton (2001: x) view mobilization to claim basic rights as an important means for poor people to shape the larger political economy.

Multiple mechanisms link micro and macro political economies to shape household assets. Deere and deJanvry (1984) identify mechanisms by which the larger economy systematically drains income and assets from farm households. These include tax in cash, kind and labor (corvée), labor exploitation, and unequal terms of trade. These processes make people vulnerable since the wealth they produce from their land and labor is siphoned off – with the systematic support of social, economic, and environmental policies. For example, forestry laws and practices in Senegal have prevented rural populations from holding onto profits from the lucrative charcoal trade (Larson and Ribot 2007) and foresters in Indonesia systematically extract labor from farmers and prevent them from trading forest products while allowing wealthy traders to profit (Peluso 1992). Scott (1976) also shows how peasant households are exploited in exchange for security. Peasants allow their patrons to take a large portion of their product or income in exchange for support during hard times.

Each household is affected by multi-scale forces that shape their assets and wellbeing. Southern African farm households contend with climate variability, AIDS, conflict, poor governance, skewed resource access and the erosion of their coping capacities. While food production support is typical of food-security interventions, household-based research shows that food purchases supported by remittances and gifts are more important in enabling households to obtain food. Donors in the region supported climate early warning systems, but these systems were found to do little to reduce vulnerability if not coupled with other measures. For example, farmers ask for guidance on specific actions to take given forecast and warning information. Many farmers lack the capacity or resources, such as credit, surplus land, access to markets or decision-making power, needed to turn climate information or specific guidance into action – these proximate factors shaped their vulnerabilities (Kasperson et al. 2005: 159–161). The analyses framed by Watts and Bohle (1993), Deere and deJanvry (1984), and Scott (1976), as well as an analysis of the power and authority hierarchies in which households are embedded (Moser and Norton 2001: 7), would give us insights into the larger political economy that would explain why credit is scarce and market access and representation are so limited.

Like entitlements analyses, livelihoods approaches (Blaikie et al. 1994; Bebbington 1999; Turner et al. 2003; Cannon et al. n.d.: 5) evaluate multi-scale factors shaping people's assets. They build on entitlements approaches, but shift the locus of analysis from the household to multi-stranded livelihood strategies that are also embedded in the larger eco-logical and political-economic environment. They also shift attention from a focus on vulnerability to hunger toward an analysis of multiple vulnerabilities, such as risk of hunger, dislocation and economic loss – a suite of factors closely related to the broader condition of poverty. In these approaches, vulnerability variables are connected with people's livelihoods, where a livelihood is "the command an individual, family or other social group has over an income and/or bundles of resources that can be used or exchanged to satisfy its needs. This may involve informa-tion, cultural knowledge, social networks, legal rights as well as tools, land, or other physical resources" (Blaikie et al. 1994: 9). Vulnerability in this framing is lower when livelihoods are "adequate and sustainable" (Cannon et al. n.d.: 5). Livelihood models also explicitly link vulner-ability to biophysical hazards by acknowledging that hazards change the resources available to a household and can therefore intensify some peo-ple's vulnerability (Blaikie et al. 1994: 21–22).

In short, entitlements and livelihoods approaches form a strong basis for vulnerability analysis. They differ in the scale of the unit of concern and analysis (exposure unit) and the scope of factors that analysts view as impinging on that unit at risk – with livelihoods approaches being much broader. When taken together they provide a powerful repertoire of ana-lytic tools for vulnerability analysts. Both approaches (1) start with the unit at risk, (2) focus on the avoidable damages it faces, (3) take the con-dition of the unit's assets to be the basis of its security and vulnerability, and then (4) analyze the causes of vulnerability in the local organization of production and exchange as well as in the larger physical, social and political-economic environment. Vulnerability analysis differs greatly from the risk-hazard approaches which start with climate events and map out their consequences across a socially static landscape. Entitlement and livelihoods vulnerability approaches put vulnerability in context on the ground, enabling us to explain why specific vulnerabilities occur at specific times in specific places.

Toward Pro-poor Climate Action

Vulnerability to hunger, famine and dislocation are correlated with poverty (Prowse 2003: 3; Cannon et al. n.d.: 5; Anderson et al. 2010; Heltberg et al. 2010). Women, minorities and other marginalized populations are also disproportionately vulnerable, sharing many vulnerabilities of the poor (Demetriades and Esplen 2010). For poor and marginalized vulnerable populations, vulnerability reduction is poverty reduction and basic development (Cannon et al. n.d.: 4; also see Prowse 2003: 3).

The weak within society tend to be of lower priority for those in power. Economically weak actors in urban slums or marginal groups far from the centers of power within semi-arid or forested zones may be of little importance to those in political office or big business. They are likely to be low priority for governments even in disaster planning (Blaikie et al. 1994: 24; ICHRP 2008). For instance, the extent to which slum dwellers are affected by extreme weather is both about settlement location and the level and quality of infrastructure and services such as water, sanitation, and drainage. These populations' lack of assets reduces their ability to adapt to changing conditions and also prevents them from making political demands for investments to reduce their risk (Moser and Satterthwaite 2010).

To counter biases against poor and marginalized, vulnerability analyses and policies must be pointedly pro-poor. This section outlines an analytic approach to pro-poor vulnerability analysis and a research agenda for the identification of vulnerability-reduction policies.

Pro-poor vulnerability analysis
Entitlements and livelihood approaches evaluate the causes of asset failure and of negative outcomes in order to identify means to counter the causes (Downing 1991; Ribot 1995; Watts and Bohle 1993; Turner et al. 2003: 8075). This focus on negative outcomes favors poor and marginalized groups because they are overrepresented in at-risk populations. This tilt in favor of the poor can also be enhanced, of course, by analytic efforts that

choose to study outcomes of most concern to the poor such as hunger, dislocation or economic losses that push people over a threshold into poverty or extreme deprivation. The focus on causality can point toward solutions.

Coping[14] and adaptation studies identify vulnerability-reduction strategies used by poor and marginalized populations and means to support those strategies. Agrawal (2010), for example, starts with household and community risk-pooling strategies and identifies institutions – civic, private and public organizations – that support these strategies. His analysis provides insights into the roles of institutions (by which he means "organizations") and therefore into potential institutional channels for coping and adaptation support. While this approach does not explain why people become vulnerable, it provides great insights into local-level vulnerability management and reduction.

While analysis of coping or adaptation strategies can also provide insights into causes of vulnerability, the entitlements and livelihoods approaches analyze the causal structure of vulnerability so as to identify a wider range of coping and adaptation opportunities (Watts 1983; Mortimore and Adams 2000; Yohe and Tol 2005; Anderson et al. 2010). Coping approaches, as well as many project-based interventions, focus on means for adapting as well as causes of adaptation and the ability to adapt. The vulnerability approach seeks to identify causes of the vulnerability – that is, causes of the risks that people need to adapt to.[15]

Tracing the causes of negative outcomes complements coping and adaptation approaches by enabling researchers and development professionals to conduct a full accounting of causality which can indicate the policy options available for reducing vulnerability at its multi-scale origins – not only coping with or adapting in the face of hazards and stress, which tends to be a response to the most-proximate factors. For example, despite laws transferring forest management to elected rural councils in Senegal, foresters force councilors to give lucrative woodfuel production opportunities to powerful urban merchants, usually leaving the rural populations destitute (Larson and Ribot 2007). Forest villagers continue to rely on low-income rain-fed farming and must cope with meager incomes. By focusing on the causes of destitution that puts forest villagers on the margins, analysts might recommend means of policy enforcement rather than, as many projects are doing, encouraging villagers to market other secondary forest products.

Vulnerability analysis most useful to policy makers starts from the outcomes we wish to avoid and works backward toward the causal factors (Turner et al. 2003: 8075; also see Blaikie 1985; Downing 1991; Füssel 2007). In addition to favoring the poor, focusing on outcomes and their causes has other advantages: (1) it best matches policy to valued attributes of the system that we wish to protect; (2) it enables policy makers to place hazards as one variable among many affecting those attributes; (3) it brings attention to the many variables at multiple scales affecting valued attributes, steering analysts toward the many possible means for reducing the probability of negative outcomes or enhancing positive ones; (4) it enables comparative analysis of the many causes of negative outcomes, helping to focus policy attention on the causes that are most important, most amenable to reforms and least costly to change – giving policy makers the biggest bang for their buck. Analyzing the "chains of causality" (Blaikie 1985), by showing how outcomes are caused by proximate factors that are in turn shaped by more distant events and processes, can tell us what kinds of interventions might stem the production of vulnerability at what scales and, where relevant, who should pay the costs of vulnerability reduction.

Vulnerability reduction measures, of course, do not only derive from understanding causes. Indeed, some causes may be (or appear) immutable, others no longer active, transient or incidental. Redressing direct causes may not always be part of the most effective solutions (Drèze and Sen 1989: 34). The objective of vulnerability analysis is to identify the active processes of vulnerability production and then to identify which are amenable to redress. Other interventions can also be identified that are designed to counter conditions or symptoms of vulnerability without attending to their causes (such as support for coping strategies or targeted poverty-reduction disaster relief). All forms of available analysis should be used to identify the most-equitable and effective means of vulnerability reduction.

Identifying multi-scale vulnerability-reduction policies

Studies of coping strategies and lessons from successful development interventions provide valuable guidance for vulnerability reduction. Large-scale causes of vulnerability, such as unequal development practices, however, are less likely to receive attention in poverty reduction, vulnerability reduction or adaptation programs. Identifying and matching solution sets or climate-related opportunities with responsive institutions at appropriate scales of social, environmental, and political-administrative organization provides an entry point into multi-scale pro-poor climate action. Such action requires a systematic understanding of both proximate and distant dynamics that place people under stress or on the threshold of disaster. This section proposes a research agenda for identifying the range of causal factors shaping various vulnerabilities for groups at risk around the world and a mapping of those causes onto solution sets for responsible and responsive institutions.

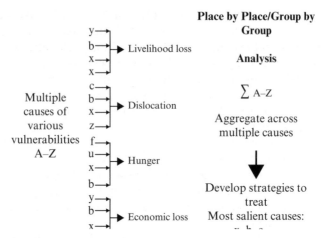

Figure 7.3   Identifying vulnerability's most salient causes

Different outcomes that we hope to avoid – such as loss of assets, livelihood, or life – are risks for different sub-groups and have different associated causal structures (Drèze and Sen 1989; Watts and Bohle 1993; Ribot 1995; Roberts and Parks 2007). Different sectors will face different stresses and risks and will have different response options (IPCC 2007: 747). Within each case, vulnerabilities of the poor, who have few resources to shield themselves or rebound from climate events and stresses, will be different from vulnerability of the rich who are able to travel to safety and draw insurance to help them rebuild. From understanding differences in the causal structures of vulnerabilities, local, national, and international policies can be developed. Explaining difference will require an analysis of the multiple causal factors for a variety of vulnerabilities of concern (see Figure 7.3).

These causal data must then be aggregated to evaluate the best point of leverage for vulnerability reduction with respect to specific vulnerabilities and overall (see Figure 7.4).

Such an analysis should reveal the frequency and importance of different causes, pointing toward strategies to address the most salient and treatable causal factors.

Identifying causal structures of vulnerability and potential policy responses can be a basis for developing a broad vulnerability-reduction strategy. It involves the aggregation of causal structures over multiple cases of vulnerability of particular groups in particular areas to specific outcomes. This aggregation may have to be broken down by sectors, by eco-zones, or by hazard areas to make such an exercise manageable. The

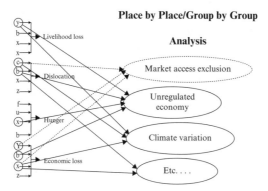

Place by Place/Group by Group

Analysis

Livelihood loss

Market access exclusion

Dislocation

Unregulated economy

Hunger

Climate variation

Economic loss

Etc. . . .

Figure 7.4   Identifying and aggregating multiple causes of vulnerability

case studies can also serve as the basis for generating recommendations for local policy. More broadly, multiple case studies can help us to understand the relative importance of different factors – both near and far – in producing and reducing vulnerability. These factors must be aggregated so as to identify the relevant scales and corresponding institutions for climate action. These steps set out a major research agenda for vulnerability reduction analysis. For this agenda to counter the biases against poorer populations, all of these steps must be consciously pro poor. For example, the cases where basic human rights such as health, livelihood, and life are at risk must take priority over the analysis of purely economic losses.

Indicators currently used to target poverty and vulnerability reduction interventions are a good starting point for identifying relevant study populations. Existing livelihoods approaches to vulnerability reduction already target the poor, strengthening their baseline nutrition, health and morale, and addressing the underlying conditions of poverty, thus reinforcing their abilities to confront stressors and bounce back after them (Cannon et al. n.d.: 6). Vulnerability studies complement successful "self-help" and "social-protection" (see Heltberg et al. 2010) coping and adaptation supports by indicating opportunities for higher-scale reforms.

Thorough vulnerability analyses would indicate the need to reform the larger political economy of institutions, policies, social hierarchies and practices that shape wellbeing, capacity for self protection, and extended entitlements. For example, while social funds, community-driven development and social safety nets are excellent means for responding to immediate stresses and needs of poor populations, examining causality through historical studies often reveals that the poverty these programs respond to is due to larger-scale uneven development investment decisions and governance policies that limit the choices available to those affected by environmental disasters (Heltberg et al. 2010; Raleigh and Jordan 2010).

Vulnerabilities and their causes are diverse. Responses to vulnerability must be developed from detailed understandings of specific problems in specific places – general principles and models are insufficient. Case studies inform us of a particular set of dynamics and opportunities for vulnerability reduction in a particular place. It is from case studies that viable solutions can follow – for specific places and more generally. To be complete, place-based approaches must take into account people's detailed knowledge of their social and production systems and the risks they face – experience with community driven development (CDD) provides this lesson (Mansuri and Rao 2003). To make results of an analysis relevant and the implication of recommendations feasible, investigations of vulnerability must consider local people's needs and aspirations and their knowledge of political-economic and social context in which any policy will have to be inscribed into law and translated into practice. Thus, while studies provide perspectives communities may not be able to generate, the steps in developing a vulnerability-reduction policy strategy must be informed and open to influence by local citizens and their representatives.

Any vulnerability case study should include an evaluation of existing vulnerability-reduction and a wide range of sectoral and regulatory policies (Burton et al. 2002: 154–157). Any given population at risk is deeply affected by existing policies. Some are aimed at assisting them. Among existing policies some may reduce vulnerability while others help produce vulnerable conditions. Policies, like institutions or organizations (*á la* Agrawal 2010) can enable coping. They can also be systematically disabling (see Larson and Ribot 2007). Policies or their unequal implementation can selectively favor some actors while making others more vulnerable. Policies from all sectors have deep distributional implications. Coudouel and Paternostro (2005) and the World Bank's Poverty and Social Impacts Analysis (TIPS) source book[16] suggest methods for poverty and social impact analysis of policies for their distributional effects. Such guidelines can also be applied to evaluating the vulnerability implications of policies and interventions.

When exploring effects of policies and practices shaping vulnerability, or when analyzing potential vulnerability-reduction measures, it is also important to account for a wide range of ancillary benefits (see Burton et al. 2002). For example, in urban areas, asset building not only reduces immediate vulnerability, but also enables poor and middle income people to make demands on their government for better services and infrastructure (Moser and Satterthwaite 2010). Most adaptation measures will go far beyond reducing of risk with respect to climate events. Hence, the set of benefits that follow from a given set of vulnerability reduction measures are also highly relevant in deciding the allocation of funds earmarked for development or for climate-related vulnerability.

Knowledge of problems and policy guidance can inform popular mobilization and policy process. Proposing policy solutions, however, is a small part of the political struggle for change. Calls for change must be backed by political voice and leverage. Bringing poor and marginalized groups into decision making through organizing or representation can reinforce their claims for justice, equity, and greater security in the face of a changing environment (Ribot 2004; Moser and Norton 2001).

Conclusion: From Climate Action Options to Institutions and Governance

While vulnerability is always experienced locally, its causes and solutions occur at different social, geographic, and temporal scales. Identifying the causes of vulnerability points toward vulnerability-reduction measures and the scales at which they can best be implemented. It also helps attribute responsibility to the polluters – providing a basis for compensation.[17] Vulnerability-reduction or compensation policies are developed, promulgated, and implemented through institutions. So are the many other sectoral, economic, and social policies that have implications for vulnerability via their effects on resource access, market access, political voice, poverty, and economic distribution. Institutions also play numerous roles in supporting people's everyday coping and livelihoods strategies (Agrawal 2010). Systematically identifying causes of vulnerability, identifying policy solutions, and mapping them to scales and appropriate institutions is a process that vulnerability-reduction analysts and activists must yet conduct.

Institutions play several important roles in wellbeing and vulnerability. Leach et al. (1999: 236) view institutions as mediating vulnerability by shaping access to resources (a part of endowment formation), the relation between endowments and entitlements (rights and opportunities with which a household can command different commodity bundles), and the relation between entitlements and capabilities (the range of things people can do or be with their entitlements). In their model, institutions enable people to obtain, transform and exchange their endowments in ways that translate into contributions to wellbeing. As such, institutions support the needs of a plurality of sub-groups, who can enter into competition and conflict when making claims to resources.

Agrawal (2010) emphasizes the role of institutions, showing how rural institutions structure risk and sensitivity in the face of climate hazards by enabling or disabling individual and collective action. Rural populations protect themselves by risk pooling via storage (over time), migration (over space), sharing assets (among households), and diversification (across assets). Exchange (via markets) can substitute for any of these risk pooling responses. Rural institutions play a role in enabling each of these risk-reducing practices. In the 77 case studies Agrawal analyzes, all of these practices depend on local institutions – mixes of public, civic, and private organizations.

Risk pooling and exchange mechanisms constitute one set of practices that shape vulnerability. Many other practices also produce or reduce climate-related vulnerabilities. Drèze and Sen (1989), for example, explored the role of media in influencing policy to prevent and respond to chronic hunger and famine. Leach et al. (1999) focus on the role of resource access, endowment formation, and entitlement mapping – the kinds of processes that might make it so the actors involved do not need to engage in risk pooling. Heltberg et al. (2010) point to social protection interventions. Cannon et al. (n.d.) examine the role of networks (akin to Sen's 1981 extended entitlements); Bebbington (1999) emphasizes social capital; Scott (1976) focuses on reciprocal relations within a moral economy; Deere and deJanvry (1984) outline mechanisms by which economic gains are coerced or extracted from peasant households; Moser and Norton (2001) emphasize the role of human rights and claim making.

Each of these enabling and disabling practices depend on different kinds of institutions – rules of the game and public, private or civic organizations – at various scales. To map vulnerability-producing and reducing practices to institutional nodes for intervention, Agrawal's (2010) analytic approach to risk-pooling could also be productively applied to each of these other vulnerability producing and reducing practices. Each can be studied for its role in the causal structure of vulnerability. Each practice – whether reciprocity or social protection – depends on institutions that, when identified, can be targeted for reform or support. But, attempting such interventions can generate social and political tension. As Leach et al. (1999) indicate, institutions and their networks can be in competition or conflict – some for enabling and others in support of disabling policies and practices.

The institutions responsible for and capable of responding to vulnerability are the locus of vulnerability governance. Governance (following World Bank 1992: 3, 1994: xiv; Leftwich 1994) is about the political-administrative, economic and social organization of authority – its powers and accountabilities. It is about how power is exercised and on whose behalf. As the global climate warms, decisions will be made at every level of social and political-administrative organization to mitigate climate change, take advantage of its opportunities, and dampen associated negative consequences – from global conventions to the decisions of local governments, village chiefs or NGOs. Multiple decisions at multiple scales affect the livelihoods of the urban and rural poor. What principles of governance should guide decisions at each of these decision-making nodes? Who will decision-making bodies represent and how? What distributions of decision-making powers and what structures of accountabilities will provide the most leverage for positive change and the checks and balances to protect poor urban and rural people's basic well being and rights? These questions remain open.

Principles to govern climate action must be designed around the processes that shape vulnerability and the actors and organizations with authority and power to make decisions that can change these processes. The first step will be aggregating case-based analyses of causality, coping and the role of institutions. This process can be tilted in favor of poor marginalized populations by analyses that explain causes of asset and entitlement failure. To translate learning into action will be a long-term iterative process to negotiate the reshaping of policies and practice. All policies change distribution and, therefore, have advocates and meet resistance. Decision-making processes that are accountable and responsive to affected populations may at least tilt policies to favor the most vulnerable – due to their sheer numbers. This means the development of and engagement with representative decision-making bodies to ensure a modicum of influence by those most in need.

For researchers, representation might mean incorporating the voice of local populations in their understanding of who is at risk, the problems they face and possible solutions, as well as sharing findings with affected populations and policy makers. For development professionals and policy makers it will mean working with representative bodies and insisting that these bodies incorporate local needs and aspirations into the design of projects and policies. In global negotiations it may mean requiring negotiators to engage in public discussions within their countries or for national groups to organize and monitor their nation's negotiators. In local and national contexts it may mean helping to mobilize the poor and marginalized to make demands and to vote. Such governance practices may help avoid negative outcomes of climate action and could make climate actions more legitimate and sustainable. Representing and responding to the needs of the most vulnerable populations might promote development that can widen the gap between climate and distress. Moving people away from the threshold of destitution by building their assets, livelihoods and options, will dampen their sensitivity, enhance their flexibility, and enable them to flourish in good times, sustain through stress and rebuild after shocks.

Acknowledgements

Many thanks to Arun Agrawal, Tom Bassett, Ashwini Chhatre, Floriane Clement, Roger Kasperson, Heather McGray, Robin Mearns, Andrew Norton, Ben Wisner and several anonymous reviewers for their challenging constructive comments on drafts and in discussions. Thanks to Tim Forsythe, Helen Epstein, Zsuzsa Gille, and Malini Ranganathan for insightful comments on the preface to this reprinting of the article.

Notes

1. For instance, if adaptations or mitigation efforts (such as reduced emissions from deforestation and decreased degradation, REDD) increase inequality within or among regions or social groups (O'Brien et al. 2007: 84).
2. This trend holds even without counting the 2004 tsunami. Twice as many people were adversely affected by climate events in the 1990s as in the 1980s, and over the past four decades great major catastrophes have quadrupled while economic losses have increased tenfold (Kasperson et al. 2005: 151–152).
3. Hurricane Katrina was a category 3 storm, as were those in Bangladesh. Katrina's surge was 4 meters. Of course, more could be done. Sidr was comparable to Katrina, which devastated New Orleans. But Katrina, despite infamous Bush-administration mismanagement, resulted in 1,300 fatalities (White House 2006).
4. The term "adaptation," although common in climate discussions, is highly problematic. It naturalizes the vulnerable populations, implying that, like plants, they should adjust to stimuli. The term implicitly places the burden of change on the affected unit – rather than on those causing vulnerability or with responsibility (e.g. government) to help with coping and enable wellbeing. "Adaptation" also suggests "survival of the fittest," which is not a desirable ethic for society.
5. The US National Research Council (September 13, 2007: 71–73), IPCC (2007: AR4-12.4, 17.2, 17.4), and 2006 Stern Review all acknowledge need for greater social science analysis.
6. For example, Leach points out that "A methodological constructivist approach can be used to understand the different perspectives of scientists, citizens and other stakeholders around the issue and to specify different roles for them in decision-making" (Leach 2008: 7). But, of course, constructivism cannot be confined to an analysis of perspectives and must be extended to an understanding of how position shapes the ways in which the world is itself apprehended and translated into meaning.
7. There is no positivist reasoning that would prevent analysis of interpretation and positionality as being part of the analytics of causality – since difference and struggles over meaning and interpretation are part and parcel of causality. In addition, discourse is no less "real" than a tree or a storm system. The causes of decisions that shape security and damage are the results of discursive battles for domination, for authority, for decision-making power and ultimately for policy and practice. Positionality shapes people's behavior and is therefore part of the material political-economic analysis of causality. These are not trivial observations of categorization. The very placing of the social-science analyses into "social constructivist" and non-"realist" categories is a means of delegitimizing these perspectives as if social, discursive, constructivist factors are not part of the causal structure of vulnerability. Indeed, they are the heart of it. Of course, any "realist" who does not understand that interpretation is multi-faceted and meaning attributed misses the point that these observations do not deny the materiality of their "science."

8. On mapping and targeting, see Downing 1991; Deressa, Hassan and Ringler 2008; Adger et al. 2004; Kasperson et al. 2005: 150.
9. For reviews of vulnerability approaches, see Kasperson et al. 2005: 148–150; Füssel and Klein 2006; and Adger 2006.
10. "Assets create a buffer between production, exchange and consumption" (Swift 1989: 11).
11. Entitlements framework is very useful, but grossly incomplete – covering only a limited set of causes. See Gasper 1993 for an analysis of its limits.
12. Household models are often limited by their failure to account for intra-household dynamics of production and reproduction – but they do not have to be. See, for example, Guyer 1981; Guyer and Peters 1987; Carney 1988; Hart 1992; Agarwal 1993; and Schroeder 1992.
13. This second innovation can be confusing since environmental claims in Sen's (1981) classic entitlements framework could be considered part of people's "rights and opportunities" and the alternative sets of utilities these can become would be part of the alternative commodity bundles people can command. Nevertheless, it is useful to view environment as contributing to people's endowments and alternative commodity bundles.
14. Coping is a temporary adjustment during difficult times, while adaptation is a permanent shift in activities to adjust to permanent change (Davies 1993; also see Yohe and Tol 2005).
15. Yohe and Tol (2005) seek to identify on causal structures – but they focus on the determinants of adaptive capacity – rather than the causes of vulnerability itself.
16. URL: http://web.worldbank.org/files/14520_PSIA_Users_Ugide_-_Chapter_1_May_ 2003.
17. Füssel (2007: 163) identifies three fundamental responses for reducing negative outcomes associated with climate change: mitigation, adaptation and compensation. Mitigation assumes climate to be the major cause of problems. Adaptation and compensation requires analysis of causality to identify a broader range of responsible factors and institutions.

References

Adger, W. Neil (2006), 'Vulnerability', *Global Environmental Change*, **16**, 268–281.
Adger, W. Neil, Nick Brooks, Graham Bentham, Maureen Agnew and Siri Eriksen (2004), 'New indicators of vulnerability and adaptive capacity', Tyndall Center for Climate Change Research Technical Paper no. 7, January.
Agarwal, Bina (1993), 'Social security and the family: coping with seasonality and calamity in rural India', *Agriculture and Human Values*, 156–165.
Agrawal, Arun (2010), 'Local institutions and adaptation to climate change', Ch. 7, pp. 173–198 in Robin Mearns and Andrew Norton (eds.), *Social Dimensions of Climate Change: Equity and Vulnerability in a Warming World*. Washington, DC, USA: The World Bank.
Anderson, Simon, John Morton and Camilla Toulmin (2010), 'Climate change for agrarian societies in drylands: implications and futures pathways', Ch. 9, pp. 199–230 in Robin Mearns and Andrew Norton (eds.), *Social Dimensions of Climate Change: Equity and Vulnerability in a Warming World*. Washington, DC, USA: The World Bank.
Batha, Emma (2008), 'Cyclone Sidr would have killed 100,000 not long ago', AlertNet, November 16, 2007, http://alertnet.org/db/blogs/19216/2007/10/16-165438-1.htm (accessed March 9, 2013).
Bebbington, A. (1999), 'Capitals and capabilities: a framework for analysing peasant viability, rural livelihoods and poverty', *World Development*, **27** (12), 2021–2044.
Bern, C. et al. (1993), 'Risk factors for mortality in the Bangladesh cyclone of 1991', *Bulletin of World Health Organization*, **73**, 72–78.

Berry, Sara (1993), *No Condition is Permanent: The Social Dynamics of Agrarian Change in Sub-Saharan Africa*. Madison, USA: The University of Wisconsin Press.

Blaikie, Piers (1985), *The Political Economy of Soil Erosion in Developing Countries*. London, UK: Longman Press.

Blaikie, Piers, T. Cannon, I. Davis and Ben Wisner (1994), *At Risk: Natural Hazards, People's Vulnerability and Disasters*. London, UK: Routledge.

Bohle, Hans-G. (2001), 'Vulnerability and criticality: perspectives from social geography', IHDP Update 2/01, 3–5.

Brooks, Nick (2003), 'Vulnerability, risk and adaptation: a conceptual framework', Working Paper 38, Tyndall Centre for Climate Change Research, Norwich UK.

Burton, I. S. Huq, B. Lim, O. Pilifosova, and E.L. Schipper (2002), 'From impact assessment to adaptation priorities: the shaping of adaptation policy', *Climate Policy*, **2**, 145–149.

Cannon, Terry (2000), 'Vulnerability analysis and disasters', in D.J. Parker (eds), *Floods*, London, UK: Routledge.

Cannon, Terry, John Twigg and Jennifer Rowell (n.d.), 'Social vulnerability, sustainable livelihoods and disasters', Report to DFID, Conflict and Humanitarian Assistance Department and Sustainable Livelihoods Support Office.

Carney, Judith (1988), 'Struggles over land and crops in an irrigated rice scheme', in J. Davidson (ed.), *Agriculture, Women and Land: The African Experience*. Boulder, USA: Westview Press.

CEDMHA (Center for Excellence in Disaster Management and Humanitarian Assistance) (2007), *Cyclone Sidr Update*, http://www.coe.dmha.org/Bangladesh/Sidr11152007.htm

Chambers, Robert (1989), 'Vulnerability, coping and policy', in Robert Chambers (ed.), 'Vulnerability: how the poor cope', *IDS Bulletin*, **20**, 2, 1–7.

Coudouel, Aline and Stefano Paternostro (2005), *Analyzing the Distributional Impacts of Reforms: A Practitioner's Guide to Trade, Monetary and Exchange Rate Policy, Utility Provision, Agricultural Markets, Land Policy, and Education*. Washington DC, USA: The World Bank.

Gasper, Des (1993), 'Entitlement analysis: concepts and context', *Development and Change*, **24**, 679–718.

Davies, S. (1993), 'Are coping strategies a cop out?', *IDS Bulletin*, **24**, (4), 60–72.

Deere, Carmine Diana and Alain deJanvry (1984), 'A conceptual framework for the empirical analysis of peasants', Giannini Foundation Paper No. 543, pp. 601–611.

Demetriades, Justina and Emily Esplen (2010), 'The gender dimensions of poverty and climate change adaptation', Chapter 5, pp. 133–144 in Robin Mearns and Andrew Norton (eds), *Social Dimensions of Climate Change: Equity and Vulnerability in a Warming World*. Washington, DC, USA: The World Bank.

Deressa, Temesgen, Rashid M. Hassan and Claudia Ringler (2008), 'Measuring Ethiopian farmers' vulnerability to climate change across regional states', IFPRI Discussion Paper 00806, Environment and Production Technology Division October. Washington DC, USA: IFPRI.

Downing, Thomas (1991), 'Assessing socioeconomic vulnerability to famine: frameworks, concepts, and applications', Final Report to the U.S. Agency for International Development, Famine Early Warning System Project, January 30, 1991.

Downing, Thomas E (1992), 'Vulnerability and global environmental change in the semi-arid tropics: modelling regional and household agricultural impacts and responses', Presented at ICID, Fortaleza–Ceará, Brazil, January 27 to February 1.

Drèze, Jean and Amartya Sen (1989), *Hunger and Public Action*. Oxford, UK: Clarendon Press.

Duarte, Mafalda, Rachel Nadelman, Andrew Peter Norton, Donald Nelson and Johanna Wolf (2007), 'Adapting to climate change: understanding the social dimensions of vulnerability and resilience', *Environment Matters* (**June–July**), 24–27.

Frank, Neil L. and S.A. Husain (1971), 'The deadliest tropical cyclone in history?', *Bulletin of American Meteorological Society*, **52** (6).

Füssel, Hans-Martin (2007), 'Vulnerability: a generally applicable conceptual framework for climate change research', *Global Environmental Change*, **17** (2), 155–167.

Füssel, Hans-Martin and Richard J.T. Klein (2006), 'Climate change vulnerability assessments, an evolution of conceptual thinking', *Climate Change*, **75**, 301–329.

Government of Bangladesh (2008), 'Cyclone Sidr in Bangladesh: damage, loss, and needs assessment for disaster recovery and reconstruction'.

Griffiths, J (1986), 'What is legal?', *Journal of Legal Pluralism*, **24**, 1–55.

Guyer, Jane (1981), 'Household and community in African studies', *African Studies Review*, **24** (2/3).

Guyer, Jane and Pauline Peters (1987), 'Introduction', to special issue on households of *Development and Change*, **18**, 197–214.

Hart, Gillian (1992), 'Household production reconsidered: gender, labor conflict, and technological change in Malaysia's Muda Region', *World Development*, **20** (6), 809–823.

Heltberg, Rasmus, Paul Bennett Siegel and Steen Lau Jorgensen (2010), 'Social policies for adaptation to climate change', Ch. 10, pp. 259–276 in Robin Mearns and Andrew Norton (eds), *Social Dimensions of Climate Change: Equity and Vulnerability in a Warming World*. Washington, DC, USA: The World Bank.

ICHRP (International Council on Human Rights Policy) (2008), *Climate Change and Human Rights: A Rough Guide*, ed. Stephen Humphreys and Robert Archer. Geneva: ICHRP.

IPCC (Intergovernmental Panel on Climate Change) (2007), 'Summary for policy makers', *Climate Change 2007: the Physical Science Basis. Contribution of Working Group I to the Fourth Assessment Report of the Intergovernmental Panel on Climate Change*, ed. S. Solomon, D. Qin, M. Manning, Z. Chen, M. Marquis, K.B. Averyt, M. Tignor and H.L. Miller. Cambridge, UK: Cambridge University Press.

Jasanoff, Sheila A. (1999), 'The songlines of risk', *Environmental Values*, **8**, 135–152.

Kasperson, R.E., K. Dow, E. Archer, D. Caceres, T. Downing, T. Elmqvist, S Eriksen, C. Folke, G. Han, K. Iyengar, C. Vogel, K. Wilson and G. Ziervogel (2005), 'Vulnerable peoples and places', pp. 143–164 in R. Hassan, R. Scholes and N. Ash (eds), *Ecosystems and Human Wellbeing: Current State and Trends*, Vol. 1. Washington, DC, USA: Island Press.

Larson, Anne and Jesse Ribot (2007), 'The poverty of forestry policy: double standards on an uneven playing field', *Journal of Sustainability Science*, **2** (2).

Leach, Melissa (2008), 'Pathways to sustainability in the forest? Misunderstood dynamics and the negotiation of knowledge, power, and policy', *Environment and Planning A*, **40**, 1783–1795.

Leach, M. and R. Mearns (1991), 'Poverty and environment in developing countries: an overview study', Report to UK ESRC (Society and Politics Group & Global Environmental Change Initiative Programme) and ODA. Brighton, UK: IDS, University of Sussex.

Leach, M., R. Mearns and I. Scoones (eds) (1997), 'Community-based sustainable development: consensus or conflict?', *IDS Bulletin*, **28** (4).

Leach, M., R. Mearns and I. Scoones (1999), 'Environmental entitlements: dynamics and institutions in community-based natural resource management', *World Development*, **27** (2), 225–247.

Leftwich, Adrian (1994), 'Governance, the state and the politics of development', *Development and Change*, **25**, 363–386.

Mansuri, Ghazala and Vijayendra Rao (2003), 'Evaluating community driven development: a review of the evidence', First Draft Report, Development Research Group, The World Bank.

McGray, Heather, Anne Hammill, Rob Bradley, E. Lisa Schipper and Jo-Ellen Parry (2007), *Weathering the Storm: Options for Framing Adaptation and Development*. Washington DC, USA: World Resources Institute.

Ministry of Food and Disaster Management of Bangladesh (2008), 'Super Cyclone Sidr 2007: impacts and strategies for interventions'.

Mortimore, Michael and W.M. Adams (2000), 'Farmer adaptation, change and "crisis" in the Sahel', *Global Environmental Change*, **11**, 49–57.

Moser, Caroline and Andy Norton (with Tim Conway, Clare Ferguson and Polly Vizard)

(2001), *To Claim our Rights: Livelihood Security, Human Rights and Sustainable Development*. London, UK: Overseas Development Institute.

Moser, Caroline and David Satterthwaite (2010), 'Toward pro-poor adaptation to climate change in the urban centers of low-and middle-income countries', Ch. 9, pp. 231–258 in Robin Mearns and Andrew Norton (eds), *Social Dimensions of Climate Change: Equity and Vulnerability in a Warming World*. Washington, DC, USA: The World Bank.

Mushtaque, A., R. Chowdhury, Abbas U. Bhuyia, A. Yusuf Choudhury and Rita Sen (1993), 'The Bangladesh cyclone of 1991: why so many people died', *Disasters*, **17** (4), 291–304.

O'Brien, K., Eriksen, S., Nygaard, L. P. and Schjolden, A. (2007), 'Why different interpretations of vulnerability matter in climate change discourses', *Climate Policy*, **7**, 73–88.

Peluso, Nancy Lee (1992), *Rich Forests, Poor People: Resource Control and Resistance in Java*. Berkeley, USA: University of California Press.

Prowse, Martin (2003), 'Toward a clearer understanding of "vulnerability" in relation to chronic poverty', CPRC Working Paper No. 24, Chronic Poverty Research Centre, University of Manchester, Manchester, UK.

Raleigh, Clionadh and Lisa Jordan (2010), 'Climate change and migration: emerging patterns in the developing world', Ch. 4, pp. 103–132 in Robin Mearns and Andrew Norton (eds), *Social Dimensions of Climate Change: Equity and Vulnerability in a Warming World*. Washington, DC, USA: The World Bank.

Rebotier, Julien (2012), 'Vulnerability conditions and risk representations in Latin-America: framing the territorializing of urban risk', *Global Environmental Change*, **22** (2), 391–398.

Ribot, Jesse (1995), 'The causal structure of vulnerability: its application to climate impact analysis', *GeoJournal*, **35**, (2).

Ribot, Jesse (2004), *Waiting for Democracy: The Politics of Choice in Natural Resource Decentralization*. Washington DC, USA: World Resources Institute.

Ribot, Jesse and Nancy Lee Peluso (2003), 'A theory of access: putting property and tenure in place', *Rural Sociology*, **68**.

Roberts, Timmons and Bradley Parks (2007), *A Climate Of Injustice: Global Inequality, North-South Politics, and Climate Policy*. Cambridge: MIT Press.

Schroeder, Richard A. (1992), 'Shady practice: gendered tenure in the Gambia's garden/orchards', paper prepared for the 88th Annual Meeting of the Association of American Geographers, San Diego, CA, April 18–20.

Scott, James (1976), *The Moral Economy of the Peasant*. New Haven, USA: Yale University Press.

Sen, Amartya (1981), *Poverty and Famines: An Essay on Entitlement and Deprivation*. Oxford, UK: Oxford University Press.

Sen, Amartya (1984), 'Rights and capabilities', in Amartya Sen (ed.), *Resources, Values and Development*. Oxford, UK: Basil Blackwell.

Smucker, Thomas A. and Ben Wisner (2008), 'Changing household responses to drought in Tharaka, Kenya: vulnerability persistence and challenge', Journal Compilation, Overseas Development Institute. Oxford, UK: Blackwell.

Swift, Jeremy (1989), 'Why are rural people vulnerable to famine?', *IDS Bulletin*, **20** (2), 8–15.

Turner II, B.L., Pamela A. Matson, James J. McCarthy, Robert W. Corell, Lindsey Christensen, Noelle Eckley, Grete K. Hovelsrud-Broda, Jeanne X. Kasperson, Amy Luers, Marybeth L. Martello, Svein Mathiesen, Rosamond Naylor, Colin Polsky, Alexander Pulsipher, Andrew Schiller, Henrik Selin and Nicholas Tyler (2003), 'Illustrating the coupled human-environment system for vulnerability analysis: three case studies', *Proceedings of the National Academy of Sciences US*, **100**, 8080–8085.

von Benda-Beckmann, K. (1981), 'Forum shopping and shopping forums: dispute processing in a Minangkabau village in West Sumatra', *Journal of Legal Pluralism*, **19**, 117–159.

Watts, Michael J. (1983), 'On the poverty of theory: natural hazards research in context', in Ken Hewitt (ed.), *Interpretations of Calamity*. London, UK: Allen Unwin.

Watts, Michael J. (1987), 'Drought, environment and food security: some reflections on peasants, pastoralists and commoditization in dryland West Africa', in Michael H. Glantz (ed.), *Drought and Hunger in Africa*. Cambridge, UK: Cambridge University Press.

Watts, Michael J. and Hans Bohle (1993), 'The space of vulnerability: the causal structure of hunger and famine', *Progress in Human Geography*, **17** (1), 43–68.

White, Andy, Jeffrey Hatcher, Arvind Khare, Megan Liddle, Augusta Molnar and William D. Sunderlin (2010), 'Seeing people through the trees and the carbon: mitigating and adapting to climate change without undermining rights and livelihoods', Ch. 11 in Robin Mearns and Andrew Norton (2010), *Social Dimensions of Climate Change: Equity and Vulnerability in a Warming World*, Washington DC: The World Bank, pp. 277–301.

White House (2006), 'The Federal response to Hurricane Katrina', February. Available at http://www.whitehouse.gov/reports/katrina-lessons-learned.pdf (accessed March 9, 2013).

Wisner, Ben (1976), 'Man-made famine in Eastern Kenya: the interrelationship of environment and development', Discussion Paper No. 96, Institute of Development Studies at the University of Sussex, Brighton, UK.

World Bank (1992), *Governance and Development*. Washington DC, USA: The World Bank.

World Bank (1994), *Governance: The World Bank's Experience*, Washington: The World Bank.

Yohe, Gary and Richard S.J. Tol (2002), 'Indicators for social and economic coping capacity – moving toward a working definition of adaptive capacity', *Global Environmental Change*, **12**, 25–40.

8. Disasters and human security: natural hazards and political instability in Haiti and the Dominican Republic
Christian Webersik and Christian D. Klose

INTRODUCTION

The discussion on the impact of natural hazards on human security has gained momentum in recent years. Given the climate change scenarios of the Intergovernmental Panel on Climate Change (IPCC), with predicted changes in tropical cyclone intensity, a growing world population, and a trend towards coastal urbanization exposing more people to natural hazards, and persistent poverty, scientists and policymakers alike are increasingly concerned with possible political ramifications of the impact of natural hazards. Apart from meteorological natural hazards, earthquakes, volcanoes and tsunamis are also believed to have destabilizing effects. But can these claims be substantiated?

In the aftermath of a disaster, the media is often quick to report on chaos and acts of violence. Some attribute this outbreak of instability to the breakdown of the social order; as Timothy Garton Ash puts it boldly: "Remove the elementary staples of organised, civilised life – food, shelter, drinkable water, minimal personal security – and we go back within hours to a Hobbesian state of nature, a war of all against all" (Ash, 2005). In Haiti, the devastating earthquake in January 2010 was followed by incidents of looting. And as earlier research confirms, this was done by "isolated individuals or pairs with the objects looted being a matter of chance or opportunity" (Quarantelli, 2008). By contrast, when a 9.0 earthquake and a subsequent tsunami hit Japan in March 2011 causing great human suffering and physical damage, there were no signs of violence or political turmoil. Yet, the political handling of the disaster created criticism and put pressure on the political leadership in Japan, eventually toppling the Japanese Prime Minister in August 2011.

In sum, the purpose of this chapter is to study the historical impacts of natural hazards on political instability in Haiti and the Dominican Republic in space and time. Both countries share the same Island of Hispaniola (thereafter Hispaniola).

POLITICAL IMPLICATIONS OF NATURAL HAZARDS

Each year, Hispaniola is affected by hurricanes and associated flooding. In addition, earthquakes affect the region. We restrict our analysis to sudden-onset hazards for practical reasons, as those hazards are more likely to show immediate effects that could be associated with riots and political violence leading to a sudden political change. Droughts and epidemics are usually long-term, potentially showing accumulating effects permitting people's grievances to cumulate. This, however, would make it more difficult to separate long-term grievances based on inequalities from long-term grievances caused by hazards.

We include earthquakes in the study though they are clearly not climate-related, as we anticipate similar socio-economic effects on affected populations, thereby increasing our sample size.

A natural hazard leads to a disaster when people lose their lives, and property is damaged or destroyed.[1] Apart from economic loss, the loss of life and disease outbreaks, what are the political consequences of natural hazards? Do they bring people together in an effort to aid the wounded and to rebuild from ruins, or do they generate grievances, unrest, political instability, and ultimately regime change? Some studies have shown that disasters have a pacifying effect rather than triggering conflict and instability (Slettebak and de Soysa, 2010; Slettebak and Theisen, 2011). This is based on the assumption that social disturbances tend to foster social integration (Durkheim, 2002, cited in Slettebak and de Soysa, 2010). Slettebak and de Soysa find a negative relation between disaster and conflict, arguing that "country-years that experienced one or more disaster in the same or previous year are less likely to have an outbreak of civil conflict than other country-years" (Slettebak and de Soysa, 2010). A more recent study using the same method in Indonesia on low-intensity conflict confirms the finding (Slettebak and Theisen, 2011). Hirshleifer explains this tendency towards cooperation following a disaster in his alliance hypothesis, "that the continuation of organized society is a collective good which benefits most of society's members" (Hirshleifer, 1987, first cited in De Alessi, 1975). Disasters can lead to cooperative behaviour, people become more generous, offering goods and services at lower than expected prices. De Alessi attributes this to utility interdependence, an economic argument of wealth maximization. In brief, individuals derive utility from "increases in the welfare of others", thus explaining the interest to respond to the suffering and victims of disasters (De Alessi, 1975). A further study on the United States has found that "societal cohesion is strengthened and community conflicts are rare" in the aftermath of disasters (Quarantelli and Dynes, 1976, cited in Slettebak and de Soysa).

American researchers following World War II conducted the first, comprehensive study on this subject. The study by Fritz written in 1961 but not published until 1996 was interested in the social response to a nuclear attack and how this might cause social disorder (Fritz, 1996). Slettebak mentions this study and points out "the initial expectation was that bombing and natural disasters would cause massive panic and a breakdown of law, order, and social norms, but this was strongly contradicted by the findings" (Slettebak, 2012: 165).

The importance of social networks in the aftermath of a natural disaster are in line with Elinor Ostrom's research on the importance of social capital in managing disasters. More specifically, according to Ostrom the effective collaboration between government and citizens builds social resilience. Ostrom calls this coproduction which is "the process through which inputs used to produce a good or service are contributed by individuals who are not in the same organisation" (Ostrom, 1996). This implies that disaster preparedness, response and recovery can be best achieved by complementarity of efforts of government officials and local citizens. As a result, cooperation is likely to occur in societies that are based on generalized reciprocity and trust. More specifically, this shows how important participatory disaster management practices can be. It may be that social capital assets in Haiti and the Dominican Republic have helped to avoid conflict and political instability in the aftermath of a disaster.

Another line of argument relates to the body of literature on disaster diplomacy. This body of research claims that disaster-related activities can trigger diplomatic efforts, ultimately bringing together enemies, such as the United States and Iran. Indeed, the relationship between the United States and Iran has been tense and to some degree hostile for many decades. Despite minimum diplomatic relations, the United States provided relief to Iranian regions that were hit by earthquakes in 1990, 2002, 2003 and 2005 (Kelman, 2012). Though the United States earthquake aid did not lead to significant diplomatic relations, it opened new channels of communication and trust. Perhaps, the provision of aid was also an opportunity for both countries to show their support for humanitarian work, claiming that humanitarian assistance is seemingly more important than politics (Kelman, 2012). In other cases, such as the peace deals of Aceh following the 2004 tsunami, disaster relief supported and fostered diplomatic processes; however, that had already started. Kelman argues, "disaster-related activities can influence, support, affect, push along, or inhibit diplomatic processes, but that does not always occur" (Kelman, 2012: 14).

By contrast and according to Slettebak and de Soysa, other studies have found a positive relationship between disasters and conflict (Drury and Olson, 1998; Brancati, 2007; Nel and Righarts, 2008). All three studies by Brancati (2007), Drury and Olson (1998), and Nel and Righarts (2008) use quantitative methods to link disasters to political instability and conflict. Whereas Brancati exclusively examines earthquakes, the other two include all types of disasters. To date, this research remains inconclusive, and more investigation is needed.

DISASTERS, VULNERABILITY AND HUMAN SECURITY

Though climate change will make extreme weather events more likely the speed and direction of this change is debatable.

There seems emerging evidence that there will be fewer but stronger tropical cyclones in the future, as argued by Professor Russell L. Elsberry of the Naval Postgraduate School[2] (see also Webster et al., 2005a; Webster et al., 2005b; Giorgi et al., 2001; Landsea et al., 2006). Recent research tends to support the thesis that a warmer climate will lead to an increase in hazard intensity (tropical cyclones) and a decrease in frequency (Knutson et al., 2010). Although this research is inconclusive as to whether the number of tropical storms has increased, the intensity and, thus, severity of category 4 and 5 hurricanes that have hit Hispaniola since the nineteenth century, has been increasing (Giorgi et al., 2001; Landsea et al., 2006).

Whether or not a natural hazard triggers a disaster depends on the vulnerability of the affected population. This is important as "natural disasters" often have a natural and human component. A natural hazard turns into a (natural) disaster when people are ill prepared for it, emphasizing the societal nature of disasters. Economic income, geographical location, natural hazard risk and historical trajectories all shape vulnerabilities. Given the economic inequalities and societal differences between individuals and groups, disasters do not affect people equally (Wisner et al., 2004). And too often, the most disadvantaged groups of the society suffer most. Research on hurricane Katrina shows that reconstruction and re-population is slowest in the poorest neighbourhoods (Mutter, 2010). These neighbourhoods are typically low-lying, swampy and on marginal lands but at greatest risk of flooding.

For some years it has been argued that the loss in human life and livelihoods, the social and physical destruction that natural hazards cause is due to the increase in human vulnerability, rather than the increase in intensity and frequency of hazards (Wisner et al., 2004). Or in Kelman's words, "the term 'natural disaster' turned out to be a poor choice, due to the connotation that the disaster is caused by nature or that these disasters are naturally what happens when society interacts with the environment" (Kelman, 2012: 11). Most likely, a combination of both factors is responsible for the disaster impact on human security. Accordingly, there has been a paradigm-shift from "seeing disasters as extreme events created by natural forces, to viewing them as manifestations of unresolved development problems" (Yodmani, 2001: vi). As a result, what is needed is a better integration of long-term development programming with disaster management. Accordingly, the impact of "natural" disasters is never determined by natural forces alone but is contingent on social, economic and demographic factors. Already in 1994, Wisner and others have argued that taking a vulnerability approach takes the focus away from the "consideration to natural events as determinants of disasters" (Wisner et al., 2004: 12).

For instance, Neumayer and Plümper found out that gender explains differences in disaster mortality (Neumayer and Plümper, 2007). Thus it may be better to omit the word "natural" in the term "natural disaster". Disasters affect human security in particular in developing countries, as Ahrens and Rudolph point out: "According to empirical evidence, it is especially the poor in developing countries who lack the administrative, organizational, financial, and political capacity to effectively cope with disasters and who are particularly vulnerable: while only 11 percent of the people exposed to natural hazards live in countries characterized by a low level of human development, they account for more than 53 percent of the total number of recorded deaths" (Ahrens and Rudolph, 2006: 208). Yet, poverty is not the only factor determining disaster vulnerability. A poor community can be poor in economic terms but endowed with disaster coping strategies rooted in social, cultural and political capacities (Yodmani, 2001).

Vulnerability is closely linked to the concept of human security, which favours a people-centred approach to security. The subject matter is not the state but the individual affected by natural and social forces. Defined by the 1994 United Nations Human Development Report as "freedom from want" or "freedom from fear", human security opposes the traditional notion of national security. The potential political impacts of natural hazards are most likely to affect local populations and will be geographically bound, and most likely will not lead to inter-state conflict. We therefore adopt a human security approach to capture the potential political instability triggered by natural hazards, as conflict is most likely to remain local and low-intensive.

But what is vulnerability in the context of natural hazards studies?

Vulnerability refers to being susceptible to disasters, argues Hans-Martin Füssel, and this is a good starting point (Füssel, 2007). This definition implies that vulnerable societies are already weakened and therefore more sensitive to external stress, such as tropical cyclones or seismic events (Webersik, 2010). Wisner et al. define disaster vulnerability as "characteristics of a person or a group and their situation that influence their capacity to anticipate, cope with, resist and recover from the impact of a natural hazard" (Wisner et al., 2004: 11). This explains why in some countries, tropical cyclones and earthquakes claim thousands of lives, such as Myanmar and Haiti, whereas in others, such as Japan[3] and the United States,[4] the biggest concern is physical damage and disruption of industry and facilities such as ports (Esteban et al., 2010). Rather than distinguishing between countries, it is more useful to look at vulnerabilities within countries, namely factors such as gender, income, educational attainment, and housing conditions to explain the number of people affected by disasters. However, Füssel's definition remains broad, and it begs the question, "vulnerable to what?", as Dilley and Boudreau argue (Dilley and Boudreau, 2001).

Bearing the above in mind, the impact of natural hazards on Haiti and the Dominican Republic is quite different. Compared to the Dominican Republic, natural hazards can cause great physical damage and, in the worst case, loss of life in Haiti. In the Dominican Republic, physical damage is the greatest concern rather than loss of life. Assessing the number of affected people per capita, Haiti is far more vulnerable to natural hazards compared to the Dominican Republic. If natural hazards can cause a "Hobbesian state of nature, a war of all against all", Haiti should show stronger signs of this relationship based on the country's greater vulnerability. By comparing more recent demographic indicators of Haiti with the Dominican Republic, the difference in vulnerability becomes apparent. In Haiti, the number of people killed by storms over the last decades is far larger compared to the Dominican Republic despite both countries' similar population size of roughly eight million people. Between 1979 and 2008, earthquakes and storms killed approximately three times as many people in Haiti compared to the Dominican Republic (Université Catholique De Louvain, 2011). When standardizing by total population, the picture remains similar: the per capita death toll remains higher in Haiti.

One explanation, apart from the political, economic and historical differences, is partly geographic. In Haiti, proportionally more people live close to the coast. In this low elevation coastal zone, defined as the contiguous area along the coast which is less than 10 meters above sea level, people are at greater risk of the adverse impact of tropical

Table 8.1   Natural hazards and socio-economic data of Hispaniola in space and time

Factor	Haiti	Dominican Republic
Area/km^{-2}	27750	48667
Population (1850)/1000 inhabitants	543	137
Population (2009)/1000 inhabitants	9036	10118
Population growth rate annual	0.02	0.03
Population growth rate/% (1850–2009)	16	73
Population density (2009)/inhabitants per km^2	326	208
Infant mortality rate/per 1000 newborn babies (2008)	72	33
GDP per capita (2009)/$US	770	4525
Number of observed storms that hit Hispaniola (1850–2009)	42	48
Storm hit rate/%	23	30
Maximum storm intensity/km h^1	130	150
Observed storm hazard/km h^1	30	45
Observed earthquake hazard (1970–2009)\ peak ground acceleration	1.52	2.72

Source: Adapted from Klose and Webersik, 2010.

cyclones as storm velocity typically decreases when storms travel over land (McGranahan et al., 2007). Still, the ratio in Haiti remains larger compared to the Dominican Republic.[5] This is despite the lower storm hit rate in Haiti compared to the Dominican Republic (see Table 8.1). Yet, using high coastal population densities as a proxy for hazard vulnerability firstly misses that there are reasons why people live at the coast, reasons that the people believe reduce their vulnerability; and secondly, Haiti has shown high vulnerability to inland freshwater flooding resulting from tropical cyclones.

One important factor of vulnerability is exposure, as described above; yet another is the level of sensitivity to natural hazards. One of the main factors shaping sensitivity and the capacity to cope with and to respond to natural hazards is income. Haiti is one of the poorest countries in the Western hemisphere: more than half of its population lives on less than a dollar per day (United Nations Development Programme, 2009a). With the lack of skilled labour and few employment opportunities, Haiti has an unemployment rate of about 70 percent. Figures for annual total gross domestic product (GDP) show that Haiti's economy, especially since the

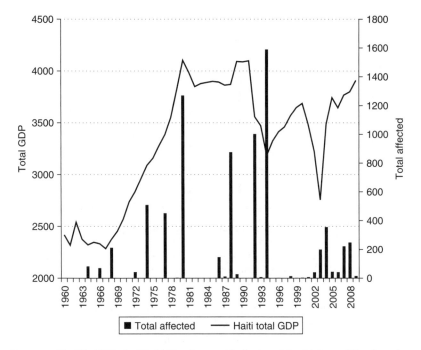

Sources: The World Bank, World Development Indicators, http://data.worldbank.org/ indicator and the OFDA/CRED International Disaster Database, Université Catholique de Louvain, Brussels, Belgium. www.emdat.be. Last accessed: 25 October 2010.

Figure 8.1   *Economic growth in Haiti (in million US$, constant 2000) and total affected people (in thousands) by meteorological hazards, 1960–2009*

1980s, neither grew nor managed to bounce back following major disasters (Figure 8.1). By contrast, total GDP of the Dominican Republic shows a steady upward trend despite disasters in 1979, 1988, and 1998 (Figure 8.2). In fact, GDP rates of Haiti and the Dominican Republic were almost equal in 1960, but since then have diverged (Figure 8.3).

Yet, GDP growth figures are not sufficient to explain vulnerability. For example, Cuba has a much lower GDP per capita but experienced far lower human loss during the 2005 Katrina hurricane. GDP growth is one indicator and other variables need to be taken into consideration. A careful in-depth vulnerability analysis of the two countries, however, would be beyond the scope of this chapter. Another caveat of total GDP is that it does not capture inequalities. With a 2001 GINI index of 60, Haiti displays great income inequality (Word Bank, 2011).[6] As a

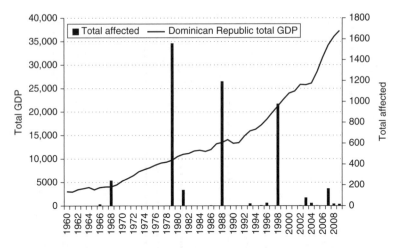

Sources: The World Bank, World Development Indicators, http://data.worldbank.org/
indicator and the OFDA/CRED International Disaster Database, Université Catholique de
Louvain, Brussels, Belgium. www.emdat.be. Last accessed: 25 October 2010.

*Figure 8.2   Economic growth in the Dominican Republic (in million US$,
constant 2000) and total affected people (in thousands) by
meteorological hazards, 1960–2009*

result, vulnerabilities also exist within a society where the poorest of the
population live in low-lying, flood-prone areas in housing with poor or
non-existent building codes. Apart from income, other social indicators
shape a society's vulnerability. Haiti's social vulnerability is captured
by indicators on health, education and income: In 2009, Haiti's human
development index of the United Nations Development Programme
(UNDP) ranks the country at 149 of 182 at the bottom of the list among
countries like Sudan, Tanzania and Ghana in sub-Saharan Africa
(United Nations Development Programme, 2009b). The levels of infant
mortality rate, a robust indicator for development, has also been fairly
low with 72 per 1,000 newborn babies dying in 2008, and this figure
remains high, compared to 33 in the Dominican Republic (World Bank,
2010).

Besides socio-economic indicators, geographical and environmental
factors also play a role. Haiti has experienced massive deforestation, and
in 2007, only 3.7 percent of the country remained forested. By contrast,
the Dominican Republic retained 28 percent forest cover in the same
year (World Bank, 2010). Deforestation remains a critical issue: between
1990 and 2007, Haiti lost 126 square kilometres whereas the Dominican

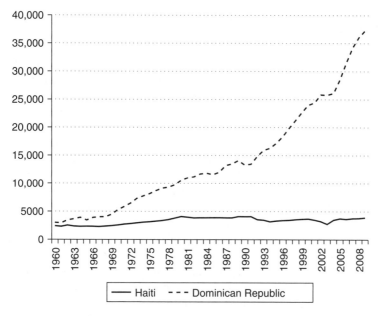

Sources: The World Bank, World Development Indicators, http://data.worldbank.org/ indicator and the OFDA/CRED International Disaster Database, Université Catholique de Louvain, Brussels, Belgium. www.emdat.be. Last accessed: 25 October 2010.

Figure 8.3   *Economic growth in Haiti and the Dominican Republic, 1960–2009 (in million US$, constant 2000)*

Republic's forest cover remained stable (World Bank, 2010). In addition, an unfavourable topography with a very hilly landscape causes water to run off unhindered.

Most important, both Haiti and the Dominican Republic had different historical trajectories in terms of resource management shaping vulnerability. In Haiti, environmental degradation (e.g., deforestation) and exploitation of natural resources (e.g., forests, water) have been accelerated since the 1920s by increasing population growth and rural poverty (Dolisca et al., 2006). Though the argument that population growth and poverty is driving environmental degradation is oversimplifying, and cannot be substantiated, the higher population densities in Haiti do expose more people to natural hazards risk (see Robbins, 2004).

DEVELOPMENT TRAJECTORIES – A BRIEF HISTORY

A discussion on economic and social vulnerabilities is incomplete without considering the differences in cultures, societies, class structures, migrations, and development trajectories, make comparison of these two extremely distinct Caribbean societies very difficult. However, we attempt the following comparison based on the similar levels of natural hazard risk over time and space. It is important to note that apart from natural hazards, domestic politics and the colonial era also played a major role in shaping the political trajectories of Haiti and the Dominican Republic.

The developmental paths of the two countries diverged in the eighteenth century when Hispaniola was divided into a French and a Spanish colony. The French colony, which was later to become Haiti, had a much larger population and a greater percentage of slaves (80 percent compared to 10 percent). In 1804, France gave up its claims to the colony and in the same year, Haiti gained independence being the first country in the "New World" in 1804. The years following independence were characterized by political coups and instability in both countries. All but one of the 22 Haitian presidents from 1843 to 1915 were either assassinated or driven out of office (Diamond, 2005). Slavery was abolished, large-scale plantations were destroyed and the land was divided into smallholder farms. Moreover, Haitians spoke Creole making it more difficult for European traders to do business with them. Consequently, agricultural productivity declined and exports dwindled. Outside occupations followed. The United States invaded Haiti in 1915 and stayed until 1934.

While the Dominican Republic was on track to economic growth and prosperity, Haiti continued to experience political instability. In 1957, Haiti came under the control of François "Papa Doc", an equally ruthless politician as the dictator Trujillo of the Dominican Republic but with no interest in developing and modernizing the country. When he died in 1971, Jean-Claude "Baby Doc" Duvalier succeeded him and ruled until 1986. The years that followed were characterized by political instability and economic decline. In 1990, former priest Jean-Bertrand Aristide was elected president. He introduced reforms and enjoyed wide popular support but was overthrown in a coup d'état in 1991. The opposition, with the involvement of the United States, targeted Aristide supporters, and in 1994, again with the military help of the United States, Aristide returned from exile and was reinstated as president. A 2004 rebellion forced Aristide into exile, and in 2006, his former prime minister, René Préval, was elected. Since 2006, politics during president Préval's second term improved Haiti's economic development slightly.

The eastern part of the island, on the other hand, remained more open towards the "Old World". Relations with Europe continued in the Spanish-speaking part of Hispaniola following independence in 1821 from Spain. Thus, economically important immigrant groups preferred to settle in the Dominican Republic, which continued to have a lower population density (Table 8.1) putting less pressure on environmental resources. It can be speculated that higher population pressures and colonial exploitation in Haiti had been largely responsible for the stripping of its forests by the mid-nineteenth century.

After its independence from Haiti in 1844, and similar to Haiti, the country experienced several political changes and then stayed under military dictatorships until 1961 (Figure 8.4). From 1916 to 1924 the United States intervened in the Dominican Republic.

Starting from 1930, Trujillo autocratically ruled the Dominican Republic until 1961 when he was assassinated. Although the country suffered under his dictatorship, it was the beginning of modernization and industrialization. This was also the beginning of environmental protection, as Trujillo began to protect forests to generate hydro-electric power and to protect his personal interest in logging. Balaguer, who shaped Dominican politics for the next three decades, continued with this path of development and environmental protection until 1996. In the last years, the countries' gross domestic product (GDP) increased though large economic inequalities persist in both countries.

NATURAL HAZARDS IN HAITI AND THE DOMINICAN REPUBLIC

One way to answer the question whether natural hazards shaped political developments on Hispaniola since 1850 is to list disaster events and compare this data with political events (regime change and political conflict) in the year[7] following a disaster. Both data sets can be compared and results interpreted.

Almost every year, disasters, caused by both tropical cyclones and earthquakes, make thousands of people on Hispaniola homeless and claim lives. Spatial and temporal data of storms and earthquakes that directly hit Haiti and the Dominican Republic between 1850 and 2009 were used in this research.

We use peak ground acceleration estimates prior the 2010 M7 earthquake event that refer only to bedrock conditions without taking into consideration variations of local effects (for example bad soil conditions and local topography) or the depth of the epicentre (a parameter as important as magnitude). Prevalent softer or harder soil conditions may significantly impact the shaking intensities and their probabilities. The United States Geological Survey (USGS) provided earthquake data.

Tropical storms have also had major impacts on Hispaniola. More important, associated precipitation can cause flash floods and trigger landslides, both important features in particular in Haiti. The United States National Oceanic and Atmospheric Administration (NOAA) provided storm data (Table 8.1).

Several major earthquakes occurred in Haiti in 1751, 1770, 1842, 1860, 1887, 1897, and most recently in 2010. Several times, these seismic events destroyed the city of Port Au Prince in Haiti's South and Cap-Haïtien in its North. A major earthquake of May 7, 1842 caused severe loss of life and destroyed the city of Cap-Haïtien. In the Dominican Republic in 1946, an earthquake and subsequent tsunami killed 2550 people. The M7 earthquake of 12 January 2010 resulted in between 46,000 and 85,000 fatalities, according to a United States government commissioned report (Associated Press 2011). The quake also destroyed most governmental buildings in Port-au-Prince (Université Catholique de Louvain, 2011).

Tropical storms also have a major impact on Hispaniola, which was directly hit by 72 storm events between 1850 and 2010. Haiti and the Dominican Republic were hit by 34 storms between 1850 and 1929 and by 38 storms between 1930 and spring 2010. When comparing the storm hazard patterns of these two periods, the intensities of extreme storm events increased by 30 percent in the Dominican Republic and by 50 percent in Haiti. The frequency of storms increased by 25 percent in the Dominican Republic and remained unchanged in Haiti.

Moreover, major storms caused human casualties and great economic damage, largely in Haiti. In 1935, a severe tropical storm killed more than 2000 people. Hurricane Georges destroyed more than 75 percent of all the crops in the country in 1998. Tropical storms Gustav and Fay directly affected more than 800,000 people (10 percent of the total population) in Haiti in August and September 2008, respectively (United Nations Development Programme, 2009a).

Spatial and temporal data of all major storms and earthquakes that occurred on Hispaniola between 1850 and 2010 are summarized in Table 8.1. The data show that Haiti and the Dominican Republic have experienced similar severe storms since 1850, although 30 percent more tropical storms hit the Dominican Republic with a maximum observed hurricane category 4 on the Saffir–Simpson Hurricane Scale. Moreover, the Dominican Republic experienced a slightly higher earthquake hazard during the last 100 years. Tropical cyclones and earthquakes, however, differently affect both countries.

POLITICAL INSTABILITY IN HAITI AND THE DOMINICAN REPUBLIC

In order to assess the impact of natural hazards on political stability, we need to measure political stability. We suggest two variants. One is to consider the rate of political instability; another is to measure the level of armed conflict. To do so, we measure political instability using a score from −10 to +10, ranging from fully institutionalized autocracies through mixed, or incoherent, authority regimes (termed "anocracies") to fully institutionalized democracies (Marshall et al., 2011). The "polity score", developed by Colorado State University and University of Maryland, is a 21-point scale measuring political regime types, from −10 (hereditary monarchy) to +10 (consolidated democracy).

Figure 8.4 illustrates regime type change over time in Haiti and the Dominican Republic. From 1850 to 1910, both countries remained relatively stable. Haiti had a brief period of minor democratic transition from 1914 to 1918. From 1946, the country entered a continued trend of negative political instability, becoming more and more autocratic. By contrast, the Dominican Republic remained almost exclusively autocratic until 1978 when it made a positive democratic transition that lasts until today. By applying the polity score, we can compare natural hazard events with political instability. Since 1850, Hispaniola experienced in total 72 storms and 5 severely damaging earthquakes that directly hit the island (Table 8.2).

In the Dominican Republic, three storms coincided with a change of political stability in the same or the following year – one in 1930, one storm in 1961 and one in 1963. Two storms (1930 and 1963) fall into a period with a negative regime change. A worsening of the autocratic rule in the Dominican Republic followed the 1930 tropical cyclone that affected the entire island. In 1963, hurricane Edith fell into a period of collapse of central political authority. Tropical storm Frances in 1961 that hit eastern part of the Dominican Republic occurred in the midst of a political transition towards a democratic regime. However, the change to a more democratic Dominican Republic between 1960 and 1962 was largely influenced by the assassination of dictator Trujillo on 30 May 1961. This decisive event happened before tropical storm Frances hit the country in October 1961.

In Haiti, in total four storms fall into a period of change in political stability in the same or the following year. Two storms coincided with a negative regime change. All these events occurred after the 1950s. In 1958, when hurricane Ella swept through southwest Haiti, the country experienced a gradual change in stronger authoritarian rule. In 1985,

Table 8.2   Major earthquakes and tropical storms/hurricanes in Haiti
(H) and the Dominican Republic (DR) since the 18th century,
political context and socio-economic consequences

Date dd.mm.yyyy	Hazard type	Location	Socio-economic damage/ regime type change/ political conflict
18.10.1751	Major earthquake	Southern Haiti	H: destruction of 75% of masonry houses in Port-au-Prince
21.11.1751	Major earthquake	Southern Haiti	H: destroyed Port-au-Prince
03.06.1770	M7.5 earthquake	Southern Haiti	H: killed 250 people; destroyed Port-au-Prince
07.06.1842	Major earthquake	Northern Haiti	H: major damages in Cap-Haïtien; 10 000 fatalities; revolts started that lead to the foundation of the Dominican Republic two years later
08.04.1860	Major earthquake	Southern Haiti	H: destruction of Port-au-Prince
23.09.1887	Earthquake	Northern Haiti	H: major damage in Moles St Nicholas
29.09.1897	Earthquake	Northern Haiti	No data available
04.08.1946	M8.0 earthquake, tsunami	Northern DR	DR: killed 2 550 people
12.01.2010	M7.0 earthquake	Southwest Haiti	H: killed 45 000–65 000; destroyed Port-au-Prince, Léogâne and Petit-Goâve
03.09.1930	Not named tropical storm	Hispaniola	DR: Political transition period 1930–33, from autocratic (−5) to autocratic (−9)
01.09.1958	Category 2 hurricane Ella	Southwest Haiti	H: 28.06.1958 (before Ella) attempted invasion of USA to arrest Duvalier, 1957–58 from −5 to −8
03.10.1961	Tropical storm Frances	Eastern DR	DR: year of political transition from autocratic (−9) to democratic (+8) 1960–62
27.09.1963	Category 2 hurricane Edith	Northeast DR	DR: interregnum period (1963–65) with complete

Table 8.2   (continued)

Date dd.mm.yyyy	Hazard type	Location	Socio-economic damage/ regime type change/ political conflict
			collapse of central political authority from democratic (+8 in 1962) to autocratic (−3 in 1966), minor armed conflict in 1965
07.10.1985	Tropical storm Isabel	Hispaniola	H: 07.02.1986 president Duvalier flees after disorder; political transition period 1985–86, from autocratic (−9) to autocratic (−8)
22.09.1998	Category 2/3 hurricane Georges	Hispaniola	H: political transition in 1999, from democratic (7 in 1998) to autocratic (−2 in 2000)
10.10.2003	Tropical storm Mindy	Hispaniola	H: The 2004 coup d'état and Aristide's
05.12.2003	Tropical storm Odette	Hispaniola	ousting; minor armed conflict in 2004, political transition period 2004–05, from autocratic (−2 in 2003) to democratic (+5 in 2006)

Source: Adapted from Klose and Webersik, 2010.

tropical storm Isabel crossed over Hispaniola along the Haitian border. In the following year on 7 February 1986, president Jean-Claude Duvalier, nicknamed "Baby Doc" fled the country forced by a popular uprising against his government. In 1998, hurricane Georges affecting all of Hispaniola was followed by a worsening of the regime type, from democratic in 1998 to autocratic in 2000. In 2003, tropical storms Mindy and Odette crossed over Haiti. In the following year, a coup d'état ousted president Aristide; the regime type turned from autocratic in 2003 to democratic in 2006.

By contrast, most earthquakes did not coincide with political change over time from 1850 to 2009. Only in 1842, a major earthquake affected

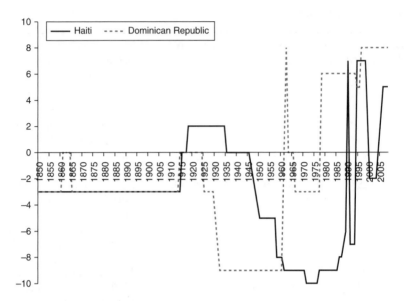

Note: Cases of foreign 'interruption' and cases of 'interregnum,' or anarchy, are converted to a 'neutral' polity score of '0.'

Source: 2008 Polity Score IV data series, a score of +10 on the polity score indicates fully developed democracies, whereas a score of −10 on the polity score indicates consistent autocracies.

Figure 8.4   *Political regime characteristics, 1850–2008, from authoritarian (−) to democratic (+)*

northern Haiti, followed by a political revolt that led into the independence of the Dominican Republic in 1844.

Apart from political instability, we plot incidents of political conflicts and natural hazards. According to the UCDP/PRIO Armed Conflict Dataset, four minor political conflicts (conflicts with more than 25 battle-related deaths per year) took place between 1946 and 2009[8] (Gleditsch et al., 2002). In the Dominican Republic in 1965, two years after hurricane Edith, this dataset reports a minor conflict. In Haiti, in 1989, 1991, and 2004, minor conflicts occurred. The 2004 conflict resulting in president Aristide's ousting occurred one year after tropical storms Mindy and Odette.

DISCUSSION

Given the few instances of political change/conflict incidence in the same or following a natural hazard event year, it is rather speculative to draw a link between the observed political changes/conflict incidence and the occurrence of natural hazards. What is possible is an effect on the capacity of the government to contain upheaval. But this assumption needs further investigation. For example, the government following the ousting of former president Aristide in 2004 was ill prepared to deal with the impact of hurricane Jeanne in September 2004 leading to severe damage and loss of lives. The focus in Haiti has been for many years to maintain peace and security. Moreover, international governmental and non-governmental agencies working in Haiti are concerned with the humanitarian situation, prioritizing emergency aid rather than long-term development assistance. What are needed most in Haiti are long-term development projects to make the country more resilient to natural hazards. One example is the lack of forest cover on steep hillsides surrounding coastal settlements that could help prevent flash floods. Already in the mid-1990s, it was estimated that Haiti retained only 1 to 3 percent forest cover (de Sherbinin, 1996). However, as expressed by a local expert, local communities have little interest in supporting re-forestation projects, as such projects do not promise an immediate return on investments.[9]

This raises the assumption that frequent regime changes weaken a country's capacity to respond and to cope with natural hazards. The fact is that, compared to the Dominican Republic, Haiti's major changes in regime type occurred fairly recently, following the end of the Cold War (Figure 8.4) Since then, Haiti has swung between autocratic and democratic regimes, only establishing a relatively stable, democratic regime in 2006. Interestingly, when plotting the regime type over time (years from 1946 to 2008), Haiti and the Dominican Republic had a fairly similar political trajectory. This statement needs clarification, as on average, Haiti remained with negative scores (autocratic regimes) following the Cold War and less developed democratic institutions over time.

In turn, it seems that stability over time, apart from economic incomes, whether the country is autocratic or democratic in nature, is important to build resilience to natural hazards. For example Cuba, with similar exposure risk to hurricanes has shown much greater disaster resilience compared to Haiti despite the absence of democratic institutions. Transitions are dangerous and destabilizing, and thereby increase vulnerability to natural hazards. This is in line with the civil war literature that argues that anocracies (regimes that are neither democratic nor autocratic) are at the highest risk of experiencing civil war (Hegre et al., 2001; Urdal, 2004). Pakistan is a good example for a country that has experienced political instability over the past years and is prone to flooding. Stable institutions are needed to build an effective and efficient disaster management. Emergencies in Haiti (earthquake), Pakistan (flooding) and Myanmar (tropical cyclone) have shown the lack of institutional capacity to respond and to manage the disaster (Bruch and Goldman, 2012).

Given the weak link between the occurrence of natural hazards and political instability (only 4 out of 72 storms and one earthquake coincided with a negative change of political regime type in the same or the following year), the hypothesis that natural hazards impact on political instability can be rejected. Moreover, often political transitions took place over a period of a couple of years in which other intervening events happened, such as global economic problems, ministers or strongmen dying or leaving their posts, changing the power balance for the leader, or alterations in development or military aid from external sources. In particular in the past decades from the 1970s onwards where most of the regime changes took place, not a single disaster was followed by a negative regime change (Figures 8.5 and 8.6).

More convincingly (but not tested in this study) it seems that political instability and political conflict results from historical socio-economic factors, such as economic inequality, political and economic exclusion, low incomes, and the influence of outsiders, all factors considered as relevant correlates of instability and war (Buhaug, 2010). Given the similar exposure to natural hazards on Hispaniola, the same explanations of political instability may shed light on explaining Haiti's vulnerability to natural hazards. As mentioned earlier, historical explanations are useful, as well as the more recent socio-economic and political trajectory. Here, future research is required.

Apart from political factors, what matters is the proportion of the population living at the coast and in river deltas exposed to climate-related hazards, such as flooding and tropical cyclones. The rationale is as follows: According to Slettebak, climate-related disasters have increased in the past decades whereas the number of earthquakes affecting populations remained somewhat stable. Despite the fact of increased reporting over time, it seems that climate-related disasters gain importance (Slettebak, 2012).

As a result, we can expect countries with large populations living in river deltas or coastal regions to face serious impacts for human security, such as Bangladesh, Vietnam, Cambodia, Indonesia, the Philippines, and China. Some of India's megacities are situated on the coast, prone to seasonal flooding, as experienced in Mumbai. What is important is

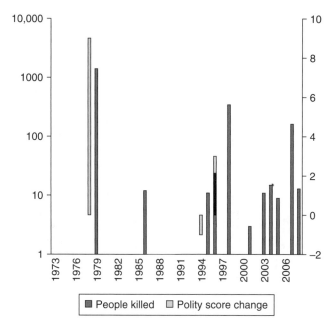

Note: Values indicate regime change from year to year. Cases of foreign 'interruption' and cases of 'interregnum,' or anarchy, are converted to a 'neutral' polity score of '0.'

Sources: People killed: EM-DAT: The OFDA/CRED International Disaster Database, Université Catholique de Louvain, Brussels, Belgium. Available at www.em-dat.net (4 April 2011).
2008 Polity Score IV data series, a score of +10 on the polity score indicates fully developed democracies, whereas a score of −10 on the polity score indicates consistent autocracies.

Figure 8.5   *Political regime change and people killed by storms and earthquakes in the Dominican Republic, 1973–2008*

that population growth and urbanization drive more people into flood-prone areas, into places where people would not have settled in the past. Low incomes prevent local communities from building according to the required building standards that may exist on paper.

When the 7.0 earthquake struck Haiti, many of the multi-story government buildings and buildings hosting aid agencies, including the United Nations, collapsed. It is estimated that approximately 40 percent of Haiti's government officials died during the quake (Bruch and Goldman, 2012). Over time, the city grew without proper urban planning. Port-au-Prince for example was planned for 250,000 people and now is home to more than two million people. This makes people settle in vulnerable areas, in flood plains or on steep hillsides, which is common in Port-au-Prince,

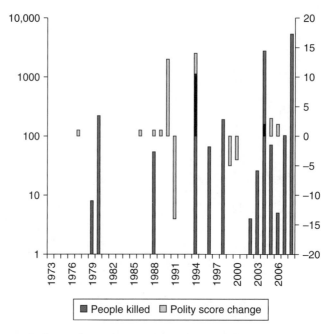

Note: Values indicate regime change from year to year. Cases of foreign 'interruption' and cases of 'interregnum,' or anarchy, are converted to a 'neutral' polity score of '0.'

Sources: People killed: EM-DAT: The OFDA/CRED International Disaster Database, Université Catholique de Louvain, Brussels, Belgium. Available at www.em-dat.net (4 April 2011).
2008 Polity Score IV data series, a score of +10 on the polity score indicates fully developed democracies, whereas a score of −10 on the polity score indicates consistent autocracies.

Figure 8.6   Political regime change and people killed by storms and earthquakes in Haiti, 1973–2008

making them more vulnerable to flooding and landslides (Bruch and Goldman, 2012).

Finally, differences in disaster vulnerability can be explained with incomes. Three countries that are most often struck by disasters, the United States, China and India have managed (climate-related) disasters more effectively due to their stable polities and economic capacity. India and China are developing rapidly with sufficient resources to protect their citizens from natural hazards (with clear limitations as experienced during the 2008 Sichuan earthquake in China when many schools collapsed killing thousands of children).

CONCLUSION

By comparing disaster events with political events, the study aimed at shedding light on the impact of disasters on political stability. Very few (four) hurricanes and (one) earthquakes were followed by a negative regime change and political conflict in the same or following year. In fact, some (three) of the regime changes were actually positive, leading to a more democratic regime type. Given the similar natural hazard risk in Haiti and the Dominican Republic, the differences in political stability, largely following the end of the Cold War, in both countries are to be found in other factors. This is perhaps not surprising, given the very different historical trajectories of the two countries, including a history of rampant deforestation in Haiti, of different patterns of United States corporate and governmental meddling in the two neo-colonial Caribbean states, of patterns of emigration bringing about remittances flows during political crises and upheavals, and organized crime.

While political instability remains unaffected, each year, people lose their lives due to flooding and earthquakes in Haiti. Over the years, the country's economy appears to be negatively affected by the impact of natural hazards. Haiti's history of political violence, weak governance, inequalities and low incomes has weakened the country's ability to prepare, cope with, resist, and recover from impacts of tropical cyclones and earthquakes. This may explain the high number of affected people, the displaced, injured and fatalities, in particular, during extreme and devastating events, such as the 2010 M7 earthquake near Port-au-Prince, which killed between 46,000 and 85,000 people. Little attention was given in Haiti to disaster preparedness and mitigation due to volatile security and political disorder. Again, Haiti's vulnerability might have been amplified, in particular, by human-influenced factors compounding the severity of natural hazards, such as massive deforestation.

Thus approaches are needed that respond to local needs through nature protection policies, including education of the public, legal measures to ensure environmental protection and participatory management for forests, fishing, agriculture, and tourism.

ACKNOWLEDGEMENTS

The authors would like to thank Tor A. Benjaminsen for his constructive comments. They also feel indebted to Halvard Buhaug and other workshop participants at the Centre for the Study of Civil War of the Peace Research Institute Oslo (PRIO). The authors also thank several anonymous reviewers whose comments substantially improved this chapter.

NOTES

1. The International Disaster Database (EM-DAT) published by the Centre for Research on the Epidemiology of Disasters (CRED) defines a disaster if at least one of the following criteria is fulfilled: Ten or more people reported killed; hundred or more people reported affected; declaration of a state of emergency; or a call for international assistance.
2. Personal communication with the authors, 12 May 2011.
3. This may change in Japan since a devastating tsunami hit the country in March 2011 and killed several thousand people.
4. With the exception of Katrina, a hurricane in 2005 that killed an estimated 1,833 people.
5. This is, of course, not relevant for seismic events, besides tsunamis.
6. The GINI index measures the degree of inequality in the distribution of family income in a country. The index is the ratio between the richest and the poorest family income of a country. If income were distributed with perfect equality, the index would be zero; if income were distributed with perfect inequality, the index would be 100.
7. Only political events within the same or following year of the disaster event were considered.
8. Unfortunately, the UCDP/PRIO Armed Conflict Dataset does not date back to 1850. A minor conflict is defined as between 25 and 999 battle-related deaths in a given year.
9. Personal communication, Port-au-Prince, 28 June 2011.

REFERENCES

Ahrens, J. and P. M. Rudolph (2006), 'The importance of governance in risk reduction and disaster management', *Journal of Contingencies and Crisis Management* **14** (4).
Ash, T.G. (2005), 'It always lies below', *The Guardian*, 8 September.
Associated Press (2011), 'US government report says far fewer people died in Haiti quake than originally estimated', *The Washington Post*, 30 May.
Brancati, D. (2007), 'Political aftershocks: the impact of earthquakes on intrastate conflict', *Journal of Conflict Resolution*, **51**, 715–743.
Bruch, C. and L. Goldman (2012), 'Keeping up with megatrends: the implications of climate change and urbanization for environmental emergency preparedness and response', Geneva: Joint UNEP/OCHA Environment Unit.
Buhaug, H. (2010), 'Climate not to blame for African civil wars', *PNAS*, 107, 16477–16482.
De Alessi, L. (1975), 'Toward an analysis of postdisaster cooperation', *The American Economic Review*, **65**, 127–138.
de Sherbinin, Alex (1996), 'Human security and fertility: the case of Haiti', *Journal of Environment and Development*, **5** (1), 28–45.
Diamond, Jared M. (2005), *Collapse: How Societies Choose to Fail or Succeed*. New York, USA: Viking.
Dilley, M. and Boudreau, T.E. (2001), 'Coming to terms with vulnerability: a critique of the food security definition', *Food Policy*, **26**, 229–247.
Dolisca, F., Carter, D.R., Mcdaniel, J.M., Shannon, D.A. and Jolly, C.M. (2006), 'Factors influencing farmers' participation in forestry management programs: a case study from Haiti', *Forest Ecology and Management*, **236**, 324–331.

Drury, C.A. and Olson, R.S. (1998), 'Disasters and political unrest: an empirical investigation', *Journal of Contingencies and Crisis Management*, 6, 153–161.

Durkheim, É. (2002), *Suicide*. London, New York: Routledge.

Esteban, M., Webersik, C. and Shibayama, T. (2010), 'Methodology for the estimation of the increase in time loss due to future increase in tropical cyclone intensity in Japan', *Climatic Change*, 102.

Fritz, Charles E. (1996), 'Disasters and mental health: therapeutic principles drawn from disaster studies', Newark, USA: University of Delaware.

Füssel, H.-M. (2007), 'Vulnerability: a generally applicable conceptual framework for climate change research', *Global Environmental Change*, 17, 155–167.

Giorgi, F., Bruce Hewitson, J. Christensen, Michael Hulme, Hans Von Storch, Penny Whetton, R. Jones, L.O. Mearns and C. Fu (2001) *Regional Climate Information-Evaluation and Projections*. Cambridge, UK: Cambridge University Press.

Gleditsch, N.P., Wallensteen, P., Eriksson, M., Sollenberg, M. and Strand, H. (2002), 'Armed conflict 1946–2001: a new database', *Journal of Peace Research*, 39, 615–637.

Hegre, H., Ellingsen, T., Gates, S. and Gleditsch, N.P. (2001), 'Towards a democratic civil peace? Democracy, political change, and civil war, 1816–1992', *American Political Science Review*, 95, 33–48.

Hirshleifer, J. (1987), *Economic Behavior in Adversity*. Chicago, USA: University of Chicago Press.

Kelman, I. (2012), *Disaster Diplomacy*. Abingdon, UK and New York: Routledge.

Klose, C.D. and Webersik, C. (2010), 'Long-term impacts of tropical storms and earthquakes on human population growth in Haiti and Dominican Republic', *Nature Precedings*, doi 10.1038/npre.2010.4737.1.

Knutson, Thomas R., John L. McBride, Johnny Chan, Kerry Emanuel, Greg Holland, Chris Landsea, Isaac Held, James P. Kossin, A.K. Srivastava and Masato Sugi (2010), 'Tropical cyclones and climate change', *Nature Geosci*, 3 (3),157–163.

Landsea, C.W., Harper, B.A., Hoarau, K. and Knaff, J.A. (2006) 'Can we detect trends in extreme tropical cyclones?', *Science*, 313, 452–454.

Marshall, M.G., Jaggers, K. and Gurr, T.R. (2011), *Polity IV Project: Political Regime Characteristics and Transitions, 1800–2009*. Center for Systemic Peace, Colorado State University, University of Maryland.

Mcgranahan, G., Balk, D. and Anderson, B. (2007), 'The rising tide: assessing the risks of climate change and human settlements in low elevation coastal zones', *Environment and Urbanization*, 19, 17–37.

Mutter, John (2010), 'Opinion: disasters widen the rich–poor gap', *Nature*, 466.

Nel, P. and Righarts, M. (2008), 'Natural disasters and the risk of violent civil conflict', *International Studies Quarterly*, 52, 159–185.

Neumayer, E. and T. Plümper (2007), 'The gendered nature of natural disasters: the impact of catastrophic events on the gender gap in life expectancy, 1981–2002', *Annals of the Association of American Geographers*, 97 (3), 551–566.

Ostrom, E. (1996), 'Crossing the great divide: coproduction, synergy, and development', *World Development*, 24, 1073–1087.

Quarantelli, E.L. (2008), 'Conventional beliefs and counterintuitive realities', *Social Research*, 75.

Quarantelli, E.L. and Dynes, R.R. (1976), 'Community conflict: its absence and its presence in natural disasters', *Mass Emergencies*, 1 (1), 139–152.

Robbins, P. (2004), *Political Ecology: A Critical Introduction*. Malden, MA, USA: Blackwell Publishing.

Slettebak, Rune T. (2012), 'Don't blame the weather! Climate-related natural disasters and civil conflict', *Journal of Peace Research*, 49 (1), 163–176.

Slettebak, R.T. and De Soysa, I. (2010), 'High temps, high tempers? Weather-related natural disasters and civil conflict', Conference on Climate Change and Security, 21–24 June 2010, Trondheim.

Slettebak, R.T. and Theisen, O.M. (2011), 'Natural disasters and social destabilization: is there a link between natural disasters and violence? A study of Indonesian districts, 1990–2003', Paper presented at the annual meeting of the International Studies Association Annual Conference, Montreal, Quebec, Canada.

United Nations Development Programme (1994), *Human Development Report 1994*, New York.

United Nations Development Programme (2009a), *Focus on Haiti: Key Statistics* (online). UNDP. Available from: http://www.undp.org/cpr/we_work/Haiti08.shtml

United Nations Development Programme (2009b), *Human Development Report 2009*. Houndmills: published for the United Nations Development Programme by Palgrave Macmillan, London, UK.

Université Catholique De Louvain (2011), *OFDA/CRED International Disaster Database*. Université Catholique de Louvain, Brussels, Belgium.

Urdal, H. (2004), *The Devil in the Demographics: The Effect of Youth Bulges on Domestic Armed Conflict, 1950–2000*. Washington DC, USA.

Webersik, C. (2010), *Climate Change and Security: A Gathering Storm of Global Challenges*. Santa Barbara, CA, USA: Praeger.

Webster, P.J., Holland, G.J., Curry, J.A. and Chang, H.-R. (2005a), 'Changes in tropical cyclone number, duration, and intensity in a warming environment', *Science*, **309**, 1844–1846.

Webster, P.J., Holland, G.J., Curry, J.A. and Chang, H.R. (2005b), 'Response to comment on "Changes in tropical cyclone number, duration, and intensity in a warming environment"', *Science*, **311**, 1713c.

Wisner, B., P. Blaikie, Terry Cannon and Ian Davis (2004), *At Risk: Natural Hazards, People's Vulnerability, and Disasters*, 2nd edition. London, New York: Routledge.

World Bank (2010), *World Development Indicators*. Washington, DC, USA.

Yodmani, S. (2001), 'Disaster risk management and vulnerability reduction: protecting the poor', Asian and Pacific Forum on Poverty.

PART III

A REGIONAL PERSPECTIVE ON CLIMATE CHANGE AND HUMAN SECURITY

9. The impact of climate change on human security in Latin America and the Caribbean

Úrsula Oswald Spring, Hans Günter Brauch, Guy Edwards and J. Timmons Roberts

1. INTRODUCTION

Climate change is projected to have multiple impacts this century on international, national and human security. If 'business as usual' levels of greenhouse gas emissions continue, a catastrophic climate change scenario becomes increasingly possible, with climate hotspots (REC, 2011), water scarcity (UNEP, 2012), decline in food production, more extreme weather events (IPCC, 2012) and environmentally induced migration, particularly within and from Latin America and the Caribbean (LAC) countries (Oswald Spring et al., 2013; Serrano Oswald et al., 2013). Physical and societal climate change impacts have been projected (IPCC, 2007, 2007a), but non-linear changes may trigger tipping points (Lenton et al., 2008), which could have geopolitical effects for international, national and human security.

LAC is particularly vulnerable to the impacts of climate change despite the fact that the region's greenhouse gas emissions represent an estimated 11 percent of the global total (IDB, 2012). Climate projections for LAC indicate that towards the end of the century, temperature increases will vary between 1 degree and 6 degrees, according to the particular emissions scenario and area concerned (Magrin et al., 2007).

Vergara et al. (2007; Vergara, 2011) state that the impacts of climate change include the potential collapse of the Caribbean coral biome, intensification of weather patterns and storms and sea level raise, increased flooding and droughts, warming of high Andean ecosystems, increased exposure to tropical diseases and risk of dieback of the Amazon rainforest ecosystem. LAC may be more vulnerable than other regions due to the diverse character of the ecosystems at risk such as the Andean mountains. The combination of significant loss of ecosystems, the major potential impacts on power and water supply, and the very likely increased costs in health and food make LAC a priority for the adaptation agenda (Vergara, 2009).

Climate change is having significant impacts on LAC's economies and these effects will become stronger as time goes on (ECLAC, 2010). The Inter-American Development Bank (IDB, 2012) comments that estimated annual damages in LAC caused by the impacts associated with a rise of 2 degrees over pre-industrial levels are estimated to be roughly US$100 billion annually by 2050. This magnitude of losses would limit development options as well as access to natural resources and ecosystem services (IDB, 2012).

Climatic variability and extreme events have been severely affecting the region (Magrin et al., 2007). LAC is extremely vulnerable to these natural disasters: the region has seen a recent increase in extreme climatic events, and with it a rise in the number of people affected. Natural disasters in the Americas in the past decade are estimated to have cost more than US$446 billion (ECLAC, 2011). Climate change also presents a challenge to hard-won development gains and is exacerbating inequality across the region and threatens to halt and then reverse advances in health and education for the most vulnerable (UNDP, 2007). Climate change also threatens the progress made in recent decades in development and in the achievement of the Millennium Development Goals (Mata and Nobre, 2006; ECLAC and IDB, 2010).

This chapter focuses on the impact of climate change on human security in Latin America and the Caribbean, particularly on climate change hotspots in Central America, the Caribbean, the Andes (Chevallier et al., 2011) and Amazonia (Figure 9.1). Here we address four research questions. First, what have been the major conceptual human security debates in LAC countries since 1990? Second, what is the state of knowledge on climate change and its possible human security impacts for LAC? What strategies for climate change adaptation are being implemented in LAC and how are they being financed? And finally, how can policies for coping with climate change be interpreted from a human security perspective?

To answer these questions, this chapter introduces two global discourses linking climate change impacts and security within LAC and discusses its environmental and social vulnerability from a human security perspective. LAC strategies for coping with climate change are assessed and policy debates on financing their adaptation measures are examined, followed by conclusions.

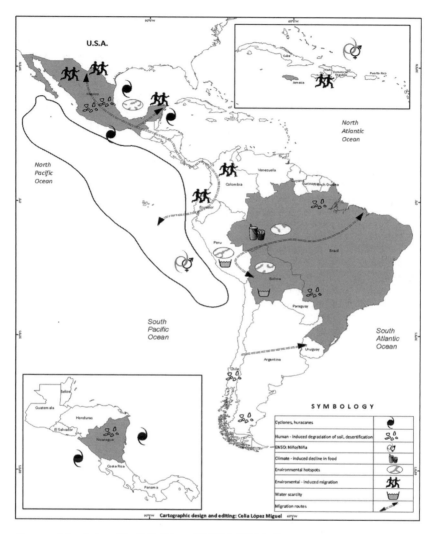

Source: Map designed by Celia López, UNAM, CRIM, Cuernavaca, Mexico.

Figure 9.1 Area of analysis and climate change hotspots

2. DISCOURSES ON CLIMATE CHANGE AND SECURITY IN LAC

Climate change has increasingly been 'securitized' (Wæver, 1995, 1997, 2008; O'Brien et al., 2010; Brauch, 2002, 2008, 2009, 2012; Brauch et al.,

Table 9.1   Basic data on Latin American and Caribbean countries

Country	UNFCCC National communications on climate change (year)					Population change and projections (millions)			
	1st	2nd	3rd	4th	5th	1950	2010	2050	2100
Mexico	1997	2001	2006	2009		27,866	113,423	143,925	127,081
Belize	2002	2011				0,069	0,312	0,529	0,555
Costa Rica	2000	2009				0,966	4,659	6,001	5,019
El Salvador	2000					2,200	6,193	7,607	6,783
Guatemala	2002					3,146	14,389	31,595	46,036
Honduras	2000					1,487	7,601	12,939	13,789
Nicaragua	2001	2011				1,295	5,788	7,846	7,261
Panama	2001					0,860	3,517	5,128	5,170
Meso America	8	4	1	1	1				
Argentina	1997	2008				17,150	40,412	50,560	49,201
Bolivia	2000	2009				2,714	9,930	16,769	20,021
Brazil	2004	2010				53,975	194,946	222,843	177,349
Chile	2000	2011				6,082	17,114	20,059	17,185
Colombia	2001	2010				12,000	46,295	61,764	58,137
Ecuador	2000					3,387	14,465	19,549	18,319
Guyana	2002					0,407	0,754	0,766	0,693
Paraguay	2002					1,473	6,455	10,323	11,364
Peru	2001	2010				7,632	29,077	38,832	35,911
Suriname	2006					0,215	0,525	0,614	0,551
Uruguay	1997	2004	2010			2,239	3,369	3,663	3,396
Venezuela	2005					5,094	28,980	41,821	40,507
South America	12	7	1						
Latin America	20	11	2	1					
Aruba	–					0,046	0,089	0,112	0,108
Bahamas	2001					0,049	0,343	0,445	0,449
Barbados	2001	2011				0,211	0,273	0,264	0,223
Cuba	2001					5,920	11,258	9,898	7,022
Dominican Republic	2003	2009				2,380	9,927	12,942	12,231
Grenada	2000					0,077	0,104	0,095	0,075
Haiti	2002					3,221	9,993	14,178	14,566
Jamaica	2000	2011				1,403	2,741	2,569	2,166
Puerto Rico						2,218	3,749	3,657	3,024
St. Vincent & Grenadines						0,067	0,109	0,113	0,096
St. Lucia						0,083	0,174	0,205	0,169
Trinidad & Tobago	2001					0,636	1,341	1,288	1,031

Table 9.1 (continued)

Country	UNFCCC National communications on climate change (year)					Population change and projections (millions)			
	1st	2nd	3rd	4th	5th	1950	2010	2050	2100
Caribbean	8	3	0	0					
LA & C	28	14	2	1					
USA (5th 2010)	1994	1997	2002	2007		157,813	310,384	403,101	478,026
Canada (5th '10)	1994	1997	2002	2007		13,737	34,017	43,642	48,290

Country	Economic data			Environmental and climate data		
	HDI 2011	GDP/capita (2010) current USD	Below poverty line (%)	Ecological Foot print 2007	CO_2 Emissions total 2008	CO_2 Emissions (tons/ cap.) 2008
Mexico	57	9,123	47.4	3.0	475,833.6	4.4
Belize	93	4,064	18.1		425.4	1.4
Costa Rica	69	7,691	21.7	2.7	8,016.1	1.8
El Salvador	105	3,426	37.8	2.0	6,112.9	1.0
Guatemala	131	2,862	51.0	1.8	11,914.1	0.9
Honduras	121	2,026	60.0	1.9	8,672.5	1.2
Nicaragua	129	1,132	46.2	1.6	4,330.7	0.8
Panama	58	7,589	32.7	2.9	6,912.3	2.0
Argentina	45	9,124	n.d.	2.6	192,378.2	4.8
Bolivia	108	1,979	60.1	2.6	12,834.5	1.3
Brazil	84	10,710	21.4	2.9	393,219.7	2.1
Chile	44	12,431	15.1	3.2	73,109.0	4.4
Colombia	87	6,225	45.5	1.9	67,700.2	1.5
Ecuador	83	4,008	36.0	1.9	26,824.1	2.0
Guyana	117	2,950	n.d.	n.d.	1,525.5	2.0
Paraguay	107	2,840	35.1	3.2	4,118.0	0.7
Peru	80	5,401	34.8	1.5	40,535.0	1.4
Suriname	104	6,254 (2009)	n.d.	n.d.	2,438.6	4.7
Uruguay	48	11,996	20.5	5.1	8,327.8	2.5
Venezuela	73	13,590	29.0	2.9	169,532.7	6.0
South America						
Latin America						
Aruba	n.d.		n.d.	n.d.		
Bahamas	53	21,985	n.d.	n.d.	2,156.2	6.4
Barbados	47	15,035	n.d.	n.d.	1,353.1	5.3
Cuba	51	5,565 (2008)		1.9	31,418.9	2.8

Table 9.1   (continued)

Country	Economic data			Environmental and climate data		
	HDI 2011	GDP/ capita (2010) current USD	Below poverty line (%)	Ecoloical Foot print 2007	CO_2 Emissions total 2008	CO_2 Emissions (tons/ cap.) 2008
Dominican Republic	98	5,215	50.5	1.5	21,617.0	2.2
Grenada	67	7,401	n.d.	n.d.	245.7	2.4
Haiti	158	671	77.0	0.7	2,434.9	0.3
Jamaica	79	5,274	9.9	1.9	12,203.8	4.5
Puerto Rico	US		US	US		US
St. Vincent & Grenadines	85	6,446	n.d.	n.d.	201.7	1.9
St. Lucia	82	6,884	n.d.	n.d.	396.0	2.3
Trinidad & Tobago	62	15,359		3.1	49,772.2	37.3
USA (5th 2010)	4	47,199	n.d.	8.0	5,461,013.7	17.3
Canada (5th '10)	6	46,236	n.d.	7.0	544,091.1	16.4

Sources: UNFCCC <http://unfccc.int/national_reports/non-annex_i_natcom/ items/2979.php> (14 January 2012); UNPD <http://esa.un.org/unpp/p2k0data.asp>; UNDP, HDI (2011) <http://hdr. undp.org/en/reports/global/hdr2011/>; World Bank, CO2 emissions (kt), at : <http://dat. worldbank.org/indicator/EN.ATM.CO2E.KT>; World Bank, GDP per capita (current US$), at: <http://data.worldbank.org/indicator/ NY.GDP.PCAP.CD>.

2008, 2009, 2011; Brauch and Scheffran, 2012) and has been taken up by scholars in LAC (Peralta, 2008; Oswald Spring, 2010, 2011). Three policy and scientific discourses have emerged on the linkage between impacts of climate change on security from the perspective of international, national and human security.

The reconceptualization of security (Buzan et al., 1998) has resulted in a 'widening' (environmental, societal, economic dimensions), 'deepening' (human, gender) and 'sectorialization' (energy, food, health, water, and livelihood) of security. The first global political discussion on human security occurred in Costa Rica in 1990 following the end of the Central American civil war crisis (Jolly and Ray, 2006, p. 4), before the human security concept was launched by the United Nations Development Programme (1994) that defined human security as 'safety from the constant threat of hunger, disease, crime and repression'. It also means 'protection from sudden and hurtful disruption in the pattern of our daily lives – whether in our homes, in our jobs, in our communities or in our environment'. By shifting the focus to a 'human-centred' perspective, the security concept was deepened and the focus on the state was complemented by adding human beings, communities and society as new referent objects.

In this chapter two pillars are examined for climate change impacts: 'freedom from hazard impacts' (environmental vulnerability: EV) and 'freedom from want' (social vulnerability and poverty eradication) that directly affect the coping capacity of people, communities and states. While the IPCC (2007) did not define the term 'coping', Brauch and Oswald Spring (2011, p. 41) use it 'to embrace the three concepts of adaptation, mitigation and resilience-building.'

2.1 Climate Change and International Security: Discourse in the United Nations and Within the EU-LAC Strategic Partnership

In April 2007, the linkage between climate change and international security was tabled by the United Kingdom in the United Nations Security Council (UNSC, 2007). In June 2009 the Small Island Developing States (SIDS) brought a resolution to the UN General Assembly calling the UN Secretary-General (UNSG, 2009) to submit a report on climate change and security linkages (Brauch and Oswald Spring, 2011; Brauch and Scheffran, 2012). On 20 July 2011, a 'Statement by the President of the Security Council' (S/PRST/2011/15) remarked on the climate change and security linkage (UNSC, 2011).

In LAC countries the debate is slowly emerging in its dialogue with Europe, discussing 'environment/climate change/energy' and the 'EU humanitarian aid to LAC'. A document of the European Commission (2008) on the EU-LAC Strategic Partnership stated that climate change 'poses a threat to economic growth and the successful implementation of poverty reduction strategies'. Prior to the Madrid EU-LAC summit in 2010 a communication of the European Commission (2009) stressed the importance of biregional cooperation on climate change. The VI EU-LAC Summit in Madrid adopted the Madrid Declaration, in which EU and LAC government representatives identified among the priority areas, 'science, research, innovation and technology; sustainable development; environment; climate change; biodiversity; energy; regional integration and interconnectivity', and the Madrid Action Plan (2010–2012) addressed cooperation on climate change.[1]

The UN Secretary-General in his second report on human security (A/66/763, 5 April, 2012) referred to UN activities on climate change and climate-related hazard events, where the human security approach can be useful (UNSG, 2012). The UN member states argued that 'climatic fluctuations and extreme weather patterns disrupt harvests, deplete fisheries, erode livelihoods and increase the spread of infectious diseases' and that the convergence with other trends 'can result in social stresses with far-reaching implications for national, regional and international stability'. The UN Secretary-General argued that the human security approach can contribute to 'improved early warning systems, more resilient coping mechanisms and better tailored adaptation strategies to the specific needs and vulnerabilities of the people.'

2.2   Emerging Latin America and Caribbean Discourses on Climate Change and Human Security

During the UN General Assembly debate on human security in 2008, Mexico, Brazil and Cuba referred to climate change as a threat to human security, while Chile pointed to natural disasters. During the UNGA debate on human security on 14 April, 2011, Costa Rica's Ambassador Sonia Picado pointed to severe human security threats in LAC:

> The [LAC] region is the most unequal region in the world [and] the most violent. . . . A human security focus allows us to appreciate the relationship between the two challenges. . . . The State is responsible for the human security of its citizens and it must be the State that takes the leading role in these projects. . . . The human security concept is particularly important for Latin America.

Brauch (2011a, 2005, 2005a) proposed a third pillar of human security as 'freedom from hazard impact' (Bogardi and Brauch, 2005), arguing that climate change[2] 'directly impacts on water, soil, food, health and livelihood security' and that climate change 'will exacerbate these sectoral security problems, if the communities and social groups fail to create mitigation and adaptation strategies with resilience-building through preventive learning and decisions'. Mexican Ambassador Claude Heller stated that human security 'involves different issues ranging from terrorism to violence in conflict all the way to climate change and natural disasters', adding that in some countries 'immigration is seen as a threat to human security'.

On 20 July, 2011, during the UN Security Council meeting on climate change and security, 14 LAC countries participated. Argentina (Group of 77 and China Chair in 2011), Barbados, Bolivia, Colombia, Costa Rica, Cuba, Peru, and Venezuela supported the G-77 and China Group that opposed the linkage between climate change and international security, arguing that climate change, as an issue of sustainable development, should be considered by the UN General Assembly, ECOSOC, UN Committee on Sustainable Development, UN Environment Programme and UNFCCC and not by the UN Security Council. The perception of high vulnerability to climate change amongst the countries of Mexico, Central America and the Caribbean may explain why their positions on security and climate change differed significantly from the South American states.

In LAC countries, the debate on climate change as a human security challenge is incipient. This is demonstrated by representatives of international organizations, mayors and social movements, and by academics at conferences (Oswald Spring, 2011). Mayors from across LAC addressed climate change impacts on human security and demanded the reduction of the vulnerability of communities and ecosystems due to various environmental, food, health and other threats that affect their sustainable development in August, 2011, in a 'Manifesto de Valparaiso'.[3] In May, 2011, the Union of South American Nation's (UNASUR) South American Defence Council inaugurated the new Defence Strategic Studies Center, which looks at various issues including the protection of strategic energy and food resources and adapting to climate change.[4]

3. VULNERABILITY TO CLIMATE CHANGE: IMPLICATIONS FOR HUMAN SECURITY, POVERTY AND INEQUALITY IN LATIN AMERICA AND THE CARIBBEAN

How vulnerable is Latin America and the Caribbean? From a human security perspective Bohle (2001, 2002) combined environmental vulnerability and social vulnerability in a dual context of global environmental change and globalization. A human security approach implies a reduction of environmental and social vulnerability by combined policy efforts to achieve 'freedom from hazard impacts' and 'freedom from want'. Here, we review briefly some indicators from the region on its physical vulnerability, its social vulnerability, likely impacts, and then turn to discussions on 'freedom from hazard impact' in the UN, and how they might be merged with the long human security discussion on 'freedom from want.'

Concepts of environmental vulnerability are widely employed in global environmental change, climate and hazard research, early warnings studies (Brauch, 2011) and in policy-making (Mesjasz, 2011). The Prevalent Vulnerability Index (PVI) offers an overview for LAC countries (Figure 9.2) that focuses on social, economic, institutional, and infrastructural capacity to recover from natural hazards (Cardona, 2007, 2011), and

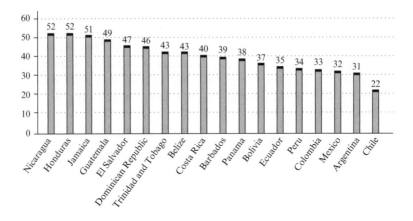

Source: IDB, at: <http://www.iadb.org/en/news/web-stories/2010-09-30/idb-natural-disaster-risks-in-latin-america-and-caribbean.8017.html>.

Figure 9.2   Prevalent Vulnerability Index (2007)

depicts predominant vulnerability conditions by measuring exposure and susceptibility in prone areas, socio-economic fragility and lack of social resilience.

Mexico and Central America have been affected by the physical and societal effects of global environmental change and are a major environmental hotspot. Mexico offers a microcosm for the analysis of complex interactions between the earth and human systems and their effects resulting in environmental scarcity, degradation and stress, natural hazard impacts (e.g. drought and hurricanes), societal outcomes, repeated domestic crises, and water and soil conflicts with public insecurity (Oswald Spring, 2010, 2011). While urbanization, internal displacement and migration to the US have been intensively studied, there are only few empirical studies on environmentally forced migration in Mexico (Alscher, 2009; Sánchez et al., 2013; Oswald Spring, 2012; Oswald Spring et al., 2013). Black et al. (2011, pp. 432–433) referred to contradictory results on the link between climate change and migration in Mexico.

According to Oswald Spring (2008, pp. 24–25) 'Social vulnerability is a concept related to unsatisfied human needs and limited access to resources which results in the loss of human security.' Mesjasz (2011) found the term 'social' ambiguous, referring to 'the characteristics of a person or group and their situation that influence their capacity to anticipate, cope with, resist and recover from the impact of a natural hazard' or 'to the inability of people, organizations, and societies to withstand adverse impacts from multiple stressors to which they are exposed'.

Pizarro (2001, p. 11) defines social vulnerability in terms of 'insecurity and defencelessness experimented by communities, families, and individuals in their livelihoods as a consequence . . . of a socio-economic event of traumatic character; and the second component is the management of resources and strategies which are utilized by these communities, families, and individuals to cope with the effects of this event'. ECLAC and IDB define vulnerability as 'the probability of a community, exposed to a natural hazard, given the degree of fragility of its elements (infrastructure, housing, productive activities, degree of organization, warning systems, political and institutional development), to suffer human and material damages' (Villagrán de León, 2006, p. 15).

Reducing poverty can reduce vulnerability. ECLAC (2011b, 2011c) analyzed the drop of poverty rate in Latin America (1990–2010) by 17 percentage points (from 48.4 to 31.4 percent), while the rate of extreme poverty fell by 10 percent (from 22.6 to 12.3 percent). The decline in both rates is mainly due to an increase in wages as well as the contribution of public money transfers, but to a much lesser extent. The PVI and poverty rate are crucial to understand how two pillars of human security may be affected.

The number and intensity of climate-related hazards in LAC increased considerably from 1970 to 2010, affecting millions of people (Figure 9.3) and resulting in huge damages and economic costs, according to ECLAC (2011, p. 1) from US$23 816000 (1971–80), to US$67 108000 (1981–90), to US$212 194000 (1991–2000) and to US$446 256000 (2001–10). The International Disaster Database (EM-DAT) has recently defined (Guha-Sapir and Vos, 2011, pp. 712–717) three relevant categories of climatological (heat/cold wave, drought, forest/land fire), hydrological (floods, landslides, avalanches, subsidence) and meteorological (tropical/winter storms) hazards (ECLAC, 2011, p. 3).

The number of these hazards and their intensity have increased (ECLAC, 2011) and may increase further in intensity (IPCC, 2007 2012), depending on GHG emission reductions and national adaptation and mitigation strategies and their effective implementation. Victims and economic costs may be reduced by resilience, early warning systems and disaster risk reduction.

A human security approach to the physical effects of climate change in LAC could help address the dual vulnerability as 'freedom from hazard impact' by reducing environmental vulnerability and social vulnerability through enhancing coping capabilities of affected societies. 'Freedom from want' implies protecting and empowering people, and reducing their social vulnerability due to poverty and socio-gender inequality. Both vulnerabilities and policy measures for addressing human security

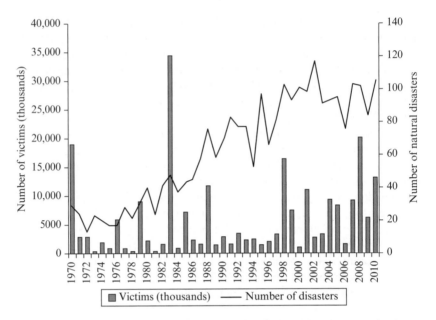

Source: ECLAC (July 2011: 3) based on EM-DAT: OFDA and CRED International Disaster Database, at: <www.emdat.be> (16 January 2012).

Figure 9.3   Number of natural disasters and number of people killed ('victims') in the Americas (1970–2010)

are closely interdependent referring to different policy agendas (Brauch, 2009a, p.983).

'Freedom from hazard impact' would imply that people can mobilize their resources to address sustainable development goals for achieving hazard-specific policies and a combination of technical, organizational and political measures in case of slow-onset (sea-level and temperature rise, droughts) and rapid-onset hazards (storms, floods, forest-fires, landslides). While hazards and environmental vulnerability/social vulnerability lead to a direct deterioration of human security, strategies must reduce these two vulnerabilities by enhancing coping capacities. To respond to these complex human security problems, primarily proactive non-military measures are needed.

The UN Development Programme (1994) introduced this human security pillar, the Commission on Human Security (CHS, 2003, 2005) developed it further, and Kofi Annan (2005) referred to a 'global partnership for development',[5] but the realization of human security goals was seen as rather disappointing (Busumtwi-Sam, 2008, p.92). A significant change would require reforms of existing power structures and institutions, especially macroeconomic structures and distribution mechanisms. 'Freedom from want' requires LAC countries to reduce poverty, social vulnerability and social inequality with gender perspective.

4. LATIN AMERICA AND CARIBBEAN COUNTRIES' ADAPTATION STRATEGIES AND INITIATIVES

All countries party to the UNFCCC are required to submit national communications that include information on emissions and the removal of greenhouse gases (GHGs), details of the activities countries are undertaking to implement the Convention, and vulnerability assessments.[6] Of 33 LAC countries (Table 9.1), 30 states tabled at least one national communication (NC), 15 submitted two, Uruguay three and Mexico is unique in having contributed five to the UNFCCC. All LAC countries have no legal quantitative emission reduction obligations under the Kyoto Protocol (1997) but a number of them such as Brazil, Mexico, Costa Rica, Ecuador and Peru and the member states of the Caribbean Community (CARICOM) have voluntarily developed targets, policies, initiatives and laws to reduce their emissions and attempt to confront the impacts of climate change.[7]

Zapata-Martí (2011, p. 1341) suggests that the increasing cost of adaptation is not being incorporated into the region's development policies, thus increasing the vulnerability of both its productive and social investments. LAC countries are required to balance concerns about climate change and adaptation with competing social, in particular poverty and inequality, and environmental priorities. In this chapter we refer to key literature citing the importance of reducing poverty and inequality as a tenet of adaptation to climate change.

According to the Intergovernmental Panel on Climate Change (IPCC, 2001) adaptation is defined as the 'adjustment in natural or human systems in response to actual or expected climatic stimuli or their effects, which moderates harm or exploits beneficial opportunities'. Adaptation is a key priority for the LAC region, as climate change impacts threaten to undermine long-term efforts in the region to achieve sustainable development (Wilk, 2010). According to a series of questionnaires conducted under the auspices of EuropeAid (2009) with national governments, Central American countries perceive adaptation as the principal problem and that agriculture and food security are among the most severely affected. South American countries also commented that adaptation is an issue of national priority (EuropeAid, 2009).

Investment in adapting to climate change must be a priority for economic and social development (ECLAC/IDB, 2010, p. 12). Mata and Nobre (2006) suggest that government interventions are needed in many Latin American countries, as various communities and sectors of the economy do not have the required financial and technical resources to bring about adaptation activities. Types of interventions vary according to national circumstances, but may include activities such as disaster prevention, integrated coastal zone management, health care planning, building codes, solid and liquid waste management and the use of traditional knowledge.

Sustainable development plans should include adaptation strategies to enhance the integration of climate change into development policies. Some LAC countries are making efforts to adapt, including through conservation of key ecosystems, early warning systems, drought and coastal management, and disease surveillance systems. However, a lack of basic information and observation and monitoring systems, insufficient capacity and appropriate political, institutional and technological frameworks, low income, and settlements in vulnerable areas undermine these efforts (Magrin et al., 2007).

Across South America, national governments have identified sectors vulnerable to climate change and corresponding adaptation priorities. They have also developed policies and strategies to manage adaptation efforts and are engaged in the implementation of country-level and multi-country adaptation projects, many of which address areas related to agriculture, fisheries, forestry and water, biodiversity, human health, coastal zones and disaster risk management. Adaptation needs in the energy and infrastructure sectors have received little attention. Most projects focus primarily on building capacity and policy research. Implementation of adaptation action is still incipient, although a few initiatives have been launched in various countries (Keller et al., 2011).

A prominent intergovernmental initiative involving all the Spanish- and Portuguese-speaking countries of Latin America and the Caribbean and also Spain, as the principal funder, and Portugal, is the Ibero-American Network of Climate Change Offices (RIOCC), which was created in 2004 by the environment ministers of each participating country. RIOCC is recognized by the Nairobi Work Programme on Impacts, Vulnerability and Adaptation to Climate Change (NWP) as an official partner organization and it also provides a regional platform for knowledge exchange, capacity building and the promotion of regional adaptation projects (RIOCC, 2008).

The Ibero-American Program on Adaptation to Climate Change (PIACC) is a leading program of RIOCC. The PIACC was created in 2005 and specifically aims to improve institutional frameworks; harmonize activities with regional adaptation initiatives and agencies; support climate research; and promote the exchange of knowledge, tools and methods for evaluating impacts, vulnerability and adaptation. Through the PIACC, Spain supports a number of regional activities such as capacity building workshops on climate scenarios as well as on the integration of adaptation into development policies and projects. A cooperation agreement was signed with UN International Strategy for Disaster Reduction in 2008 in order to promote the exchange of know-how and experiences between disaster risk reduction and adaptation to climate change, and to integrate both issues within regional and UNFCCC processes. Under this agreement, capacity building, institutional strengthening, coordination and communication activities have been pursued in various RIOCC countries (RIOCC, 2008).

The 2008 Lima Declaration agreed to launch the joint EU-LAC Environment programme, Euroclima, with the aim to provide LAC decision-makers and the scientific community with better knowledge of climate change and its consequences. The three-year programme was approved at the end of 2009 and started in May 2010 with EuropeAid at the helm with a budget of €5 million ($6.9 million). It co-financed the 'Review of the Economics of Climate Change in South America' project. Among the programme's objectives include: reducing people's vulnerability to the effects of climate change and reducing social inequalities and the socio-economic impact of climate change through cost-efficient adaptations. Expected results include enhanced policy dialogue on climate change issues, and improved sharing of information and data on scientific and socio-economic matters related to climate change.[8]

The UNDP Community-Based Adaptation Programme's activities in LAC focus on Bolivia, Guatemala and Jamaica with the objective of enhancing adaptive capacity, which allows communities to reduce their vulnerability to adverse impacts of future climate hazards. The programme is funded by the Global Environment Facility Small Grants Programme as part of Strategic Priority on Adaptation. The funding budget is US$5.5 million from GEF and US$4.5 million in co-financing for the whole of the Community-Based Adaptation Program includes 10 countries in total. The implementation period was 2007–11. In Bolivia, the program focuses on rural livelihoods and ecosystems in the context of water, agriculture and health; in Guatemala on community-based natural disaster risk reduction activities in rural communities; and in Jamaica on support for adaptation both in coastal regions and in the agricultural sector, focusing on improved natural resource management.[9]

Despite the progress made, the gap between current adaptation actions and what is urgently required remains problematic. These major gaps include insufficient attention in the areas of human health, water, coastal zone management, the maintenance of ecosystem services, forestry and fisheries. Adaptation action related to the hydroelectric sector requires greater attention, as well as urban adaptation for the region's vast urban areas and the gender implications of climate change. A substantial amount of the ongoing adaptation programming in the region remains focused on assessment, research, capacity building and knowledge communication, and there is a need to direct more attention toward practical action (Keller et al., 2011).

5. FINANCING ADAPTATION MEASURES IN LATIN AMERICA AND THE CARIBBEAN

Protecting the health and livelihoods of Latin American and Caribbean citizens requires efforts to assist recovery from previous natural disasters, and planning to reduce the impact of future events and the adverse effects of climate change impacts. Probably most adaptation to climate change is autonomous and self-financing, being done by households, institutions, and local and national governments. However specialists, NGOs and developing country negotiators argue that much adaptation work will require support from international agencies. Estimates of the possible global cost for adaptation in developing countries range above US$80 billion per year to 'climate proof' international development efforts (World Bank, 2009).

In the UNFCCC negotiations, since 1992 there have been promises of new and additional funding from the wealthier countries to the poorer ones who need to address the problem of climate change which they did not create. The 1997 Kyoto Protocol established the Clean Development Mechanism (CDM), which was designed to bring sustainable development funding and technology transfer to developing countries by allowing wealthier nations to purchase credits from emissions-reduction projects funded there (its success in this regard is questionable and its future is in doubt). Then in 2001 in Marrakesh, UN negotiators created two funds for voluntary contributions (the Least Developed Countries Fund or LDCF, and the Special Climate Change Fund), and the Adaptation Fund under the Kyoto Protocol, which was to be funded by a 2 percent levy on the proceeds from the CDM. This fund promised significant revenues and a governance structure more balanced to the desires of developing country Parties.

The LDCF was limited to the world's 48 poorest nations (of which only Haiti is a member from LAC). Under the Marrakesh Accords, the LDCs were each funded with US$300000 to prepare 'National Adaptation Programmes of Action,' and to produce a list of top priority projects required to adapt. Most countries produced their NAPAs in 2006–08; as the region's only LDC, Haiti's was submitted in December 2006 and defined eight priority activities in its NAPA. As its first project under the LDCF, Haiti developed a project addressing the adaptive capacities of coastal communities consisting of four components: 1. Systemic, institutional and individual capacity development; 2. A sustainable financial framework for Climate Risk Management (CRM) in coastal areas; 3. Piloting of on-the-ground coastal adaptation measures; 4. Knowledge management, codification of best practices and dissemination.[10] After the success of the NAPA programme, the idea was extended to other nations above the LDCs in income, such as the LAC countries. Most adaptation projects funded by international agencies have been in the area of planning, not in concrete adaptation actions, a trend identified earlier (Roberts et al., 2008) but which appears to be continuing today.

The Global Environment Facility (GEF), created to support the efforts of the 1992 Rio Earth Summit, was slow in initiating funding for adaptation because its charter required that it only pay for 'global public goods.' For adaptation, this suggested that only the 'incremental costs' of making a development project climate-proof – above what would take place during normal implementation – would be paid by international climate funds. For example, if flooding was expected to become more frequent or higher because of climate change, then the incremental cost of adaptation would be the cost of raising the road the few feet above current flooding levels to deal with expected new impacts. Donor nations resisted spending climate funds for anything but the costs that were clearly new and due to climate change. Developing countries argued that proving such additionality is nearly impossible (Roberts and Parks, 2007), and subsequently the rules have been somewhat relaxed.

LAC has received US$544 million from the GEF for climate change related activities via organizations such as the World Bank, IDB, UNDP and UNEP, while US$2514 million of GEF co-financing has been mobilized. About half of these totals were allocated to capacity-building, national communications and adaptation, while the other half was earmarked for mitigation and energy efficiency projects (Samaniego, 2009).

Other multilateral and bilateral agencies have begun to fund adaptation projects in the LAC region. Among multilaterals the Inter-American Development Bank, the International Development Research Centre, Oxfam, MDG Achievement Fund, UNDP and the World Bank are listed as top funders in South America by Keller et al. (2011, p. 32). Top bilateral donors were the European Union, Germany, Spain, Switzerland and the

US (both of these lists are in alphabetical order). Medeiros et al. (2011) report that national, regional and global adaptation projects being implemented in the Caribbean region have been funded by the GEF, the IDB, the UNDP and the governments of Spain, the UK and US.

The Brazilian National Climate Change Fund provides direct financial support to finance research that addresses adaptation priorities. Moreover, the second Working Group of the Brazilian Panel on Climate Change will provide a synthesis of climate change vulnerability, impacts and adaptation options in the First National Assessment Report, which is anticipated to be released to the public in 2012 (Keller et al., 2011). Redesigning and rebuilding infrastructure is expected to account for the largest share of adaptation costs.

The most vulnerable sectors that require special attention in the adaptation agenda are agriculture and forest resources, water resources, energy and transportation infrastructure, tourism, health, urban development, and disaster risk management (Wilk, 2010). According to estimates, to close these gaps and build infrastructure able to withstand the rigors of climate change, Latin America needs to spend the equivalent of 6 percent of its GDP on infrastructure (Moreno, 2011). One estimate is that the financing requirement for adaptation in LAC's agricultural sector is on the order of US$1.2 billion per year from the present to 2050 (IFPRI, 2009).

The Inter-American Development Bank adopted on March 2011 its 'Integrated Strategy for Climate Change Adaptation and Mitigation, and Sustainable and Renewable Energy' which will enable IDB to meet its lending target (25 percent) for activities related to climate change, renewable energy and environmental sustainability under the Bank's Ninth General Capital Increase. Priorities under the Plan of Action include adaptation.

The IDB and ECLAC came together to argue that LAC governments will have to work with public and private banks and especially with the private sector and civil society to optimize the impact of the international funds obtained for their main initiatives (ECLAC/IDB, 2010, p. 9).

6.   CONCLUSION

Latin America and the Caribbean are highly vulnerable to climate change. Even in the event that an ambitious global climate change agreement can be achieved by 2015 and implemented by 2020, the negative impacts of climate change will be unavoidable. At present, adaptation efforts in the region are currently dispersed, considerably underdeveloped and unfunded. LAC countries are not sufficiently prepared to confront the impacts of climate change due to deficits in technical and scientific knowledge. The region needs to continue exploring how best to advance national debates linking climate change to national interests while working on adaptation plans, and building on reliable data about potential losses in capital stock, infrastructure, food security, trade and the country's natural resources, all of which could have adverse effects on the human security of LAC societies (Garibaldi et al., 2012). Without greater commitment to reducing its vulnerability, the region's sustainable development will be seriously compromised, adversely affecting, among other things, its ability to reach all the Millennium Development Goals (Magrin et al., 2007).

Given the level of vulnerability in the region, and the strong imbalance between private investment, loans and grants going towards climate change mitigation rather than adaptation, LAC countries will require greater attention to their adaptation needs at the regional, national and international levels. The IDB's leadership in committing a substantial portion of its funding to go to climate change reflects an important shift, as do regional and national assessments on vulnerability and adaptation options; and incipient adaptation policies and actions. However, the next steps will require launching protective adaptation measures. National contexts matter, as do broader conceptions of what adaptation actually is in practice.

Significant effects from climate change are observable in LAC, making it essential to reduce risks to those most exposed. These will be most effective if combined with efforts to reduce poverty levels and inequality in the region (UNEP/ECLAC, 2010). Human development is the most secure foundation for adaptation to climate change, and policies that promote inclusive and equitable growth, expand opportunities in health and education, provide insurance for vulnerable populations, improve disaster management, and support post emergency recovery all enhance the resilience of poor people facing climate risks. That is why adaptation planning should be seen not as separate from, but rather as an integral part of wider strategies for poverty reduction and human development (UNDP, 2007).

Eakin and Lemos (2006) argue that the benefits of globalization for the adaptive capacity of Latin American governments are unlikely to be easily secured. They suggest the inability of the reconfigured national state to tackle the growing social and political inequality is central to the vulnerability problem, and that as long as inequality persists, it is unlikely that increased vulnerability and low adaptive capacity among the poor in Latin America will change. Attempts to facilitate adaptation to climate change and build adaptive capacity among the most vulnerable groups can only be secured through policy reform and by implementing re-distributive policy such as eliminating bureaucratic excesses, and enhancing access to information, knowledge and technology (Eakin and Lemos, 2006).

From a human security perspective, adaptation policies for LAC require a 'sustainability transition' (Grin et al., 2010). The National Communications of LAC countries have outlined adaptation projects that may enhance the human security of people by contributing to 'freedom from want' and 'freedom from hazard impact'. However, the best human security strategy includes reduction of greenhouse gas emissions, principally by the world's largest emitters, and building resilience based on bottom-up initiatives reinforced by top down governmental support.

The G-20 Leaders' Summit in Los Cabos in 2012 adopted a declaration that avoided any specific obligations. Leaders committed to maintain a focus on inclusive green growth as part the G-20 agenda, reiterated their commitment to fight climate change and expressed their commitment to the full implementation of the outcomes of Cancun and Durban climate talks and supported the operationalization of the Green Climate Fund.[11] In the 'Policy Commitments by G-20 Members' climate change was not even mentioned. Within the financial track a study group had discussed 'the most effective ways to mobilize resources for the fight against climate change' but no binding commitments were included in the G-20 Leaders Declaration.

Therefore, besides these policy declarations, the political will is lacking to replace the prevailing business-as-usual policies with a true 'transition towards sustainability'. This requires a fundamental change from carbon intensive production and consumption to a drastically reduced and in some cases carbon free and dematerialized system, including a transformation of the worldview and dominant mindset of policy-makers away from short-sighted economic interests. The later this fundamental turn in policies occurs, the higher the costs and disruptions will become (Stern, 2006). LAC countries need to be making long-term plans with climate change in mind, even as the global community sends mixed signals about its commitment to address this mounting problem.

To avoid catastrophic climate change impacts for LAC countries, a business-as-usual strategy will not solve the 'climate paradox' wherein those countries suffering most are those that did the least to cause the problem. Policy declarations without willingness and obligation to implement them, and the postponement of creating an ambitious, fair and legally binding climate treaty until 2015 and implemented by 2020, will probably result in more intensive climate-related hazards, causing further human losses and economic damage.

To facilitate a fundamental transformation towards a fourth 'sustainability revolution' (Oswald Spring and Brauch, 2011) a new 'social contract for sustainability' (Clark et al., 2004; WBGU, 2011) is required relying on 'freedom from want' and 'freedom from hazard impact' for human beings. Such a humanitarian policy and societal vision requires linking the long-term goals of 'sustainable development' and human security that may contribute to significantly reduce the impacts of climate change and enhance human security in Latin America and the Caribbean.

NOTES

1. See Council of the European Union, VI EU-LAC Madrid Summit, 18 May, 2010, Madrid Declaration, at: <http://www.consilium.europa.eu/uedocs/cms_Data/docs/pressdata/en/er/114535.pdf> and the Madrid Action Plan, 18 May, 2010 at: <http://www.consilium.europa.eu/uedocs/cms_Data/docs/pressdata/en/er/114540.pdf>.
2. See at: <http://www.un.org/News/Press/docs//2011/ga11072.doc.htm>.
3. See at: <http://www.eclac.cl/rio20/noticias/paginas/5/43755/Manifiesto_Valparaiso_2011_rev.pdf>.
4. MercoPress 'Unasur defence strategic studies centre opens this week in Buenos Aires', 23 May, 2011: <http://en.mercopress.com/2011/05/23/unasur-defence-strategic-studies-centre-opens-this-week-in-buenos-aires>. Accessed 15 May, 2012.
5. See Annan (2005): Executive Summary, at: <http://www.un.org/largerfreedom/summary.html>.
6. See UNFCCC National Reports: <http://unfccc.int/national_reports/items/1408.php>.
7. The LAC region is a relatively minor contributor to global GHG emissions accounting for approximately 11% of the global total in 2011. However, the fact that in absolute terms the region accounts for a small amount of emissions does not relieve it of its global responsibilities. On a per capita basis and in proportion to the size of its economies, the region contributes more GHG emissions than do other developing countries, such as China and India (ECLAC/IDB, 2010). In this chapter, we focus on adaptation taking into account LAC's modest contribution to global GHG emissions, which render the countries of the region considerably less responsible for global warming than the rich industrialized countries. Although all countries should embark on a pathway towards low carbon resilient growth and reduce emissions, mitigation actions in LAC are having little impact given the region's low contribution to global emissions. As a result, the relevance of adaptation actions in the region is considerably more pertinent in relation to human security than mitigation actions.
8. EuropeAid (2012), 'Euroclima', at: <http://www.euroclima.org/> accessed 25 April 2012.
9. UNDP Climate Change website: <http://www.undp.org/content/undp/en/home/ourwork/environmentandenergy/strategic_themes/climate_change.htmlkm=1>>
10. See Haiti experiences with the NAPA process: <http://unfccc.int/cooperation_support/least_developed_countries_portal/items/6500.php>.
11. See G20 Leaders' Declaration. <http://g20.org/images/stories/docs/g20/conclu/G20_Leaders_Declaration_2012_1.pdf>.

REFERENCES

Alscher, S. (2009), 'Environmental factors in Mexican migration: the cases of Chiapas and Tlaxcala', EACHFOR EU Project, Mexico Case Study Report, Bielefeld: University Bielefeld, 30 January.

Annan, Kofi A. (2005), *In Larger Freedom: Towards Security, Development and Human Rights for All. Report of the Secretary General for Decision by Heads of State and Government in September 2005*, A/59/2005, New York: United Nations, Department of Public Information, 21 March.

Black, Richard, Kniveton, Dominic, Schmidt-Verkerk, Kerstin (2011), 'Migration and climate change: towards an integrated assessment and sensitivity', *Environment and Planning A*, **43**, 431–450.

Bogardi, J., and H.G. Brauch (2005), 'Global environmental change: a challenge for human security – defining and conceptualising the environmental dimension of human security', in: Rechkemmer, A. (ed.), *UNEO – Towards an International Environment Organization – Approaches to a Sustainable Reform of Global Environmental Governance*, Baden-Baden: Nomos, pp. 85–109.

Bohle, H. G. (2001), 'Vulnerability and criticality: perspectives from social geography', in: *IHDP Update*, 2/01, 3–5, <http://www.ihdp.uni-bonn.de/html/publications/update/IHDPUpdate01_02.html>.

Bohle, H. G. (2002), 'Land degradation and human security', in: Plate, Erich (ed.), *Environment and Human Security, Contributions to a Workshop in Bonn*, Bonn, Germany.

Brauch, H. G. (2002), 'Climate change, environmental stress and conflict – AFES-PRESS Report for the Federal Ministry for the Environment, Nature Conservation and Nuclear Safety', in: Federal Ministry for the Environment, Nature Conservation and Nuclear Safety (eds), *Climate Change and Conflict. Can Climate Change Impacts Increase Conflict Potentials? What is the Relevance of this Issue for the International Process on Climate Change?*, Berlin: BMU, pp. 9–112.

Brauch, H. G. (2005), *Environment and Human Security: Freedom from Hazard Impacts*, InterSecTions 2, Bonn, Germany: UNU-EHS.

Brauch, H. G. (2005a), *Threats, Challenges, Vulnerabilities and Risks of Environmental and Human Security*, UNU-EHS, Source 1, Bonn, Germany: UNU-EHS.

Brauch, H. G. (2008), 'Conceptualising the environmental dimension of human security in the UN', in: *Rethinking Human Security. International Social Science Journal Supplement*, 57, Paris, France: UNESCO, pp. 19–48.

Brauch, H. G. (2009), 'Securitizing global environmental change', in: H. G. Brauch, Ú. Oswald Spring, J. Grin, C. Mesjasz, P. Kameri-Mbote, N. Chadha Behera, B. Chourou and H. Krummenacher (eds), *Facing Global Environmental Change: Environmental, Human, Energy, Food, Health and Water Security Concepts*, Berlin: Springer, pp. 65–102.

Brauch, H. G. (2009a), 'Human security concepts in policy and science', in: H. G. Brauch, Ú. Oswald Spring, J. Grin, C. Mesjasz, P. Kameri-Mbote, N. Chadha Behera, B. Chourou and H. Krummenacher (eds), *Facing Global Environmental Change: Environmental, Human, Energy, Food, Health and Water Security Concepts*, Berlin: Springer, pp. 965–990.

Brauch, H. G. (2011), 'Security threats, challenges, vulnerabilities and risks in US national security documents (1990–2010)', in H. G. Brauch, Ú. Oswald Spring, C. Mesjasz, J. Grin, P. Kameri-Mbote, B. Chourou, P. Dunay and J. Birkmann (eds) (2011), *Coping with Global Environmental Change, Disasters and Security: Threats, Challenges, Vulnerabilities and Risks*, Hexagon Series on Human and Environmental Security and Peace, vol. 5, Berlin: Springer-Verlag, pp. 249–274.

Brauch, H. G. (2011a), 'The environmental dimension of human security: freedom from hazard impacts', Presentation to the Informal Thematic Debate of the 65th Session of the United Nations General Assembly on Human Security, 14 April 2011, <http:// www.un.org/en/ga/president/65/ initiatives/Human%20Security/DrBrauch.pdf>.

Brauch, H. G. (2012), 'Policy responses to climate change in the Mediterranean and MENA region during the Anthropocene', in: J. Scheffran, M. Brzoska, Brauch, H. G., P. M. Link and J. Schilling (eds), *Climate Change, Human Security and Violent Conflict: Challenges for Societal Stability*, Berlin: Springer, pp. 719–794.

Brauch, H. G. and J. Scheffran (2012), 'Introduction: climate change, human security, and violent conflict in the Anthropocene', in: J. Scheffran, M. Brzoska, Brauch, H. G., P. M. Link and J. Schilling (eds), *Climate Change, Human Security and Violent Conflict: Challenges for Societal Stability*, Berlin: Springer, pp. 3–40.

Brauch, H. G. and Ú. Oswald Spring (2011), 'Introduction: coping with global environmental change in the Anthropocene', in H. G. Brauch, Ú. Oswald Spring, C. Mesjasz, J. Grin, P. Kameri-Mbote, B. Chourou, P. Dunay and J. Birkmann (eds) (2011), *Coping with Global Environmental Change, Disasters and Security: Threats, Challenges, Vulnerabilities and Risks*, Hexagon Series on Human and Environmental Security and Peace, vol. 5, Berlin: Springer-Verlag, pp. 31–60.

Brauch, H. G., Ú. Oswald Spring, C. Mesjasz, J. Grin, P. Dunay, N. Chadha Behera, B. Chourou, P. Kameri-Mbote and P.H. Liotta (eds) (2008), *Globalisation and Environmental Challenges: Reconceptualising Security in the 21ˢᵗ Century*, Berlin: Springer.

Brauch, H. G., Ú. Oswald Spring, J. Grin, C. Mesjasz, P. Kameri-Mbote, N. Chadha Behera, B. Chourou and H. Krummenacher (eds), *Facing Global Environmental Change: Environmental, Human, Energy, Food, Health and Water Security Concepts*, Berlin: Springer.

Brauch, H.G., Ú. Oswald Spring, C. Mesjasz, J. Grin, P. Kameri-Mbote, B. Chourou, P. Dunay and J. Birkmann (eds) (2011), *Coping with Global Environmental Change, Disasters and Security: Threats, Challenges, Vulnerabilities and Risks*, Hexagon Series on Human and Environmental Security and Peace, vol. 5, Berlin: Springer-Verlag.

Busumtwi-Sam, J. (2008), 'Menschliche Sicherheit und Entwicklung', in: Cornelia Ulbert and Sascha Werthes (eds), *Menschliche Sicherheit. Globale Herausforderungen und regionale Perspektiven*, Baden-Baden, Germany: Nomos, pp. 81–93.

Buzan, B., O. Wæver and J. de Wilde (1998), *Security. A New Framework for Analysis*, Boulder – London: Lynne Rienner.

Cardona, O. D. (2007), *Indicators of Disaster Risk and Risk Management*, Summary Report, Updated Version, June, Washington DC, USA: Inter-American Development Bank.

Cardona, O. D. (2011), 'Disaster risk and vulnerability: concepts and measurement of human and environmental insecurity', in H. G. Brauch, Ú. Oswald Spring, C. Mesjasz, J. Grin, P. Kameri-Mbote, B. Chourou, P. Dunay and J. Birkmann (eds) (2011), *Coping with Global Environmental Change, Disasters and Security: Threats, Challenges, Vulnerabilities and Risks*, Hexagon Series on Human and Environmental Security and Peace, vol. 5, Berlin: Springer-Verlag, pp. 107–122.

Chevallier, P., Pouyaud, B., Suarez, W. and Condom, T. (2011), 'Climate change threats to environment in the tropical Andes: glaciers and water resources', *Regional Environment Change*, **11**, Supplement 1, March: S179–S188.

CHS (Commission on Human Security) (2003, 2005), *Human Security Now, Protecting and Empowering People*, New York, USA: Commission on Human Security, <http://www.humansecurity-chs.org/finalreport/>.

Clark, W. C., P. J. Crutzen and H. J. Schellnhuber (2004), 'Science and global sustainability: toward a new paradigm', in: H. J. Schellnhuber, P. J. Crutzen, W. C. Clark, M. Claussen and H. Held (eds), *Earth System Analysis for Sustainability*, Cambridge, MA – London: MIT Press, pp. 1–28.

Eakin, H and M. C. Lemos (2006), 'Adaptation and the state: Latin America and the challenge of capacity-building under globalization', *Global Environmental Change*, **16** (1), 7–18.

ECLAC (2010), *Economics of Climate Change in Latin America and the Caribbean. Summary 2010*. Santiago, Chile: ECLAC.

ECLAC (2011a), *Natural Disaster Prevention and Response in the Americas and Financing and Proposals*, Santiago, Chile: ECLAC.

ECLAC (Sanchez, Marco V. and Pablo Sauma, eds) (2011b), *Vulnerabilidad economica externa, proteccion social y pobreza en America Latina*, Santiago, Chile: CEPAL.

ECLAC (2011c), *Social Panorama of Latin America 2011*, Briefing Paper, Santiago, Chile: CEPAL.

ECLAC/IDB (2010), *Climate Change: A Regional Perspective*, Santiago, Chile: CEPAL.

EuropeAid (2009), *Climate Change in Latin America*, Brussels, Belgium: European Union.

European Commission (2008), 'The strategic partnership between the European Union, Latin America and the Caribbean: a joint commitment', Brussels, Belgium: European Communities.

European Commission (2009), *Communication from the Commission to the European Parliament and the Council, The European Union and Latin America: Global Players in Partnership, Brussels, 30 September 2009, COM(2009) 495 final*, Brussels, Belgium: European Communities.

Friedman, L. (2012), 'Mexico approves groundbreaking climate bill', *ClimateWire*, April 20.

Garibaldi, J. A., M. Araya and G. Edwards (2012), 'Shaping the Durban platform: Latin America and the Caribbean in a future high ambition deal', *CDKN Policy Brief*, March, Climate and Development Knowledge Network, London.

Grin, J., J. Rotmans and J. Schot (2010), *Transitions to Sustainable Development. New Directions in the Study of Long Term Transformative Change*, New York – London: Routledge.

Guha-Sapir, D. and F. Vos (2011), 'Quantifying global environmental change impacts: methods, criteri and definitions for compiling data on hydro-meteorological disasters', in H. G. Brauch, Ú. Oswald Spring, C. Mesjasz, J. Grin, P. Kameri-Mbote, B. Chourou, P. Dunay and J. Birkmann (eds) (2011), *Coping with Global Environmental Change, Disasters and Security: Threats, Challenges, Vulnerabilities and Risks*, Hexagon Series on Human and Environmental Security and Peace, vol. 5, Berlin: Springer-Verlag, pp. 693–717.

IDB (Inter-American Development Bank) (2012), 'The climate and development challenge for Latin America and the Caribbean: options for climate resilient low carbon development: executive summary', Inter-American Development Bank.

IFPRI (International Food Policy Research Institute) (2009), *Climate Change: The Impact on Agriculture and Costs of Adaptation*, Washington DC, USA.

IPCC (Intergovernmental Panel on Climate Change) (2001), *Climate Change 2001: Impacts, Adaptation and Vulnerability*, IPCC Third Assessment Report, Cambridge, UK: Cambridge University Press.

IPCC (2007), *Climate Change 2007. Impacts, Adaptation and Vulnerability*, Working Group II Contribution to the Fourth Assessment Report of the IPCC, Cambridge, UK: Cambridge University Press, December.

IPCC (2007a), *Climate Change 2007. Synthesis Report*, Geneva, Switzerland: IPCC.

IPCC (2012), *Special Report on Managing the Risks of Extreme Events and Disasters to Advance Climate Change Adaptation*, Geneva, Switzerland: IPCC, February.

Jolly, R. and D. Basu Ray (2006), *National Human Development Reports and the Human Security Framework: A Review of Analysis and Experience*, Falmer, UK: University of Brighton, Institute of Development Studies, April.

Keller, M., D. Medeiros, D. Echeverría, J.E. Parry (2011), *Review of Current and Planned Adaptation Action: South America*, Winnipeg, Canada: International Institute for Sustainable Development, November.

Lenton, T., H. Held, E. Kriegler, J. W. Hall, W. Lucht, S. Ramstorf and H. J. Schellnhuber (2008), 'Tipping elements in the Earth's climate system', in: *Proceedings of the National Academy of Science* (PNAS), **105** (6), 12, 1786–1793.

Magaña, V., C. Conde, Ó. Sánchez and C. Gay (2000), *Evaluación de Escenarios Regionales de Clima Actual y de Cambio Climático Futuro Para México 2000*, México, D.F.

Magrin, G., C. Gay García, D. Cruz Choque, J.C. Giménez, A.R. Moreno, G.J. Nagy, C.

Mata, Luis Jose and Carlos Nobre (2006), 'Background paper: impacts, vulnerability and adaptation to climate change in Latin America', Bonn, Germany: United Nations Framework Convention on Climate Change.

Medeiros, D., H. Hove, M. Keller, D. Echeverría and J.E. Parry (2011), *Review of Current and Planned Adaptation Action: The Caribbean*, Winnipeg, Canada: International Institute for Sustainable Development, November.

Mesjasz, C. (2011), 'Economic vulnerability and economic security', in H. G. Brauch, Ú. Oswald Spring, C. Mesjasz, J. Grin, P. Kameri-Mbote, B. Chourou, P. Dunay and J. Birkmann (eds) (2011), *Coping with Global Environmental Change, Disasters and Security: Threats, Challenges, Vulnerabilities and Risks*, Hexagon Series on Human and Environmental Security and Peace, vol. 5, Berlin: Springer-Verlag, pp. 123–156.

Moreno, L. A. (2011), 'The decade of Latin America and the Caribbean: a real opportunity', Remarks by the President of the IDB at the book launch of *The Decade of Latin America and the Caribbean: A Real Opportunity*, Buenos Aires, Argentina, May 27.

O'Brien, K., A. Lera St. Clair and B. Kristoffersen (2010), *Climate Change, Ethics and Human Security*, Cambridge, UK: Cambridge University Press.

Oswald Spring, Ú. (2008), *Gender and Disasters. Human, Gender and Environmental Security. A HUGE Challenge*, Source 8/2008, Bonn: UNU-EHS.

Oswald Spring, Ú. (2010), 'El cambio cambio climatico, conflictos sobre recursos y vulnerabilidad social', in: G. C. Delgado, C. Gay, M. Imaz, and M. Amparo Martinez (eds), *Mexico Frente Al Cambio Climatico*, Mexico, D.F.: UNAM: Coleccion El Mundo Actual, pp. 51–82.

Oswald Spring, Ú. (2011), 'Reconceptualizar la seguridad ante los riesgos del cambio climático y la vulnerabilidad social', in: Daniel Rodríguez (ed.), *Las dimensiones sociales del cambio climático: una agenda para México*, México D.F.: Instituto Mora.

Oswald Spring, Ú. (2012), 'Environmentally-forced migration in rural areas: security risks and threats in Mexico', in: J. Scheffran, M. Brzoska, H. G. Brauch, P. M. Link and J. Schilling (eds), *Climate Change, Human Security and Violent Conflict: Challenges for Societal Stability*, Berlin et al.: Springer, pp. 315–350.

Oswald Spring, Ú. and H.G. Brauch (2011), 'Coping with global environmental change – sustainability revolution and sustainable peace', in H. G. Brauch, Ú. Oswald Spring, C. Mesjasz, J. Grin, P. Kameri-Mbote, B. Chourou, P. Dunay and J. Birkmann (eds) (2011), *Coping with Global Environmental Change, Disasters and Security: Threats, Challenges, Vulnerabilities and Risks*, Hexagon Series on Human and Environmental Security and Peace, vol. 5, Berlin: Springer-Verlag, 1487–1503.

Oswald Spring, Ú., I. Sánchez Cohen, R. Pérez, A. Martín, J. Garatuza, E. Gómez, C. Watts and M. Miranda (eds.) (2010), *Retos de la Investigación del Agua en México*, Cuernavaca, Mexico: CRIM-UNAM/CONACYT.

Oswald Spring, Ú., S. E. Serrano Oswald, A. Estrada Álvarez, F. Flores Palacios, M. Ríos Everardo, H. G. Brauch, T. E. Ruíz Pantoja, C. Lemus Ramírez, A. Estrada Villareal, M. Cruz (2013), *Vulnerabilidad Social y Género entre Migrantes Ambientales*, Cuernavaca, México: CRIM-DGAPPA-UNAM, in press.

Peralta, O. (2008), *Cambio climático y seguridad nacional*, México, D.F.: Centro Mario Molina para Estudios Estratégicos sobre Energía y Medio Ambiente.

Pizarro, R. (2001), 'La Vulnerabilidad Social y sus Desafíos: Una Mirada desde América Latina', *Estudios Estadísticos y Prospectivos*, 6, CEPAL.

REC (Regional Environmental Change) (2011), 'Climate hotspots: key vulnerable regions; climate change and limits to warming', *Regional Environmental Change*, 11, Supplement 1, March.

Red Iberoamericana de Oficinas de Cambio Climático (The Ibero-American Network of Climate Change Offices) (RIOCC) (2008), Contribution of a regional cooperative structure to the objectives of the UNFCCC, Nairobi Work Programme (NWP). RIOCC Pledges to Support the NWP.

Roberts, J. T. and B. Parks (2007), *A Climate Of Injustice: Global Inequality, North-South Politics, and Climate Policy*, Cambridge, USA: MIT Press.

Roberts, J. T., K. Starr, T. Jones and D. Abdel-Fattah (2008), 'The reality of official climate aid', *Oxford Energy and Environment Comment*, November, Oxford: Oxford Institute for Energy Studies.

Samaniego, J. L. (Coord.) (2009), *Climate Change and Development in Latin America and the Caribbean Overview 2009*, Santiago, Chile: UN ECLAC.

Sánchez Cohen, I., Ú. Oswald Spring, G. Díaz Padilla, J. Cerano Paredes, M. A. Inzunza Ibarra, R. López López and J. Villanueva Díaz (2013), 'Forced migration, climate change, mitigation and adaptive policies in Mexico: Some functional relationships', *International Migration*, **51**, 4 (August).

Scheffran, J., M. Brzoska, H. G. Brauch, P. M. Link and J. Schilling (eds) (2012), *Climate Change, Human Security and Violent Conflict: Challenges for Societal Stability*, Berlin et al.: Springer.

Serrano Oswald, S. E., H. G. Brauch and Ú. Oswald Spring (2013), 'Teorías de Migracíon', in Oswald Spring, Úrsula et al. (eds), *Vulnerabilidad Social y Género entre Migrantes Ambientales*, Cuernavaca, México: CRIM-DGAPPA-UNAM, in press.

Stern, N. (2006), *The Economics of Climate Change – The Stern Review*, Cambridge, UK: Cambridge University Press.

United Nations Development Programme (UNDP) (1994), *Human Development Report 1994. New Dimensions of Human Security*, New York: UNDP – Houndmills, UK: Palgrave Macmillan.

United Nations Development Programme (UNDP) (2007), *Human Development Report 2007/2008. Fighting Climate Change: Human Solidarity in a Divided World*, Basingstoke, UK: Palgrave Macmillan.

United Nations Environment Programme (UNEP) (2012), *Global Environmental Outlook, GEO 5*. Nairobi – New York: UNEP.

UNEP/ECLAC (2010), *Vital Climate Change Graphics for Latin America and the Caribbean*. Panama City, Panama: United Nations Environment Programme.

United Nations General Assembly (UNGA) (2009), 'Climate change and its possible security implications', Resolution adopted by the General Assembly, A/RES/63/281, New York: United Nations General Assembly, 11 June, <http://www.un.org/News/Press/ docs/2007/ sc9000.doc.htm>.

United Nations General Assembly (UNGA) (2011), 'Informal thematic debate on human security', <http://www.un.org/en/ga/president/ 65/initiatives/HumanSecurity.html>.

UNSC (United Nations Security Council) (2011), 'Security Council, in statement, says "contextual information" on possible security implications of climate change important when climate impacts drive conflict', UN Security Council, 6587th Meeting (AM & PM), 20 July.

UNSC (United Nations Security Council) (2007), 'Security Council holds first-ever debate on impact of climate change on peace, security, hearing over 50 Speakers', UN Security Council, 5663rd Meeting, 17 April.

UNSG (United Nations Secretary-General) (2009), *Climate Change and its Possible Security Implication*, New York: UN, 11 September.

UNSG (2012), *Follow-up to General Assembly Resolution 64/291 on Human Security. Report of the Secretary-General* A/66/763, 5 April.

Vergara, W. (ed.) (2009), 'Assessing the potential consequences of climate destabilization in Latin America Latin America and Caribbean', Region Sustainable Development Working Paper 32, Washington DC, USA: The World Bank.

Vergara, W. (2011), 'The economic and financial costs of climate change in regional economies in Latin America', Presented at The Economic and Financial Costs of Climate Change in Regional Economies in Latin America and the Caribbean Event, Conference of Parties (COP 17), Durban Exhibition Centre, Durban, South Africa, 7 December.

Vergara, W., H. Kondo, E. Pérez Pérez, J. M. Méndez Pérez, V. Magaña Rueda, M. C. Martínez Arango, J. F. Ruíz Murcia, G. J. Avalos Roldán and E. Palacios (2007), 'Visualizing future climate in Latin America: results from the application of the Earth Simulator', Latin America and Caribbean Region Sustainable Development Working Paper 30, November, Washington DC, USA: The World Bank.

Villagrán de León, J. C. (2006), *Vulnerability: A Conceptual and Methodological Review*. Source, 4/2006, Bonn, Germany: UNU-EHS.

Wæver, O. (1995), 'Securitization and desecuritization', in: R. D. Lipschutz (ed.), *On Security*, New York, USA: Columbia University Press, pp. 46–86.

Wæver, O. (1997), *Concepts of Security*, Copenhagen, Denmark: Department of Political Science.

Wæver, O. (2008), 'The changing agenda of societal security', in: H. G. Brauch, Ú. Oswald Spring, C. Mesjasz, J. Grin, P. Dunay, N. Chadha Behera, B. Chourou, P. Kameri-Mbote and P.H. Liotta (eds), *Globalization and Environmental Challenges: Reconceptualizing Security in the 21st Century*, Berlin: Springer, pp. 581–593.

WBGU (2011), *World in Transition – A Social Contract for Sustainability*, Berlin, Germany: German Advisory Council on Global Change, July, <http://www.wbgu.de/fileadmin/templates/dateien/veroeffentlichungen/hauptgutachten/jg2011/wbgu_jg2011_kurz_en.pdf>.

Wilk, D. (Coordinator) (2010), *Analytical Framework for Climate Change Action*, Inter-American Development Bank, Washington DC.

World Bank (2009), *World Development Report 2010: Development and Climate Change*, Washington, DC, USA: The World Bank.

Zapata-Martí, R. (2011), 'Strategies for coping with climate change in Latin America: perspective beyond 2012', in H. G. Brauch, Ú. Oswald Spring, C. Mesjasz, J. Grin, P. Kameri-Mbote, B. Chourou, P. Dunay and J. Birkmann (eds) (2011), *Coping with Global Environmental Change, Disasters and Security: Threats, Challenges, Vulnerabilities and Risks*, Hexagon Series on Human and Environmental Security and Peace, vol. 5, Berlin: Springer-Verlag, pp. 1341–1354.

10. Human security and climate change in the Mediterranean region
Marco Grasso and Giuseppe Feola

1. INTRODUCTION

This chapter investigates human security and its intersections with climate change in the Mediterranean Region (MR). It does so by measuring human security at national level, and by critically discussing an ethical approach for improving human security in the MR. Adopting a regional perspective is particularly useful to account for the traits of human security, in that it favours the recognition of its interregional dynamics, its environmental, cultural and governance dimensions, and its multifaceted relationship with climate change (Liverman and Ingram 2010).

There are different and controversial definitions of the MR (Brauch 2001, 2003). We adopt a medium concept that includes all the twenty countries with Mediterranean coastlines, plus Portugal, Serbia, Macedonia and Jordan. All these countries have a number of shared features that, for the purposes of this chapter, make it possible to consider them as forming a fairly homogenous region: (i) a common history; (ii) relatively similar cultures; (iii) a distinct Mediterranean economy; (iv) comparable natural and climatic characteristics, and environmental threats (Brauch 2010).

Climate change is expected to have, and to some extent already has, had, multiple physical and socio-economic impacts on the MR. Physical impacts are characterized by the high variability of projected effects in different sub-regions. Coastal areas, large deltas and semi-arid zones will suffer the most (UNEP/MAP 2009). Physical impacts mainly consist in the increase of drought periods that will further limit the already constrained water availability, accelerate desertification processes, and threaten biodiversity. In general, physical effects coupled with anthropogenic activities are expected dramatically to increase environmental pressure in the entire Mediterranean area (IPCC 2007). As far as societal and economic aspects are concerned, it is widely agreed that these outcomes of changing climatic dynamics will severely affect Mediterranean agriculture (Iglesias et al. 2011; Giannakopoulos et al. 2009), fishing, tourism, coastal zones and infrastructure, and that they will ultimately endanger public health (UNEP/MAP 2009).

However, human security in the MR is not threatened by climate change alone. In fact, the MR has been facing several socio-demographic and economic challenges, ranging from the economic crisis to demographic changes, to political and social tensions. While these challenges are far from being settled, so that full appreciation of their effects on human security is hardly possible, it seems clear that they are already having, and possibly will have long in the future, effects on the Mediterranean peoples' capacity to "end, mitigate, or adapt to threats to their human, environmental, and social rights; have the capacity and freedom to exercise these options; and actively participate in pursuing these options" (Barnett et al. 2006 p. 18).

Owing to this overlap between exceptional environmental and social change trends, the MR is an especially interesting case for investigation of the intersections between climate change and human security. Because Parts 1 and 2 of this Handbook provide an exhaustive overview of the notion of human security and contextualize its features in the milieu of climate change, we neither analyze nor review these aspects here. Rather, we propose a measure of human security at national level that can serve to gain better understanding of human security, to scrutinize its connection with climate change in the MR, and ultimately to develop an ethical approach for raising its level in this region.

The chapter is structured as follows. It first conducts a review of current environmental and social change trends in order to highlight the most severe threats to human security in the MR, or sub-regions within it. It then carries out a quantitative indicator-based analysis of human security, the purpose being to identify and quantify the relevant determinants of human security in the region. In particular, the indicators are used to cluster Mediterranean countries (based on their similarity with respect to the indicators) in order to explore how countries differ in terms of human security dimensions. Finally, on the basis of the emerging evidence, the chapter proposes an ethical approach for the improvement of human security in the region where states assume the role of mediators of the factors that most significantly influence human security.

2. ENVIRONMENTAL AND SOCIAL CHANGE IN THE MEDITERRANEAN REGION

The MR is undergoing several environmental and social changes with the potential to change the environmental, institutional, demographic and cultural configurations of the region, and significantly to affect human security in it.

Table 10.1   Summary of environmental and social change trends

Trends		Sub-region	
		Southern Europe	MENA
Environmental change		Increased frequency of extreme events (especially droughts, heat waves, storms and wind storms) Higher average temperatures Reduced precipitations Sea level rise Increased variability	
Social change	Demography	Shrinking and ageing population	Growing and still young population
	Financial crisis and economic globalization	Serious economic crisis and slow recovery; Strong integration into the global economy	Economic crisis but recovery; Limited integration into the global economy
	Political and social trends	Persistence of conflict (passive, active for Balkan region); Social mobilization for democratic reforms	Persistence of conflict (active); Social mobilization for democratic reforms

While it is necessary analytically to distinguish among the different environmental and social change trends (Table 10.1), it is important to bear in mind that they are closely interlinked, and that threats to human security often arise from the interplay among these trends in specific contexts. In fact, as made clear below, several environmental changes largely determined by climate dynamics interact with social ones, and they have the potential jointly to affect human security. More in detail, it is necessary to note that climate change, owing to its ramified and overarching impacts in the MR, can be seen as a *threat multiplier* (Brklacich et al. 2006; Dokos et al. 2008) because of its many potential interactions with other factors of human insecurity. For example, climate change is believed to be linked with violent conflict (Homer-Dixon 1994; Scheffran and Battaglini 2010; Hsiang et al. 2011), although the evidence is contradictory, given that several studies do not find confirmation of such a link (Barnett 2009; Koubi et al. 2012; Buhaug 2010; Tol and Wagner 2010). Climate change is also linked with migration (Afifi and Warner 2008; Feng et al. 2010; Warner 2010; Black et al. 2011), whereby migration often results from a closely intertwined set of factors, among which are also demographic (e.g. overpopulation), economic (e.g. unemployment), and political (e.g. conflict, lack of human rights and freedom) ones. Some authors also hypothesize a causal, though indirect, connection between climate change and the Arab Spring (Johnstone and Mazo 2011).

2.1   Environmental Change

The MR is considered to be a climate change "hot spot" because of its high environmental vulnerability (Giorgi 2006; Scheffran and Battaglini 2010). Beside other sources of environmental stress, such as earthquakes, desertification and volcanic eruptions, that have traditionally character-ized this region, in recent decades climate change has moved to centre stage by bringing ecosystems and human populations under increasing strain. With reference to the period between 1860 and 1995 and the central and western MR, for example, Piervitali et al. (1997) found several signs of climate change, among which a significant increase in heat waves, a rise in surface air temperature, and a reduction in cloudiness and precipita-tion. Similarly, on studying the risk of climate change for agriculture, Moonen et al. (2002) recorded change trends in extreme temperature and rainfall events, frost, flooding, and drought risks. Hoerling et al. (2011) also recorded a climate-related increase in Mediterranean droughts. These trends are already affecting ecosystems (e.g. animal and life cycles, Penuela et al. 2002) and human activities (e.g. agriculture, Ben Mohamed et al. 2002), and they are projected to continue and increase in magnitude in the next decades. Higher average temperatures are expected in the MR (Gibelin and Deque 2003; Giorgi et al. 2004), and, also in combination with reduced precipitations, are projected to cause longer droughts and increased or accelerated desertification (Gibelin and Deque 2003; Arnell 2004; Puigdefabregas and Mendizabal 1998; Black 2009; Bou-Zeid and El-Fadel 2002; UNDP 2009), and to reduce water availability in urban areas as well (Bigio 2009). Surprisingly, despite its importance, water scarcity is seldom on the agenda of political leaders in the MENA (Middle-Eastern and North African) countries (Sowers et al. 2010). The magnitude and frequency of such extreme events as heat waves, storms and windstorms is also expected to increase (Giannakopoulos et al. 2009; Maracchi et al. 2005; Schwierz et al. 2010). Furthermore, sea level rise is expected to affect coastal areas through increased soil salinity, coastal inundation and increased coastal erosion (Sánchez-Arcilla et al. 2011; Iglesias et al. 2011; Bigio 2009; Nicholls and Hoozemans 1996). Both marine and terrestrial ecosystems are expected to undergo significant modifications as a conse-quence of climate change (Turley 1999; Schroeter et al. 2005; Thuiller et al. 2005; Scarascia-Mugnozza et al. 2000; Metzger et al. 2006). While climate and environmental models tend to differ in terms of the projected magni-tudes of the above-mentioned climate change effects, they are consistent in showing that these trends are likely to be the ones that most affect ecosystems and human activities in the MR (Giorgi and Lionello 2008).

An important feature of climate change is induced variability in both its temporal and spatial dimensions. Interannual variability is projected to increase as a consequence of less regular meteorological patterns (Giorgi et al. 2004). The spatial variability of climate change effects is likely to be high in that it depends on the complex and diverse interactions of environmental trends with local physical and environmental conditions, and on the sensitivity, vulnerability and adaptive capacity of different natural and social systems (Ferrara et al. 2009; Grasso and Feola 2012).

The Environmental Vulnerability Index (EVI) (UNEP 2005) confirms the severity of climate change threats for the MR. The EVI, which also considers other sources of environmental threat such as pollution, biodiversity loss, and natural disasters, shows that climate change is the main driver of environmental vulnerability in the region (see Table 10.2). It also suggests that no clear sub-regional pattern can be identified, because almost all countries are highly vulnerable to climate change and related threats (e.g. desertification).

2.2   Social Change

We focus in what follows on four interlinked demographic, economic, political/institutional and social trends characterizing the MR that we consider to require closer attention from the relevant literature. A *caveat* is in order. While these challenges are far from settled, so that full appreciation of their effects on human security is hardly possible, it seems clear that they are already having, and possibly will have long into the future, effects on human security in the MR.

Demography

Demographic dynamics in the MR are characterized by a shrinking and ageing population in large parts of southern European countries, and by a growing and still young population in the MENA ones. Population is ageing especially in Spain, Italy, Portugal and Greece, where a very low or negative population growth rate and an ageing of both the entire population and the working-age population are observed (Gesano et al. 2009). The former Yugoslavian countries (i.e. Croatia, Serbia, Bosnia-Herzegovina, Montenegro, and Macedonia) are projected to experience a steadily ageing working population and a rapid increase in the elderly population (Gesano et al. 2009). Population ageing is projected to occur to a significant extent also in Albania and Turkey, where, however, the working-age population is expected to continue to increase at least until

2020 (Gesano et al. 2009). Although the MENA countries have begun the demographic transition, they are experiencing, and are projected to experience, a rapidly increasing population, with a growing but still limited elderly population and a young and increasing working-age population (Tabutin and Schoumaker 2005; Gesano et al. 2009; UNDP 2009).

These different trends are expected to create opposite socio-economic problems in different parts of the MR (Gesano et al. 2009). Southern Europe is facing, and will increasingly face, the challenge of funding its welfare and health care systems, which may prove to be particularly difficult under conditions like the present financial crisis (see below). The MENA countries, on the other hand, are facing, and will increasingly face, the challenge of creating jobs for a young and growing working-age population so as to fight unemployment, and consequently poverty and frustration (Laipson 2002; Assaad and Roudi-Fahimi 2007), which many argue are among the triggers of the so-called Arab spring (Al-Momani 2011; Warf 2011). Given the scant openness of the MENA countries to globalization (Noland and Pack 2004, see below), this may prove particularly difficult, and it may exacerbate other phenomena traditionally linked with population growth, such as urbanization (Cohen 2004), and international migration towards the MR's northern shore (which in turn could influence human security in North Mediterranean countries (i.e. European).

Financial crisis and economic globalization

Also with regard to economic globalization trends, the MR is characterized by a cleavage between countries on its northern and southern shores. Southern European countries have been strongly hit by the recent financial and economic crises (Verney 2009; Filippetti and Archibugi 2010). Greece, but also Portugal, Spain and Italy, have been among the European countries most seriously affected, although in different ways because of the different structures of their economic and financial systems (Escribiano 2010; Armingeon and Baccaro 2011). Recovery is expected to be slow, especially amid current EU economic governance difficulties and market speculations.[1]

The economic crisis has less severely impacted on Eastern European countries and the MENA ones. The latter, in particular, have begun a significant recovery (IMF 2010), although its future is uncertain (IMF 2012). As noted by Dabrowski (2010), economies relying little on external financing and trade have suffered less than more sophisticated economies. This relative isolation from the globalized economy, however, has significant drawbacks in terms of economic development. The MENA countries generally have a difficult relationship with globalization (Noland and Pack 2004) and an inward-looking model of development. The Index of Globalization (Dreher et al. 2008), for example, shows that MENA countries with the exception of Israel are consistently lower than southern European ones in the ranking of globalized economies (Table 10.2). The economies of MENA countries are traditionally based on support by the public sector, the oil economy, remittances, and a limited openness to international markets (World Bank 2003; UNDP 2009), although sub-regional international integration has improved in the past two decades (Romagnoli and Mengoni 2009). Such a model of development is widely considered to be unsustainable due to changing conditions, among which the decline of oil resources and the volatility of oil prices, the increase in international competition, the pressure of internal labour markets (i.e. a rapidly growing working-age population), and the slowing of migration opportunities as a reaction of European countries to growing migration flows (World Bank 2003; Noland and Pack 2004; UNDP 2009). Despite the looming financial crisis, greater engagement in the global markets and translating the gains of economic growth into poverty reduction and human development are therefore considered critical factors for the MENA countries in the next decades (Amin et al. 2012; Aryveetey et al. 2011).

Political/institutional and social trends
In recent decades, the MR has seen the persistence of violent conflict (both intra- and inter-state) in several sub-regions, such as the Balkans, Cyprus, the Middle East and North Africa. Several, if not all, Mediterranean countries were involved in such conflicts at different times after World War II – although some countries experienced conflict on their territory (e.g. Balkan countries, Israel, Egypt, Syria, Jordan) while others did not (e.g. Spain, France, Italy). Conflicts are ongoing in some parts of the Mediterranean region (e.g. the Middle East). The Peace Index (IEP 2011), for example, shows that Mediterranean countries perform rather poorly on a global level. According to this index, which considers a variety of issues including internal and external conflict, military expenditures, criminality, and weapons, no Mediterranean country (with the exception of Portugal and Slovenia) ranks among the most peaceful 35 countries in the world. Moreover, several MENA countries perform particularly poorly (e.g. Israel, Turkey, Syrian Arab Republic, Libyan Arab Jamahiriya and Lebanon).

The MENA countries have been traditionally ruled by authoritarian regimes, and some scholars have argued that Arab culture and Islamic religion are not compatible with democracy (Al-Momani 2011; Warf 2011). Common explanations point out that democratic institutions in the MENA countries did not develop because of a combination of factors, among them oil-affected economies and social interests, foreign aid, the Arab-Israeli conflict, and a sort of clustering effect in the sub-region (El-Badawi and Makdisi 2007; UNDP 2009; Diamond 2010). The recent wave of social mobilizations in favour of democratic reforms, commonly referred to as the Arab Spring (Sorenson 2011; Warf 2011), is expected to affect this political and social configuration (Al-Momani 2011; Warf 2011). By and large, the Arab Spring was fuelled by relatively highly educated young cohorts, making widespread use of the new media (Warf 2011), which rebelled against the restriction of political freedoms, corruption, and the incapacity of the ruling elites to deal with persistent social and economic issues (e.g. inequality, unemployment) (UNDP 2009; Al-Momani 2011; Sorenson 2011; Warf 2011). However, serious doubts exist concerning its potential to produce lasting and widespread change. The "possibility for the creation of failed states or international interventions, and the necessity of governments to deal with large numbers of refugees, sectarian tensions, and deeply rooted economic problems" (Al-Momani 2011, p.159), as well as the possible radicalization of religious forces (Sorenson 2011), are among possible backlashes. Moreover, there is the risk that Arab countries may undergo an incomplete democratic construction which ultimately leads to disappointment and a possible democratic rejection (Sorenson 2011).

3.   MEASURING HUMAN SECURITY AND CLUSTERING THE MEDITERRANEAN REGION

Brklacich et al. (2006) provide a reference framework for the analysis of human security in the MR. Environmental and societal change refer to the two types of trends (i.e. double exposure) that influence the context for human security. As seen in the previous section, the main environmental change threat in the Mediterranean is climate change, which also interacts with other trends such as desertification and water availability. The societal change trends considered here are demographic, economic, social and institutional trends. These contextual factors influence the exposure and the responsive capacity via influence on the institutional context, control over and access to assets, and distribution of rights and resources (Brklacich et al. 2006).

Responsive capacity, finally, is linked to a set of fundamental functionings, in that the weaker a country is in these domains of well-being (or achievable functionings) that specify human security, the less are its institutional and social capacities and possibilities to carry out effective adaptation actions. Accordingly, adopted here is the notion of human security

Table 10.2 Environmental Vulnerability Index, Index of Globalization and Peace index for the Mediterranean countries

Country	Environmental Vulnerability Index[1]	Environmental Vulnerability Index (3 most policy relevant sub-indices)	Index of globalization (economic)[2,3]	Peace index [2,4]
Albania	Highly Vulnerable	Water, Desertification, Climate Change	86	63
Algeria	Vulnerable	Desertification, Water, Climate Change	108	129
Bosnia and Herzegovina	Vulnerable	Water, Desertification, Climate Change	65	60
Croatia	Highly Vulnerable	Water, Agriculture and fisheries, Human health aspects	41	37
Cyprus	Vulnerable	Human health aspects, Water, Agriculture and fisheries	11	71
Egypt	Vulnerable	Water, Human health aspects, Climate Change	109	73
France	Highly Vulnerable	Climate Change, Agriculture and fisheries, Desertification	25	36
Greece	Highly Vulnerable	Human health aspects, Water, Climate Change	39	65
Israel	Extremely Vulnerable		27	145
Italy	Extremely Vulnerable	Human health aspects, Climate Change, Exposure to natural disasters	46	45
Jordan	Vulnerable	Water, Human health aspects, Desertification	52	64
Lebanon	Extremely Vulnerable	Human health aspects, Water, Climate Change	na	137
Libyan Arab Jamahiriya	At risk	Desertification, Water, Agriculture and fisheries	na	143

Macedonia, FYR	Highly Vulnerable	Human health aspects, Water, Agriculture and fisheries	64	78
Malta	Extremely Vulnerable	Human health aspects, Water, Climate Change	4	na
Montenegro	na	na	18	89
Morocco	Vulnerable	Water, Climate Change, Desertification	107	58
Portugal	Highly Vulnerable	Climate Change, Water, Agriculture and fisheries	20	17
Serbia	na	na	73	84
Slovenia	Highly Vulnerable	Human health aspects, Water, Agriculture and fisheries	23	10
Spain	Highly Vulnerable	Agriculture and fisheries, Desertification, Water	31	28
Syrian Arab Republic	Highly Vulnerable	Water, Human health aspects, Desertification	131	116
Tunisia	Vulnerable	Water, Human health aspects, Desertification	71	44
Turkey	Highly Vulnerable	Water, Desertification, Human health aspects	95	127

Notes:
1. UNEP (2005).
2. Position in the global ranking. Year 2011.
3. Dreher (2006), Dreher et al. (2008) (see also: http://globalization.kof.ethz.ch/). Year 2008.
4. Institute for Economics and Peace (2011). Year 2009–2010.

put forward by Alkire, who views it as the protection and promotion of a limited number of aspects of well-being which constitute its "vital core" (Alkire 2003, p. 2), the "central component of human well-being" (UNU 2007, p. 6). On this view, the protection of human security does not include all aspects of human well-being, but only the crucial ones. It is worth noting that, in our opinion, as thus understood the notion of vital core does not have a precise theoretical meaning; rather, it "may be specified in terms of human rights or capabilities related to absolute poverty" (Alkire 2003, p. 25). Therefore, the basis and epistemological foundation of this notion of human security consist mainly in practical reasons; and, in fact, the expression "human security" is employed in this chapter according to this perspective.

3.1   Methodology

Following King and Murray (2002), and Brklacich et al. (2006), eight indicators were selected to cover the economic, social, institutional and environmental dimensions of human security (see Table 10.3).

Consistently with the definition of human security provided above, an indicator set was selected to cover the vital core aspects of human security. This resulted in fewer indicators than proposed, for example, by Lonergan et al. (2000). Finally, a balanced number of indicators per dimension were adopted.

By and large, we argue that these indicators represent the key factors whereby countries and their peoples can "end, mitigate, or adapt to threats

Table 10.3   Correspondence between the indicator set used in this study and three relevant frameworks

Indicator	King and Murray (2006)	Brklacich et al. (2006)	Lonergan et al. 2000
GNI	Income	Assets	Economy
GINI	Income	Distribution of resources	Economy
HALE at birth	Education	Assets	Social
School life expectancy ISCED 1–6	Health	Assets	Social
Voice and accountability	Democracy	Institutions	Institutions
Government effectivenes	Democracy	Institutions	Institutions
IRWR per capita	–	Ecological context	Environment
Agricultural area per Capita	–	Ecological context	Environment

to their human, environmental, and social rights; have the capacity and freedom to exercise these options; and actively participate in pursuing these options" (Barnett et al. 2006, p. 18).

Data for the indicators were collected from secondary sources such as United Nations (UN), World Bank (WB), Food and Agriculture Organization (FAO) and UNICEF. It should be warned that the latest data available may be – given the extreme dynamic of the MR pointed out above – relatively out-dated: if fact, they refer largely to 2009–10.

The groups of countries were identified by means of a cluster analysis based on their similarity with respect to the human security dimensions (indicators) considered.[2]

Finally, the statistical difference between the clusters for each of the indicators of human security was tested using the Kolmogorov-Smirnov Z non-parametric test.

3.2 Results

Indicators of human security

Table 10.4 shows the indicator values for each of the 24 Mediterranean countries considered.

Figure 10.1 graphically represents the distribution of the Mediterranean countries in relation to pairs of selected indicators, i.e. economic, social, institutional and environmental. These graphs show that the MENA countries, with the exception of Israel, tend to perform worse than other Mediterranean countries. The performance of Eastern European countries varies, with Slovenia often showing values close to Western European ones.

Three clustering options were generated, i.e. with two, three and four clusters. The option with two clusters corresponded consistently (i.e. independently from the clustering method) with two groups of almost equal size (i.e. 11 and 10 countries). The options with three and four clusters consistently yielded the same group of 11 countries and also consistently set France aside from the rest of the countries. The option with four clusters, however, generated two groups with very few countries. For this reason, this option was dropped in the following analysis, and the one with three clusters was preferred. Moreover, in order to facilitate the statistical analysis, France was grouped with the cluster to which it was most similar (cluster 3).

Interestingly, the MENA countries plus Albania, Macedonia FYR, Bosnia and Herzegovina, and Turkey all clustered together independently of the clustering method adopted. This suggests a very strong similarity in terms of the factors determining human security among these countries.

Table 10.4   Indicator values

Country	GNI	GINI	HALE at birth	School life expectancy ISCED 1-6	Voice and account-ability	Government effectivenes	Internal Renewable Water Resources IRWR per capita	Agricultural area per capita
Albania	3950	35.3	64	11.3	0.158	-0.204	8504.6	0.357
Algeria	4420	33.0	62	12.8	-1.044	-0.591	332.2	1.219
Bosnia and Herzegovina	4700	36.3	67	13.2	-0.048	-0.646	9056.1	0.569
Croatia	13810	29.0	68	13.9	0.559	0.639	8276.6	0.271
Cyprus	26940	29.0	70	14.2	1.062	1.320	913.3	0.184
Egypt	2070	32.1	60	11.0	-1.118	-0.300	23.4	0.044
France	43990	32.7	73	16.1	1.260	1.442	2929.1	0.477
Greece	28630	34.3	72	16.5	0.882	0.608	5197.1	0.740
Israel	25740	39.2	73	15.4	0.580	1.095	107.7	0.070
Italy	35080	36.0	74	16.3	1.040	0.517	3137.2	0.234
Jordan	3740	37.7	63	13.1	-0.849	0.281	114.0	0.165
Lebanon	7970	45.0	62	13.5	-0.334	-0.675	1314.0	0.168
Libyan Arab Jamahiriya	12020	na	64	na	-1.889	-1.118	98.6	2.526
Macedonia, FYR	4400	42.8	66	13.3	0.129	-0.136	2647.1	0.528
Malta	16690	26.0	72	14.4	1.205	1.110	124.7	0.023

	GNI per capita	GINI	School life expectancy	HALE at birth	Voice and accountability	Government effectiveness	IRWR per capita	Agricultural area per capita
Montenegro	6550	36.9	65	na	0.299	−0.031	na	0.828
Morocco	2790	40.9	62	10.5	−0.791	−0.109	894.7	0.971
Portugal	20940	38.5	71	15.8	1.211	1.207	3587.3	0.330
Serbia	5990	28.2	65	na	0.318	−0.154	na	0.685
Slovenia	23520	31.2	71	16.8	0.987	1.163	9501.3	0.248
Spain	31870	34.7	74	16.4	1.187	0.936	2550.2	0.639
Syrian Arab Republic	2410	42.0	63	11.3	−1.633	−0.609	350.2	0.678
Tunisia	3720	40.8	66	14.5	−1.269	0.414	406.5	0.961
Turkey	8730	41.2	66	11.8	−0.119	0.352	3020.2	0.540

Notes:

na = Data not available. Units: USD (GNI per capita), Years (HALE at birth, School life expectancy ISCED 1–6), dimensionless (GINI, Voice and accountability, Government effectiveness), hectares (Agricultural area per capita), cubic meters per year (IRWR per capita). Reference years: 2007 (HALE at birth, IRWR per capita, Agricultural area per capita), 2004–2008 (School life expectancy ISCED 1–6), 2009 (GNI, Voice and accountability, Government effectiveness), 2000–2010 (GINI).

Sources: WB (GNI per capita, GINI, Voice and accountability, Government effectiveness), UN (HALE at birth, IRWR per capita), UNESCO (School life expectancy ISCED 1–6), FAO (agricultural area per person).

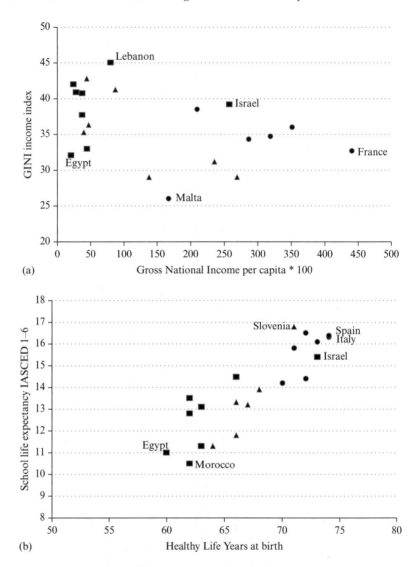

Notes: Squares stand for MENA countries (Algeria, Egypt, Israel, Jordan, Lebanon, Morocco, Syrian Arab Republic, and Tunisia); triangles stand for Eastern European countries (Albania, Bosnia and Herzegovina, Croatia, Macedonia FYR, Slovenia and Turkey); circles stand for Western European countries (Cyprus, France, Greece, Italy, Malta, Portugal, Spain). Serbia and Montenegro are not plotted due to missing data.

Figure 10.1   Distribution of the Mediterranean countries in relation to economic (a), social (b), institutional (c), and environmental indicators (d)

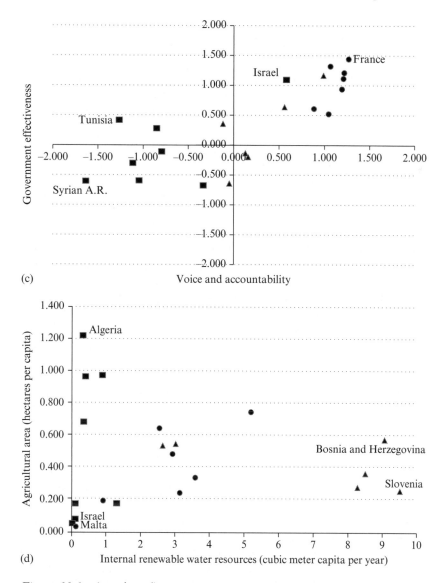

Figure 10.1 (continued)

It is also remarkable that France set aside from all other countries in the clustering options with three or four clusters (although, as said, France was allocated to the most similar cluster in order to carry out the analysis). This suggests that France might actually be in a different condition from all other countries, EU countries included. Serbia and Montenegro, which

Table 10.5   Clusters generated (two- and three-cluster options)

Cluster	Two clusters	Three clusters
Cluster 1	Albania, Algeria, Bosnia and Herzegovina, Egypt, Jordan, Lebanon, Macedonia FYR, Morocco, Syrian Arab Republic, Tunisia, Turkey	Albania, Algeria, Bosnia and Herzegovina, Egypt, Jordan, Lebanon, Macedonia FYR, Morocco, Syrian Arab Republic, Tunisia, Turkey
Cluster 2	Croatia, Cyprus, France, Greece, Israel, Italy, Malta, Portugal, Slovenia, Spain	Croatia, Malta, Portugal, Slovenia
Cluster 3	–	Cyprus, France, Greece, Israel, Italy, Spain

were not inputted in the clustering process due to missing data, presented a profile similar to that of countries in cluster 1 for the two-cluster option, and of cluster 2 for the three-cluster option (Table 10.5).

In regard to the remaining countries, the option with three clusters gives somewhat deeper insight into the potential differences among countries at a higher level of human security. Among those, Croatia, Malta, Portugal and Slovenia seem to differ from Cyprus, Greece, Israel, Italy (and France).

Table 10.6 shows the mean indicator values and standard deviations for different clusters of countries. The Kolmogorov-Smirnov Z test was performed to explore the difference between clusters. This table shows that the clusters are indeed statistically significantly different for most indicators. The indicators belonging to the environmental dimension constitute the sole, but interesting, exception. The differences are particularly clear among the 11 countries in cluster 1 and all the other countries. In the three-cluster option, the only statistically significant difference between clusters 2 and 3 is that of GNI.

These results confirm those obtained using a different index, and indeed for a different but somehow akin dimension, i.e. adaptive capacity, by Grasso and Feola (2012) in the same region. In particular, the results show a North-South divide whereby MENA countries seem to perform significantly worse than EU countries. Surprisingly, Turkey, Bosnia and Herzegovina, Macedonia and Albania are consistently clustered together with MENA countries. The most marked differences between MENA and other Mediterranean countries are observed with respect to social, economic and institutional factors determining human security (see tables and figures).

Table 10.6 *Mean and standard deviation of indicators for different clusters of countries*

Indicator	2-cluster option				3-cluster option						All countries	
	Cluster 1 (n = 11)		Cluster 2 (n = 10)		Cluster 1 (n = 11)		Cluster 2 (n = 4)		Cluster 3 (n = 6)		(n = 21)	
	Mean	Stdv	Mean	Stdv	Mean	Stdv	Mean	Stdv	Mean	Stdv	Mean	Stdv
GNI (USD)	4445	2113***	26721	8883	4445	2113***	18740	4328**	32042	6773	15053	12950
GINI	38.8	4.2*	33.1	4.3	38.8	4.2*	31.18	5.33	34.3	3.4	36.1	5.1
HALE at birth (years)	63.7	2.2***	71.8	1.9	63.7	2.2**	70.5	1.7	72.7	1.5	67.6	4.6
School life expectancy ISCED 1-6 (years)	12.4	1.3***	15.6	1.1	12.4	1.3*	15.2	1.3	15.8	0.9	13.9	2.0
Voice and accountability	−0.629	0.615***	0.957	0.254	−0.629	0.615**	0.991	0.306	1.002	0.244	0.145	0.954
Government effectivenes	−0.202	0.410***	1.004	0.318	−0.202	0.410**	1.030	0.264	0.986	0.373	0.372	0.714
IRWR per capita (cubic meter per year)	2424.0	3298.2	3632.5	3205.8	2424.0	3298.2	5372.5	4328.5	2472.4	1794.6	2999.4	3232.1
Agricultural area per captia (hectares)	0.564	0.374	0.322	0.233	0.564	0.374*	0.218	0.135	0.391	0.269	0.448	0.331

Note: Significant difference between clusters (cluster 1 vs. cluster 2, and cluster 2 vs. cluster 3): *** 1%, ** 5%, * 10% (Kolmogorov-Smirnov Z).

4.   AN ETHICAL APPROACH FOR THE IMPROVEMENT OF HUMAN SECURITY IN THE MEDITERRANEAN REGION

The stark North-South divide made clear by the clustering analyses, whereby EU Mediterranean countries consistently rank better than the others in the region, especially in terms of non-environmental factors determining human security, prompts significant considerations and implications regarding the role of states as mediators of human security, and more generally in regard to the relationship between human security and climate change in the MR. The existence of countries (by and large those belonging to cluster 1 in the two-cluster analysis of section 3) in the region with an insufficient/low level of human security urges identification of subjects or groups of subjects that are responsible for remedying a morally unacceptable situation otherwise likely to continue unabated.

To this end, we adopt here an international political theory perspective that endorses a statist focus. Accordingly, states are agents of justice (Erskine 2008) and have moral obligations that they are eventually reasonably compelled to observe (Nardin 2006, 2008). Furthermore, we consider states in a regional context. A regional approach for addressing human security is feasible only under the following simultaneous conditions: (i) the physical vulnerability of the region is high, so that it is possible to envision a process of "bounding" which requires that states share (a) relevant characteristic(s) which leads to definition of a "community of place" irrespectively of national boundaries (Newman 2003); (ii) within this community development, physical vulnerability and the means to deal with it are unevenly allocated. The MR, as the clustering unambiguously makes clear, fulfils both these conditions: the sharing of the same likely climate impacts and of similar physical vulnerabilities facilitates closeness among countries of the region which might be the foundation for the emergence of a regional community of place. At the same time, there is great heterogeneity in the levels of social vulnerability and adaptive capacity. As a matter of fact, two relevant groups of countries can be identified in the MR with regard to human security: (i) those that have sufficient/higher levels of human security; (ii) those that have an insufficient/lower level of human security. Belonging to the first group (cluster 3, three-cluster analysis) are Cyprus, France, Greece, Israel, Italy and Spain; to the second (cluster 1, two-cluster analysis) Albania, Algeria, Bosnia and Herzegovina, Egypt, Jordan, Macedonia FYR, Morocco, Syrian Arab Republic, Tunisia, Turkey.

Given these two features, the "regionalization" of human security makes it, in our opinion, much easier to consider its − immoral – distribution across the MR. Indeed, in principle, every country in the region has at the same time, yet with very different proportions, a responsibility to promote human security, and to some, yet greatly differentiated, extent suffers from inadequate levels of human security. However, it can be argued that, on ethical grounds: (i) there are countries that have the possibility and the responsibility to promote human security; and (ii) others that, because of their higher social vulnerability, should be supported in attaining sufficient/higher levels of human security. On practical grounds, for instance, Grasso and Feola (2012) have shown that a similar approach based on responsibility and (lack of) adapative capacity can be usefully applied to agricultural adaptation in the MR.

4.1   Remedial Responsibility

To justify the former claim, it is first necessary to frame and vindicate a suitable notion of responsibility. It might be said that a subject can be responsible for a certain action that has already occurred. But it is also possible to argue that a subject is responsible for an outcome that she/he/it is expected to achieve in the future. The first kind of responsibility is defined as retrospective responsibility (Miller 2007; Erskine 2003): it pertains to the subject that created a bad situation, even unintentionally (Miller, 2007), and entails that the benefits and burdens fall on her/him/it (Miller 2004). The second kind of responsibility is instead defined as remedial or prospective (Miller 2007; Erskine 2003) and relates to the subject in charge of remedying the bad situation. With this distinction in mind and acknowledging the nature of human security and the results of the clustering analyses carried out, it is possible to argue that retropective responsibility is ruled out both by the no-harm principle (Shue 1996), since there is no subject that directly brought about harm to others by act or by omission, and by the principle of historical justice (Gardiner 2004) since an insufficient/lower level of human security does not depend on any specific actions. On the other hand, instead, higher-level human security EU Mediterranean countries should be held remedially responsible in the first instance. In fact, according to Miller's (2001) analysis of remedial responsibility, two principles expressly demand that EU Mediterranean countries support and promote human security in the weaker areas of the region. First, the principle of community, which states that when people are linked together by ties arising from shared activities and commitments, by common identities and history, as in the case of the Mediterranean region (Brauch 2010), they have special mutual responsibilities greater than those that they have towards outsiders; second, the principle of capacity, which states that those with a greater capacity to act have a special obligation to do so.

Consequently, states belonging to the MR that have the ability to pay and capacity to act should be held remedially responsible towards other Mediterranean countries characterized by scant human security.

4.2   Social Vulnerability

The ethical relevance of insufficient/low human security states – that is, of the second claim advanced above – is fortunately more straightforward to vindicate. The degree of vulnerability to climate impacts ought to be considered when characterizing such states ethically. It is useful to refer to a starting-point notion of vulnerability, which as regards social systems is also termed "social vulnerability" (Brooks et al. 2005; Kelly and Adger 2000). This perspective underlines the centrality of the human dimension, in that it focuses on prior damages and not on future stresses, and makes social vulnerability broadly understandable as a state of well-being pertaining directly to individuals and social groups, and whose causes are related not only to climate impacts but also to social, institutional and economic factors such as poverty, class, race, ethnicity, gender (Paavola and Adger 2006). In the context of climate change, the social vulnerability produced by climate impacts can be assumed to endanger a number of critical aspects of well-being: those which, in fact, constitute its vital core and which ultimately define the notion of human security adopted here (see above). The latter thus becomes the metric of social vulnerability itself. Hence it is the compromising of this notion of human security, which encompasses the ability to convert resources into valuable actions against climate impacts, that characterizes the ethical significance of low-level human security countries. What, therefore, is ultimately the moral imperative for putting the most socially vulnerable first, or more precisely, for using human security as the criterion for defining the ethical status of Mediterranean states? Several general constructs of justice validate this imperative. For instance, many liberal theories of justice show particular concern for the weakest parties with least human security. However, there are also universal principles of justice which postulate that people have a moral right not to suffer from the adverse effects of climate change. More specifically, Shue's third general principle of equity of guaranteed minimum (Shue 1999) states, from a sufficientarian standpoint, that those who have less than enough for a decent human life should be given enough, and that being socially vulnerable means being deprived and having far less than enough. Hence these subjects should be given the amount of assistance needed to reach a level of human security sufficient for them to cope with, and to recover from, climate impacts.

In light of these ethical considerations, we argue that it is possible to divide Mediterranean countries into two groups: the group of countries ethically bound to support human security; and that of countries with insufficient/low level of human security. Practically, we adopt a prudential and inclusive perspective – that is, the three-cluster one in the first case – in order to include only the highest-ranking countries, and the two-cluster one to include the largest possible number of countries characterized by insufficient/lower human security. Belonging to the first group (cluster 3, three-cluster analysis) are Cyprus, France, Greece, Israel, Italy and Spain; to the second (cluster 1, two-cluster analysis) Albania, Algeria, Bosnia and Herzegovina, Egypt, Jordan, Macedonia FYR, Morocco, Syrian Arab Republic, Tunisia, Turkey.

5. CONCLUSIONS

The review of the relevant literature has unambiguously shown that climate change has numerous and closely interlinked physical and socio-economic impacts that magnify the threats to human security in the MR. Our empirical analysis has shown that there exists a significant divide within the area under scrutiny whereby EU countries have much higher levels of human security than MENA ones. Importantly, the weakness factors of MENA countries are to be ascribed mainly to social, economic and institutional factors producing human security. The unbalanced levels of human security and of its determinants that characterize the MR move the discourse towards the ethical domain, and more precisely to the development of a regional ethical approach. This we have framed in terms of remedial responsibility and social vulnerability, where a group of states have the responsibility of promoting human security in the weaker states of the region. In particular, in light of the empirical analysis carried out in section 3 and of the theoretical considerations put forward in section 4, it seems possible to claim that cluster 3, three-cluster countries, are ethically responsible for the promotion of the social, economic and institutional factors determining human security in more socially vulnerable cluster 1, two-cluster countries.

We believe that the inclusion of the ethical dimension in a regional approach to human security in the MR may promote its overall increase, and it may mitigate the consequent conflicts among interests, so that the impacts inflicted by climate change on the factors determining human security can be effectively addressed. We argue that, in regions characterized by high degrees of inequality, such as the Mediterranean basin, ethical considerations may also provide reasoned elements for debate among regional stakeholders with regard to the development of a successful strategy for improving human security that devises agreed and unified initiatives. Otherwise, the emerging haphazard approach, in which the notion of human security itself is fragmented and unclear, and even more so its determinants, will lead to the ineffective use of resources to increase the well-being of the entire region.

NOTES

1. See, for instance, the view of the European Central Bank President Mario Draghi in a Q&A session with the *Wall Street Journal* on 23 February 2012 (Internet: http://blogs.wsj. com/eurocrisis/2012/02/23/qa-ecb-president-mario-draghi/, accessed 12 March 2012).
2. Clusters were first created by using ten different methods, i.e. resulting from the combination of two measure calculations (Minkowski at cubic power, and Squared Euclidean Difference) and five clustering methods (Furthest neighbour, Ward's method, Within-group linkage, Between-group linkage, Centroid clustering). The reason for using several methods was the need to achieve reliable data, since clustering analysis is often sensitive to the method adopted. For each method, three solutions were generated, i.e. with 2, 3 and 4 clusters of countries. The consistency between clusters generated with different methods was verified through Spearman's test (see Johnson and Wichern (1998) for further details on the clustering methods).

REFERENCES

Afifi, T., and Warner, K. (2008), 'The impact of environmental degradation on migration flows across countries', Working Paper No. 5/2008, Bonn, Germany: UNU Institute for Environment and Human Security.
Al-Momani (2011), 'The Arab "youth quake": implications on democratization and stability', *Middle East Law and Governance*, **3** (1), 159–170.
Alkire, S. (2003), *A Conceptual Framework for Human Security*. Oxford, UK: University of Oxford, Centre for Research on Inequality, Human Security and Ethnicity (CRISE).
Amin, M., Assaad, R., al-Baharna, N., Dervis, K., Desai, R.M., Dhillon, N.S., Galal, A., Ghanem, H., Graham, C., and Kaufmann, D. (2012), *After the Spring. Economic Transitions in the Arab World*. Oxford: Oxford University Press.
Armingeon, K., and Baccaro, L. (2011), 'The sorrows of young euro: policy responses to the sovereign debt crisis', University of Geneva, Switzerland, http://www.unige.ch/ses/socio/ rechetpub/dejeuner/dejeuner2011-2012/thesorrowsofyoungeuro.pdf
Arnell, N.W. (2004), 'Climate change and global water resources: SRES emissions and socio-economic scenarios', *Global Environmental Change*, **14**, 31–52.
Aryveetey, E., Shantayanan, D., Kanbur, R., and Kasekende, L. (2011), 'The economics of Africa' (Working paper No. WP 2011-07), Charles H. Dyson School of Applied Economics and Management, Cornell University.
Assaad, R. and Roudi-Fahimi, F. (2007), 'Youth in the Middle East and North Africa: demographic opportunity or challenge?', Population Reference Bureau.
Barnett, J. (2009), 'The prize of peace (is eternal vigilance): a cautionary editorial essay on climate geopolitics', *Climatic Change*, **96** (1–2), 1–6.
Barnett, J., Matthew, R.A. and O'Brien, K.L. (2006), 'Global environmental change and human security: an introduction', in Matthew, R.A., Barnett, J., McDonald, B., O'Brien, K. (eds), *Global Environmental Change and Human Security*. Cambridge, USA: MIT Press, pp. 3–32.

Ben Mohamed, A., van Duivenbooden, N., and Abdoussallam, S. (2002), 'Impact of climate change on agricultural production in the Sahel – Part 1. Methodological approach and case study for millet in Niger', *Climatic Change*, **54** (3), 327–348.

Bigio, A.G. (2009), 'Adapting to climate change and preparing for natural disasters in the coastal cities of north Africa', Washington DC, USA: The World Bank.

Black, E. (2009), 'The impact of climate change on daily precipitation statistics in Jordan and Israel', *Atmospheric Science Letters*, 10, 192–200.

Black, R., Adger, W.N., Arnell, N.W., Dercon, S., Geddes, A., and Thomas, D. (2011), 'The effect of environmental change on human migration', *Global Environmental Change*, **21** (Supplement 1), 3–11.

Bou-Zeid, E., and El-Fadel, M. (2002), 'Climate change and water resources in Lebanon and the Middle East', *Journal of Water Resources Planning and Management*, **128** (5), 343–355.

Brauch, H.G. (2001), 'Environmental degradation as root causes of migration: desertification and climate change. Long-term causes of migration from North Africa to Europe', in P. Friedrich and S. Jutila (eds), *Policies of Regional Competition*. Schriften zur öffentlichen Verwaltung und öffentlichen Wirtschaft, Vol. 161, Baden-Baden, Germany: Nomos, 102–138.

Brauch, H.G. (2003), 'Natural disasters in the Mediterranean (1900–2001): from disaster response to disaster preparedness', in H.G. Brauch, P.H. Liotta, A. Marquina et al. (eds), *Security and Environment in the Mediterranean: Conceptualising Security and Environmental Conflicts*, Berlin/Heidelberg, Germany: Springer-Verlag, 863–906.

Brauch, H.G. (2010), *Climate Change and Mediterranean Security: International, National, Environmental and Human Security Impacts for the Euro-Mediterranean Region during the 21st Century. Proposals and Perspectives*. Barcelona, Spain: IEMed.

Brklacich, M., Chazan, M., and Bohle, H.G. (2006), 'Human security, vulnerability, and global environmental change', in Matthew, R.A., Barnett, J., McDonald, B., and O'Brien, K. (eds), *Global Environmental Change and Human Security*. Cambridge, USA: MIT Press, pp. 35–51.

Brooks, N., Adger, W.N., and Kelly, P.M. (2005), 'The determinants of vulnerability and adaptive capacity at the national level and the implications for adaptation', *Global Environmental Change*, **15**, 151–163.

Buhaug, H. (2010), *Climate Not To Blame For African Civil Wars*. Proceedings of the National Academy of Sciences Early edition.

Cohen, B. (2004), 'Urban growth in developing countries: a review of current trends and a caution regarding existing forecasts', *World Development*, **32**, 23–51.

Dabrowski, M. (2010), 'The global financial crisis and its impact on emerging market economies in Europe and the CIS: evidence from mid-2010', Warsaw, Poland: CASE – Centre for Social and Economic Research.

Diamond, L. (2010), 'Why are there no Arab democracies?', *Journal of Democracy*, **21**, 93–112.

Dokos, T., Afifi, T., Bogardi, J., Dankelman, I., Dun, O., Goodman, D.L., Huq, S., Iltus, S., Pearl, R., Pettengell, C., Schmidl, Sl., Stal, M., Warner, K., and Xenarios, S. (2008), *Climate Change: Addressing the Impact on Human Security*. Athens, Greece: Hellenic Foundation for European and Foreign Policy and Hellenic Ministry of Foreign Affairs.

Dreher, Axel (2006), 'Does globalization affect growth? Evidence from a new Index of Globalization', *Applied Economics*, **38** (10), 1091–1110.

Dreher, A., Gaston, N., and Martens, P. (2008), *Measuring Globalisation – Gauging its Consequences*. New York, USA: Springer.

El-Badawi, I., and Makdisi, S. (2007), 'Explaining the democracy deficit in the Arab world', *The Quarterly Review of Economics and Finance*, **46**, 813–831.

Erskine, T. (2003), 'Introduction. Making sense of responsibility', in T. Erskine (ed.), *International Relations: Key Questions and Concepts. Can Institutions have Responsibilities? Collective and Moral Agency*, Basingstoke, UK: Palgrave MacMillan, pp. 1–18.

Erskine, T. (2008), 'Locating responsibility: the problem of moral agency in international relations', in D. Snidal, and C. Reus-Smit, *The Oxford Handbook of International Relations*. Oxford, UK: Oxford University Press, pp. 699–707.

Escribano, G. (2010), 'Southern Europe's economic crisis and its impact on Euro-Mediterranean relations', *Mediterranean Politics*, **15**, 453–459.

Feng, S., Krueger, A.B., and Oppenheimer, M. (2010), 'Linkages among climate change, crop yields and Mexico–US cross-border migration', *Proceedings of the National Academy of Sciences*, **107** (32), 14257–14262.

Ferrara, R.M., Trevisiol, P., Acutis, M., Rana, G., Richter, G.M., and Baggaley, N. (2009), 'Topographic impacts on wheat yields under climate change: two contrasted case studies in Europe', *Theoretical and Applied Climatology*, **99**, 53–65.

Filippetti, A., and Archibugi, D. (2010), 'Innovation in times of crisis: the uneven effects of the economic downturn across Europe', MPRA Paper No. 22084.

Gardiner, S.M. (2004), 'Ethics and global climate change', *Ethics*, **114**, 555–600.

Gesano, G., Heins, F., and Naldini, A. (2009) 'A report to the Directorate-General for Regional Policy Unit Conception, forward studies, impact assessment. Background paper on: Demographic Challenge', ISMERI Europa.

Giannakopoulos, C., Le Sager, P., and Bindi, M. et al. (2009), 'Climatic changes and associated impacts in the Mediterranean resulting from a 2°C global warming', *Global and Planetary Change*, **68** (3), 209–224.

Gibelin, A.-L.G., and Déqué, M.D. (2003), 'Anthropogenic climate change over the Mediterranean region simulated by a global variable resolution model', *Climate Dynamics*, **20** (4), 327–339.

Giorgi, F. (2006), 'Climate change hot-spots', *Geophysical Research Letters*, **33** (L08707), 4 pp.

Giorgi, F., and Lionello, P. (2008), 'Climate change projections for the Mediterranean region', *Global and Planetary Change*, **63** (2–3), 90–104.

Giorgi, F., Bi, X., and Pal, J. (2004), 'Mean, interannual variability and trends in a regional climate change experiment over Europe. II: climate change scenarios (2071–2100)', *Climate Dynamics*, **23**, 839–858.

Grasso, M., and Feola, G. (2012), 'Mediterranean agriculture under climate change: adaptive capacity, adaptation, and ethics', *Regional Environmental Change*, **12** (3), 607–618.

Hoerling, M. et al. (2011), 'On the increased frequency of Mediterranean drought', *Journal of Climate*, **25**, 2146–2161.

Homer-Dixon, T.F. (1994), 'Environmental scarcities and violent conflict: evidence from cases', *International Security*, **19** (1), 5–40.

Hsiang, S.M., Meng, K.C., and Cane, M.A. (2011), 'Civil conflicts are associated with the global climate', *Nature*, **476**, 438–441.

Iglesias, A., Mougou, R., Moneo, M., and Quiroga, S. (2011), 'Towards adaptation of agriculture to climate change in the Mediterranean', *Regional Environmental Change*, **11** (Suppl 1), 159–166.

Institute for Economics and Peace (IEP) (2011), *Global Peace Index*. Sydney, Australia: IEP.

Intergovernmental Panel on Climate Change (IPCC) (2007), *Climate Change 2007 – Impacts, Adaptation and Vulnerability: Working Group II Contribution to the Fourth Assessment Report of the IPCC*. Cambridge, UK: Cambridge University Press.

International Monetary Fund (IMF) (2010), *World Economic Outlook April 2010*. Washington DC, USA: IMF.

International Monetary Fund (IMF) (2012), *World Economic Outlook April 2012*. Washington DC, USA: IMF.

Johnson, R.A., and Wichern, D.W. (1998), *Applied Multivariate Statistical Analysis (4th ed.)*. Upper Saddle River, USA: Prentice Hall.

Johnstone, S., and Mazo, J. (2011), 'Global warming and the Arab Spring', *Survival*, **53**, 11–17.

Kelly, P.M., and Adger, W.N. (2000), 'Theory and practice in assessing vulnerability to climate change and facilitating adaptation', *Climatic Change*, **47**, 325–352.

King, G., and Murray C.J.L. (2002), 'Rethinking human security', *Political Science Quarterly*, **116** (4), 585–610.

Koubi, V., Bernauer, T., Kalbhenn, A., and Spilker, G. (2012), 'Climate variability, economic growth, and civil conflict', *Journal of Peace Research*, **49**, 113–127.

Laipson, E. (2002), 'The Middle East demographic transition: what does it mean?', *Journal of International Affairs*, **56** (1), 175–188.

Liverman, D., and Ingram, J. (2010), 'Why regions?', in Ingram, J., Ericksen, P., and Liverman, D. (eds), *Food Security and Global Environmental Change*. London-Washington: Earthscan, pp. 203–211.

Lonergan, S., Gustavson, K., and Carter, B. (2000), 'The Index of Human Insecurity', *Aviso*, **6**, 1–7.

Maracchi, G., Sirotenko, O., and Bindi, M. (2005), 'Impacts of present and future climate variability on agriculture and forestry in the temperate regions: Europe', in Salinger, J., Sivakumar, M.V.K., and Motha, R.P. (eds), *Increasing Climate Variability and Change*. Springer: Netherlands, pp. 117–135.

Matthew, R.A., Barnett, J., McDonald, B., and O'Brien, K. (eds) (2009), *Global Environmental Change and Human Security*. Cambridge, USA: MIT Press.

Metzger, M., Rounsevell, M., Acostamichlik, L., Leemans, R., and Schroter, D. (2006), 'The vulnerability of ecosystem services to land use change', *Agriculture, Ecosystems and Environment*, **114**, 69–85.

Miller, D. (2001), 'Distributing responsibilities', *The Journal of Political Philosophy*, **9**, 453–471.

Miller, D. (2004), 'Holding nations responsible', *Ethics*, **114**, 240–268.

Miller, D. (2007), *National Responsibility and Global Justice*. Oxford, UK: Oxford University Press.

Moonen, A.C. et al. (2002), 'Climate change in Italy indicated by agrometeorological indices over 122 years', *Agricultural and Forest Meteorology*, **111** (1), 13–27.

Nardin, T. (2006), 'International political theory and the question of justice', *International Affairs*, **82** (3), 449–465.

Nardin, T. (2008), 'International ethics', in D. Snidal and C. Reus-Smit, *The Oxford Handbook of International Relations*. Oxford, UK: Oxford University Press, pp. 594–610.

Newman, D. (2003), 'Boundaries', in Agnew, J., Mitchell, K., and Toal, G. (eds), *The Companion to Political Geography*. Malden, USA: Blackwell, pp. 122–36.

Nicholls, R.J., and Hoozemans, F.M.J. (1996), 'The Mediterranean: vulnerability to coastal implications of climate change', *Ocean and Coastal Management*, **31** (2–3), 105–132.

Noland, M., and Pack, H. (2004), 'Islam, globalization, and economic performance in the Middle East', Policy Brief No. PB-04-4, Institute for International Economics.

Paavola, J., and Adger, W.N. (2006), 'Fair adaptation to climate change', *Ecological Economics*, **56**, 594–609.

Penuelas, J., Filella, I., and Comas, P. (2002), 'Changed plant and animal life cycles from 1952 to 2000 in the Mediterranean region', *Global Change Biology*, **8**, 531–544.

Piervitali, E., Colacino, M., and Conte, M. (1997), 'Signals of climatic change in the Central-Western Mediterranean basin', *Theoretical and Applied Climatology*, **58**, 211–219.

Pratt et al. (2004), *Vulnerability Index (EVI)*. SOPAC Technical Report 383.

Puigdefabregas, J., and Mendizabal, T. (1998), 'Perspectives on desertification: western Mediterranean', *Journal of Arid Environments*, **39**, 209–224.

Romagnoli, A., and Mengoni, L. (2009), 'The challenge of economic integration in the MENA region: from GAFTA and EU-MFTA to small scale Arab Unions', *Economic Change and Restructuring*, **42** (1), 69–83.

Sánchez-Arcilla, A., Mösso, C., Sierra, J.P., Mestres, M., Harzallah, A., Senouci, M., and El Raey, M. (2010), 'Climatic drivers of potential hazards in Mediterranean coasts', *Regional Environmental Change*, **11** (Supplement 3), 617–636.

Scarascia-Mugnozza, G., Oswald, H., Piussi, P., and Radoglou, K. (2000), 'Forests of the Mediterranean region: gaps in knowledge and research needs', *Forest Ecology and Management*, **132**, 97–109.

Scheffran, J., and Battaglini, A. (2010), 'Climate and conflicts: the security risks of global warming', *Regional Environmental Change*, **11**, 27–39.

Schroeter, D., Cramer, W., Leemans, R., Prentice I.C., Arnell, N.W., Bondeau, A., Bugmann, H., Carter, T.R., Gracia, C.A., de la Vega-Leinert, A.C., Erhard, M., Ewert, F., Glendining, M., House, J.I., Kankaanpaa, S., Klein, R.J.T., Lavorel, S., Lindner, M., Metzger, M.J., Meyer, J., Mitchell, T.D., Reginster, I., Rousenvell, M., Sabate, S., Sitch, S., Smith, B., Smith, J., Smith, P., Sykes, M.T., Thonicke, K., Thuiller, W., Tuck, G., Zaehle, S., and Zierl, B. (2005), 'Ecosystem service supply and vulnerability to global change in Europe', *Science*, **310**, 1333–1337.

Schwierz, C., Köllner-Heck, P., Zenklusen Mutter, E., Bresch, D., Vidale, P.-L., Wild, M., and Schär, C. (2010), 'Modelling European winter wind storm losses in current and future climate', *Climatic Change*, **101**, 485–514.

Shue, H. (1996), *Basic Rights. Subsistence, Affluence and U.S. Foreign Policy – Second Edition*. Princeton, USA: Princeton University Press.

Shue, H. (1999), 'Global environment and international inequality', *International Affairs*, **75**, 531–545.

Sorenson, D.S. (2011), 'Transitions in the Arab World. Spring or Fall?', *Strategic Studies Quarterly*, **5** (3), 22–49.

Sowers, J., Vengosh, A., and Weinthal, E. (2010), 'Climate change, water resources, and the politics of adaptation in the Middle East and North Africa', *Climatic Change*, **104** (3–4), 599–627.

Tabutin, D., and Schoumaker, B. (2005), 'The demography of the Arab World and the Middle East from the 1950s to the 2000s', *Population*, **60** (5/6), 505–591 and 593–615.

Thuiller, W. et al. (2005), 'Climate change threats to plant diversity in Europe', *Proceedings of the National Academy of Sciences*, **102**, 8245–8250.

Tol, R.S.J., and Wagner, S. (2010), 'Climate change and violent conflict in Europe over the last millennium', *Climatic Change*, **99** (1–2), 65–79.

Turley, C.M. (1999), 'The changing Mediterranean Sea – a sensitive ecosystem?', *Progress In Oceanography*, **44**, 387–400.

UNDP (2009), *The Arab Human Development Report. Challenges to Human Security in the Arab Countries*. New York, USA: UNDP.

UNEP (2005), *Building Resilience in SIDS The Environmental Vulnerability Index*. United Nations Environment Programme (UNEP), South Pacific Applied Geoscience.

UNEP/MAP (2009), *State of the Environment and the Mediterranean*. Athens, UNEP/MAP.

UNU (United Nations University) (2007), *Measuring Human Well-Being: Key Findings and Policy Lessons*. UNU WIDER Policy Brief 3. Helsinki, Finland: UNU WIDER.

Verney, S. (2009), 'Flaky fringe? Southern Europe facing the financial crisis', *South European Society and Politics*, **14**, 1–6.

Warf, B. (2011), 'Myths, realities and lessons of the Arab Spring', *The Arab World Geographer*, **14** (2), 166–168.

Warner, K. (2010), 'Global environmental change and migration: governance challenges', *Global Environmental Change*, **20** (3), 402–413.

World Bank (WB) (2003), *Trade, Investment, and Development in the Middle East and North Africa*. Washington DC, USA: World Bank.

11. Climate change and human security in the Arctic
Mark Nuttall

INTRODUCTION

It is no longer the case that climate change can be viewed as an environmental issue to be researched, discussed and potentially managed without consideration of the human dimensions. Indeed, the scientific consensus that climate change (as opposed to climate variability) is anthropogenic in its origins draws attention not only to its underlying human causes but to its social, cultural and economic impacts and consequences. This underscores the reality that climate change is a fundamentally human issue. This opening statement may seem obvious, given what we know about, how we respond to, and how we anticipate the impacts of climate change on societies across the world. However, the study of climate change is still often considered the prerogative of the physical and natural sciences and, to some extent, is still dominated by work on climate scenarios and global circulation models, with social scientists called upon by scientists and policy-makers to explain and justify exactly what they have to contribute to their refinement. As such, the social sciences are presented with theoretical and methodological challenges in how they study and evaluate climate impacts on society, culture, economy, politics, and everyday human life, as well as reflecting on the advocacy and policy positions social scientists should take. For the most part, this has led to uncertainty and disagreement in disciplines such as sociology and anthropology as to what an appropriate social science approach to climate change should be (e.g., Lever-Tracy 2008; Grundman and Stehr 2010).

Yet things are changing and social scientists do not (even if they ever did) necessarily see their role in climate change research as something that is an add-on to the science of global circulation models, greenhouse gas emissions scenarios and projections of future climate. In particular, anthropological engagement with climate change in a variety of ethnographic settings and methodological and theoretical areas has been increasing in recent years (e.g. Crate and Nuttall 2009; Hastrup 2009). Anthropologists are also finding roles to play in regional and global assessments of climate change, theorizing and conceptualizing how to bridge temporal and spatial scales, as well as illuminating ways of understanding how to disentangle the effects of natural variability and change on the environment from those of human action. Such work involves dialogue within and across the social sciences and the natural sciences and is advancing understanding of the ways in which the study of anthropogenic climate change allows for an interdisciplinary focus on the mutuality of nature and society without necessarily subscribing to a systemic approach that privileges ecological perspectives over an understanding of the richness and complexity of the human world.

In recent work, arguments have been put forward by anthropologists and others that understanding the human dimensions of climate change necessitates a discussion of how it poses threats to, and undermines, human security (Barnett and Adger 2007; Crate and Nuttall 2009). This places emphasis on socio-economic conditions and situations, political settings, rights, ownership and entitlements, resource dependence, and vulnerability and resilience. Climate change not only raises questions about human security, however, but highlights the precarious nature of human insecurity. As Barnett and Adger (2007: 641) put it,

> The vulnerability (potential for loss) of people to climate change depends on the extent to which they are dependent on natural resources and ecosystem services, the extent to which the resources and services they rely on are sensitive to climate change, and their capacity to adapt to changes in these resources and services. In other words, the more people are dependent on climate sensitive forms of natural capital, and the less they rely on economic or social forms of capital, the more at risk they are from climate change.

Barnett and Adger point out the importance of recognizing that environmental change does not undermine human security in isolation from a broader range of social factors, including, among other things, poverty, the degree of support (or conversely discrimination) communities receive from the state, their access to economic opportunities, and the effectiveness of decision-making processes. All of this affects the abilities of communities to respond and adapt to climate change effectively.

This chapter considers climate change and human security in the high latitudes of the circumpolar North. Both polar regions—the Arctic and Antarctica—play crucial roles in global climate dynamics, and both regions are experiencing rapid physical changes in response to global warming. Scientists have long understood the importance of studying the polar regions because of the ways they influence the Earth's weather systems—for example, the sea ice in both the Arctic and Antarctic is a major element in the global climate system, while the Southern Ocean plays a significant role in processes of biogeochemical cycling and the exchange of gases between the ocean and the atmosphere. Meteorological stations show the Antarctic Peninsula has experienced strong and significant warming over the last 50 years, while Antarctic sea ice extent has also been affected, although satellite data since 1978 show regional trends rather than a ubiquitous trend in sea ice duration—for example, sea ice duration has increased in the Ross Sea, but decreased in the Bellingshausen Sea (Chapin et al. 2005; Lemke et al. 2007). In the Arctic, observations of the lowest recorded extent of sea ice in September 2007 and September 2012 have provoked concerns from scientists that the previous distribution cannot be retrieved if scenarios of ice-free summers are realized (Wadhams 2012).

The Arctic and Antarctica also contain the world's largest ice sheets. In the circumpolar North, the melting of the Greenland inland ice has captured scientific and public attention in recent years as one of the starkest examples of global climate change, placing Greenland at the epicenter of discussion about processes of ecological reconfiguration and topographical reshaping (Nuttall 2009). Scientific climate models suggest that average temperatures in Greenland will rise by more than 3°C this century, which would mean large-scale melting of the inland ice. Even if future conditions were to stabilize somewhat, the more widely accepted scenario is still one where the inland ice will eventually melt completely, even though it will take the next two to three centuries to do so. The entire Greenland ice sheet contains enough water to raise global sea levels by seven metres (AMAP 2011).

Although they are often assumed to be regions worthy of comparative study, the Arctic and Antarctic are polar opposites in several important ways. They are fundamentally different geographically in that a large portion of the Arctic consists of the ice-covered Arctic Ocean, which is surrounded by many islands and archipelagos, and the northern parts of the mainland areas of the North American and Eurasian continents, whereas Antarctica is an ice-covered landmass—a continent in its own right—surrounded by an ocean. The Arctic and Antarctica also have different environmental patterns, climatic systems, wildlife habitats, and social and political settings. For instance, there is much lower terrestrial biodiversity in the Antarctic than in the Arctic—and there are no indigenous peoples living in Antarctica. Indeed, Antarctica has no permanent human population—although the operations of scientific research bases and the seasonal visits of tourists and others means there is a year-round human presence. In Antarctica, issues of human security and climate are largely about governments' concepts of security, often defined in terms of sovereignty, natural resources and the continuation of national Antarctic science programmes. Concern over ice-shelves breaking away or sea ice distribution frames human security in Antarctica in different ways from how it is usually envisaged in other parts of the world. Antarctica has also been subject to an international framework for environmental management and conservation under the Antarctic Treaty System for the past fifty years, whereas no such regime exists in the northern circumpolar region. The Arctic, however, does have a permanent human population of around four million (although this figure is closer to ten million depending on the broader definition of the circumpolar North – see Nuttall 1998). Apart from some areas of the Arctic Ocean, the Arctic is encompassed in the territories of eight sovereign nation states—the United States of America, Canada, Denmark/Greenland, Iceland, Norway, Sweden, Finland, and the Russian Federation. Indigenous peoples also comprise a minority of northern populations in Arctic countries (except in Greenland where they are the majority) and so the concerns of human security become apparent and immediate for many communities in a region undergoing rapid climate change.

There is significant evidence of threshold changes in northern hemisphere climate patterns and temperature-induced Arctic ecosystem changes. Many scientists now argue that the climate of the Arctic may be approaching a "tipping point" beyond which will lie a future shaped by dramatic, far-reaching, and irreversible climatic, environmental, economic, political and social change (e.g., Arctic Council 2013; Nuttall 2012a; Wadhams 2012; Wassmann and Lenton 2012). Although climate change affects all northern residents, the Arctic's indigenous peoples and local communities dependent on fishing, hunting, herding and agriculture are facing special challenges to traditional knowledge and understanding of the environment, that make prediction, travel safety, and access to customary food sources more difficult (e.g., Krupnik and Jolly 2002; Nuttall et al. 2005).

In this chapter, I discuss how the local and regional consequences of ecosystem transformation arising from global climate change have profound and far-reaching implications for human security and human-environment relations. Because of the diversity of human populations across the eight states of the Arctic, I limit myself to discussion that makes particular reference to research carried out in Inuit communities in Greenland, Canada and Alaska. However, I point to the need to understand the consequences of climate change in context—rapid social, economic and demographic change, resource management and resource development, animal-rights and anti-whaling campaigns, trade barriers and conservation policies have significant implications for human security in the Arctic. In many cases, climate change merely magnifies existing societal, political, economic, legal, institutional and environmental challenges to human security that northern peoples living in resource dependent communities experience in their everyday lives (Heikkinen et al. 2011; Lynge 1992; Nuttall et al. 2005; Wenzel 1991).

CLIMATE CHANGE AND THE ARCTIC

Recent regional and global climate change assessments carried out under the auspices of the Arctic Council and the United Nations Intergovernmental Panel on Climate Change (IPCC) confirm that the Arctic has been warming more rapidly than any other region of the globe. For several decades, surface air temperatures have warmed in the Arctic at approximately twice the global rate. Sea ice extent is diminishing, the area of discontinuous permafrost is undergoing a thaw, there are significant reductions in seasonal snow patterns and to the mass balance of glaciers, terrestrial and marine animals are being affected, and the cultures and live-lihoods of northern residents are under threat (ACIA 2005; AMAP 2011; Anisimov et al. 2007).

A number of seas border the northern rims of the North American and Eurasian continents. Each of these stretches of water, which include the Beaufort, Barents, Kara and Laptev seas, has its own distinctive marine ecosystem but they all form part of the Arctic Ocean. As much as 90 percent of the surface of the entire Arctic Ocean is usually covered by ice in winter and 70 percent year round, although scientists have observed changes in the area, extent and thickness of multi-year (or perennial) sea ice cover over the last thirty years. An increase of average Arctic air temperatures by 1.5°C since 1980 has reduced summer sea ice extent from 8 million km^2 in the 1970s to 3.4 million km^2 in 2012. This dramatic summer ice retreat has been accompanied by a significant decrease in sea ice extent in other seasons (Stroeve et al. 2012), and to changes in the types of ice, especially a reduction in multi-year ice (Wadhams 2012). Furthermore, the mean thickness of sea ice has decreased by 40 percent (Wadhams 2012), and oceanographers have reported a reduction of 73 percent in the frequency of ridges and have observed major changes in ice dynamics (Wadhams and Davis 2000). Sea ice is highly dependent on the temperature difference between ocean and atmosphere, and on near-surface oceanic heat flow, and reacts quickly to changes in atmospheric conditions. These rapid changes are set to continue with some atmosphere-ocean climate models predicting a reduction in sea ice of around 60 percent in the next 50–100 years, while other coupled models predict an "ice-free" Arctic summer by 2040 (e.g. Wang and Overland 2009). Still others (e.g. Maslowski et al. 2012) predict an ice-free September within the next five to eight years.

The likely disappearance of multi-year ice in the Arctic Ocean and other northern seas will be immensely disruptive to ice-dependent micro-organisms and will have tremendous consequences for Arctic wildlife. The effects of climate change on Arctic ecosystems have significant implications for the distribution and trophic dynamics of wildlife populations and entire food webs—as the amount of sea ice decreases, for example, seals, walrus, polar bears and other species that depend on it would suffer drastically—although quantitative predictions are made difficult because of the interaction of climate with other processes and variables which results in complex responses at the population level (Krebs and Bertaux 2006). Nonetheless, there are increasing concerns that, in the immediately foreseeable future, climate change will have a potentially devastating and transformative impact on the Arctic environment and on the resource-dependent livelihoods of northern communities, particularly the indigenous peoples of the region such as Inuit in Greenland, Canada, Alaska and Chukotka, Sámi of northern Fennoscandia, and peoples of the Russian North such as Evenki and Nenets, whose livelihoods, societies and cultures are inextricably bound up with the Arctic environment and its wildlife (Nuttall et al. 2005).

Animals are not viewed merely as economic resources by northern peoples, and "wildlife" is not an indigenous categorization informing local understandings of commensalism that best describes the intricate human-animal relations that infuse circumpolar places (Anderson and Nuttall 2004; Kalland and Sejersen 2005; Nuttall 1992). Nor is it appropriate to describe and reduce northern landscapes and marine environments to ecosystems. The circumpolar North is a human world in which the movement of animals, the advance and retreat of glaciers, the dripping and trickling sounds of meltwater, the texture and distribution of thinning ice, the increasing frequency and ferocity of storms and the forces of coastal erosion are considered to mean something powerful and extraordinary, as well as troubling and worrying, to those who live there, indicating the sentience, agency and unpredictability of a world of becoming (Nuttall 1992, 2009).

Because many northern species of terrestrial and marine mammals as well as freshwater and ocean fish are a cornerstone of local community and regional economies, climate change poses a threat to food security in northern regions because it influences animal availability, human ability to access wildlife, and the safety and quality of wildlife for consumption (Meakin and Kurvits 2009). For example, in northern Canada, residents of First Nations communities in both Yukon Territory and the Northwest Territories have been witness to changes in climate that are affecting the availability of species, but they are also experiencing difficulties in their abilities to access and harvest them and there has been a corresponding decline in the nutrient intake from traditional foods (Guyot et al. 2006).

Research on the actual effects of climate change and climate variability on animals hunted and consumed in indigenous communities is, however, still limited. What we do know—with a great degree of certainty—is that migration routes of caribou and reindeer, seals, whales, fish and geese will be disrupted and likely shift, with consequences for the hunting, herding, trapping and fishing economies of many small, remote Arctic settlements and communities (Nuttall et al. 2005). Most Arctic mammals and fish species depend upon the presence of sea ice and, in turn, many Inuit coastal communities in Greenland, Canada, Alaska and Chukotka depend on harvesting these species. Climate variability appears to have already caused relatively rapid shifts in the composition of Arctic marine ecosystems, especially in the Bering and Barents seas. It is difficult for climate scientists, though, to determine whether such changes are happening because of natural environmental fluctuations or human activities. The eastern Bering Sea, for example, has experienced significant changes in the past 100–150 years, in part due to changes in climate but also to commercial exploitation of marine mammals, fish and invertebrates. Climate change may have contributed to animal population changes, but the impacts of commercial whaling in the nineteenth and twentieth centuries on the ecosystems of northern seas have also taken a significant toll (Nuttall et al. 2005). Higher ocean temperatures and lower levels of salinity, changes in seasonal sea ice extent, rising sea levels and many other (as yet undefined) effects are certain to affect marine species significantly, with implications for Arctic coastal communities dependent on hunting and fishing. Furthermore, climate change could also result in serious consequences for the health and well-being of indigenous peoples, in particular from indirect impacts on the epidemiology of infectious diseases in both humans and wildlife and, given the traditional diet based on products of hunting and fishing activities, from the occurrence of zoonotic food borne and waterborne diseases.

Concern over the effects of climate change on water resources has also been expressed in recent years (IPCC 2008). Some of the longest rivers in the world are found in the Arctic, and they include the Yukon, Mackenzie, Ob, Lena and Yenisei. A significant amount of Arctic water, especially flowing into the Arctic Ocean, originates in the headwater basins of these rivers. Huge deltas and wetlands are also found in the far northern reaches, the largest being Russia's Lena River Delta. The second-largest Arctic delta is Canada's Mackenzie Delta which is fed by the Mackenzie River. Some 1,800 km in length, the Mackenzie River is the main branch of the second-largest river system in North America (after the Mississippi-Missouri river system). Its watershed, the Mackenzie Basin, drains approximately 20 percent of Canada. Glacial melt from much of the Canadian Rockies eventually ends up in the Arctic Ocean or other northern waters. For example, the North Saskatchewan River winds its way through the boreal and aspen parkland landscapes of northern Alberta. Its source, like those of several other great Alberta rivers, is high in the Rockies, and it begins as a small trickle of meltwater from the Saskatchewan Glacier which extends from the Columbia Icefield in Jasper National Park. Eventually, the waters of the North Saskatchewan flow into Hudson Bay. The Athabasca River likewise originates from an outflow glacier of the Columbia Icefield—the Athabasca Glacier—and, carving its way through Alberta's north country to Lake Athabasca, empties its waters into the Slave River which joins the Mackenzie River via Great Slave Lake, ultimately ending up in the Beaufort Sea. The water systems of the Arctic and Subarctic are particularly susceptible to current and predicted climate change. Virtually all major hydrological processes and related aquatic ecosystems are affected by the presence and absence of snow and ice.

CHANGING ARCTIC ECONOMIES

Climate change will bring economic benefits as well as mean social and economic costs to the Arctic. Climate warming may enhance biological production in some fish species, for example, with positive results for commercial fisheries. Extractive industries have been a significant driving force for environmental and socio-economic change in parts of the Arctic for over a century and today much of the circumpolar North is on the verge of major energy and mineral exploration and development (Nuttall 2010a). One recent study by the United States Geological Survey suggests that 25 percent of the world's untapped reserves in areas known to contain oil are found in the Arctic, making the Arctic attractive to the oil and gas industry as the final frontier for hydrocarbon extraction. Current developments in Alaska, Canada, Norway and Russia and further exploration in Greenland will likely continue as climate change contributes to reductions in sea ice, opening new sea and river routes and reducing exploration, development and transportation costs.

Global political changes, issues of sovereignty and energy security, and growing global energy needs and demands strongly influence the patterns, rates and economics of resource exploration and extraction at high latitudes. Countries such as China, India, South Korea and European Union member states are looking to the Arctic as a crucial source of oil, gas and minerals that could meet their energy needs in a future characterized by resource scarcity. Increased interest in the potentially resource-rich circumpolar continental shelf raises the prospect of international disputes over Arctic sovereignty. As the world looks northwards for its supplies of oil and gas, territorial challenges are provoking nations like Canada and Russia to reassert their claims over their northern hinterlands (e.g., see Byers 2009). Canada's Northwest Passage, for example, is not recognized as Canadian waters by the United States. Instead, the US argues the Northwest Passage is an international strait. Furthermore, there is an unresolved maritime boundary dispute between the US and Canada, extending from the Alaska-Yukon border into the Beaufort Sea. Russia faces similar issues over the future use of the Northern Sea Route. Once part of a major Arctic transportation system, Russian shipping in the Northern Sea Route has declined in recent years, but many other countries including the US, China and Japan consider it a potentially important transportation artery and increased maritime traffic is anticipated.

Opportunities to develop new global trade links may arise as shipping routes open up across an ice-free Arctic Ocean.The oil, gas and mining industries would benefit from lower operational costs, and tourists will find that access to previously remote and inaccessible places will be easier. The danger, however, is that these benefits will accrue largely to powerful transnational corporations and foreign companies rather than northern indigenous and local communities. New shipping routes and global interest in developing extractive industries in the Arctic also raise issues concerning resource governance and environmental protection. True, increased development of non-renewable resources may promise jobs, money and prosperity, but there is urgent need for indigenous peoples to be able to participate effectively in the planning processes for such development and to assess whether development in a climate-changed Arctic will ensure economic, cultural and environmental sustainability (Nuttall 2010a; Sirina 2009). If climate change is to alter the social, cultural and physical environments of the Arctic, then indigenous people must be assured that they will play a key role in the regional and global dialogues that will determine the kind of economic development that will take place in their homelands. This is reinforced by the Inuit Circumpolar Council's (ICC) declaration on principles for resource development (ICC 2009). For the most part, however, as climate change becomes apparent in the nature of increasingly frequent extreme weather events, the scenario is an uncertain, unpredictable and unfamiliar future for indigenous peoples. Their livelihoods and cultures—indeed the very future of human security in the Arctic—may well depend on their ability to adapt to—but also to anticipate—climate change and to participate in and benefit from, the shaping of the new forms of economies, governance, livelihoods, and sustainable development necessary to meet the challenge of climate change.

INUIT AND CLIMATE CHANGE

In the contemporary Arctic, indigenous peoples in many small, often remote communities, as well as those living in larger towns in some places, maintain a strong connection to the Northern environment through customary activities of hunting seals, whales, or caribou, by fishing, and by herding reindeer, which provide the basis for livelihoods and food production in a way that usually marks them off from non-indigenous people who do not live in such resource dependent communities. Indeed, one of the defining attributes of being indigenous in the Arctic today is that it refers to the quality of a specific people relating their identity and their cultural and economic well-being to a particular place (be it a local community or to the broader idea of belonging to the Arctic), but also to their traditional cultural and economic dependence on local resources, thus distinguishing them culturally from other peoples such as more recent settlers and immigrant communities. Traditional food-related activities have crucial economic and dietary importance; they are also important for maintaining social relationships and cultural identity (Nuttall et al. 2005).

At the risk of generalizing, it is nonetheless fair to say that from east Greenland, across to Greenland's west coast, throughout northern Canada and northern and western Alaska, and to the Bering Sea coast of Siberia, Inuit have often responded creatively to changes in the environment and to the movement of terrestrial and marine species. Their ability to thrive by hunting and fishing in the extreme environment of the planet's far northern reaches has often been influenced by fluctuations in climate, the changing ecology of the Arctic, and the availability of animals in specific habitats. But such human presence in the North over thousands of years dispels the scientific description of the biodiversity of northern places as low, meager and sparse. Seasonal variations in climate and extreme and unusual weather events have often greatly altered the abundance and availability of animals, with implications for the ability of hunters and fishers to travel on winter sea ice, open water in the summer, across northern terrestrial landscapes, or through forests to secure animals and fish for food, clothing and other uses. Many of the animals Inuit rely on, for instance, such as seals and whales, are migratory, with some species only available seasonally and in localized areas. Yet Inuit societies have developed the capacity and flexibility, as well as the skills and technology, to harvest a wide variety of animal and plant species, in all seasons and practically any weather conditions. Like other hunting peoples, Inuit have not perceived or considered the environment and animals as distinct from human society and human-environment relations and human-animal relations are best described as a relational ontology of mutual engagement (Bird-David 1990; Fienup-Riordan 1983; Nuttall 1992).

Climate change, however, now appears to be testing Inuit in ways they have not experienced in living memory, with implications for the disruption of human-environment relations. According to Inuit reports, contemporary environmental changes in climate and local ecosystems can be noticed not only in the altered migration routes and behaviour of animals such as seals or caribou, but in the very *taste* of animals. I recall an elderly Inuk from western Nunavut at a community workshop on climate change that was held several years ago in Iqaluit, the capital of Canada's Nunavut territory, once describe the difficulty she found in sewing caribou skins to make clothing. There was something not quite right with the fur of the caribou now, she remarked. She had also noticed that caribou were leaner and the meat was not easy to cook. Nor did it quite taste the same. She wondered what was the reason for this difference, but cautioned the other participants at the workshop, all of whom had their own stories to tell about the changes they were seeing and experiencing, that she did not know whether climate change was to blame—it was just that she was observing changes in the caribou near her community and that it was something to worry about.

The dependence of Arctic indigenous societies on traditional resource use continues for several critically important reasons. One is the obvious economic and dietary importance of being able to secure access to customary, local foods. Many of these local foods—fish, and meat from marine mammals or caribou and birds, for instance, as well as berries and edible plants—are nutritionally superior to the foods which are presently imported into remote Arctic communities (and which are often expensive to buy—and it is the case that more perishable foods such as fresh vegetables are often unavailable in remoter communities). Another reason is the cultural value and social importance of hunting, herding, and gathering animals, fish, and plants, as well as bringing animals home, processing, distributing, consuming, and celebrating them. Animals, therefore, not only provide indigenous peoples with the economic and nutritional bases for survival, they remain important for social identity and cultural survival. Dependence on animals for food and social, cultural and economic well-being is reflected in community hunting and fishing regulations (for example as evident in seal hunting and whaling communities in Greenland, Alaska and northern Canada), in herding practices (in reindeer-dependent societies in northern Fennoscandia and Siberia) and, across the Arctic, in patterns of sharing and reciprocity based on social relatedness (Caulfield 1997; Dahl 2000; Kalland and Sejersen 2005; Nuttall 1992). As I have witnessed and experienced during fieldwork in northern communities in Greenland and Finland, these aspects of culture, of social relatedness between persons, and between persons, animals and the environment form a fundamental basis for how communities think about resilience (Heikkinen et al. 2011; Nuttall 1992). At the same time they are essential for bodily and cultural health and well-being (Nuttall et al. 2005).

Having no access to traditional foods is a serious threat to human security in northern communities. Fergurson (2011) has shown how food insecurity affects Inuit communities throughout Canada's Arctic. In particular, she argues that those who are most vulnerable are often the most impoverished. Furthermore, while the duality of Inuit food systems has the potential to support a diverse supply of food, various pressures threaten these food systems. Fergurson argues that when both traditional and western foods are available they are not necessarily accessible or acceptable as a result of overlapping social, economic, environmental and political factors, such as socio-economic change, climate change and geography, the impact of specific policies and legislation, and the influence of the international community and environmental organizations. For central west Greenland, Ford and Goldhar (2012) also point to how constrained access and availability of key wildlife resources and increased harvesting dangers are affecting individuals and households closely linked to the traditional subsistence economy.

The impacts of climate change, however, run deeper. In Greenland, for example, people do not generally experience or talk about the environment, *nuna*, in temporal and spatial terms as something set apart from the human world, or as an ecosystem as characterized, described and defined by science, but as a fundamental part of the human world, a place in which people dwell, move around, and in which they engage in social relations of morality and exchange with animals and other constituent aspects of it, such as sea ice, glaciers, mountains, rivers, and so on. Inuit describe such shared relations as *nunaqatigiit*. The epistemological and ontological qualities of *nunaqatigiit* are also expressed in perceptions and understandings of *pinngortitaq*. Often simply translated from Greenlandic as "nature" or "creation", the literal meaning of *pinngortitaq*, however, is "to come into being". The word *pinngorpoq* refers to a continuous process of "becoming", "to come into existence", indicating the unfolding of possibility and opportunity in the world and indicating the difficulty of seeing a separation of nature and society in the Greenlandic worldview (Nuttall 2009). *Pinngortitaq* is not a mere description of an observed material, tangible nature and its properties, but an expression of a world that is experienced as one of movement, becoming, and emergence. Over the years, I have worked extensively in many parts of Greenland, in small villages as well as in larger towns such as the capital Nuuk, and I have been struck by the fact that people do not necessarily talk of the environment around them as *changing*, or being transformed, but of it being in a constant process of *becoming* (Nuttall 2009).

The maintenance of human security and the avoidance of human insecurity depends largely on how people perceive and conceptualize change and such processes of becoming—in short, people's world view goes some way to determine the kinds of reactive, adaptive or anticipatory strategies they utilize and it shapes and influences the nature of community perspectives on resilience to change. In Greenland, for instance, *nuna*—and everything in it—is said by people to have its own essence (*inua*); for example, *qaqqap inua*, "the essence of mountains", while the inland ice has *sermersuap inua*, "the essence of the great ice". Weather, or climate, or "the outside", is known as *sila*, and *silap inua* is "the essence of *sila*"—but its meaning is deeper, and people understand and talk about *sila* as the breath of life, the reason things become and how they move and change. *Sila* is also the word for "intelligence/consciousness", or "mind" and it integrates the self with the natural world (Nuttall 1992, 2009). Inuit talk about this integration of weather, climate, the outside, and of the individual mind as *silaqatigiit*. As *sila* links the individual, *nuna* and *pinngortitaq*, a person who lacks *sila* is said to be separated from an essential relationship with the environment that is necessary for human well-being and balance. Lack of *sila* can be explained and observed as a temporary disorientation often evident when someone is said to experience *silaqaraluarneq*, the state of being out of one's mind (*silaqaraluarneq* can also mean "the weather is out of its mind"). In this way climate change is perceived as potentially disruptive to *silaqatigiit*.

Given this context, when some people in Greenland experience a change in the weather, this change is experienced in a deeply personal way and as an unraveling of the ontological relations inherent in *silaqatigiit*. And when they talk about their concerns about climate change, they articulate them in terms of how their own sense of self, personhood and well-being is changing in relation to external climatic fluctuations and their effects on ice, currents, wind direction and the movement and behaviour of animals. Yet, at the same time climate change is also often understood as being consistent with the constant making of the world as one of becoming, with its inherent uncertainty, and with the environment coming into existence, moving and reshaping through continuous actualization and realization. However, the current rate of such change is faster than many people recall having experienced it in living memory (Nuttall 2009).

How Greenland Inuit respond to change and remain resilient in a world of movement and becoming is dependent, in part, on them continuing to learn how to grow up and dwell in an environment where one is always prepared for surprise, where one is constantly challenged by uncertainty, and where one can never take anything for granted. But as I have written elsewhere, based on extensive fieldwork in northwest and south Greenland, being resilient in the face of such change also depends on the strength of a sense of community, kinship and relatedness, and close social association (including the relational ontologies of engagement entangled in *nunaqatigiit* and *silaqatigiit*). In a world of flux, uncertainty and unpredictability, social relationships are a source of constancy (Dahl 2000; Nuttall 1992, 2009). If a person breaks from networks of kin and social relationships, they are set adrift from the human security of their social world. Loss of community is thus a threat to individual and social identity and to human security and, combined with loss of livelihood, exposes people to the impacts of climate change in a way that makes it difficult for them to respond effectively, if at all. To become a stranger in one's own land does not happen solely because the environment has changed, but also because political, economic and social changes threaten the social cohesion of community, endanger one's livelihood and separate one from a fundamental relationship with people, animals and place, and from *nuna*, *pinngortitaq* and *sila* (Nuttall 1992, 2009).

COMMUNITIES ON THE EDGE

Recent scientific research reveals that there has been a distinct warming trend in the lowland permafrost of northern regions of 2–4°C over the last hundred years, with thawing, thermokarst formation and severe erosion disturbing animal and human activities. Further warming is likely to continue this trend and increase the hazards and risks people face when involved in travelling across the land in permafrost regions. It will also result in pressure on buildings, communication and transportation links, industrial infrastructure and oil and gas pipelines and production facilities, posing challenges to architectural and engineering design (ACIA 2005). The widespread thawing of discontinuous permafrost in Alaska illustrates some of these hazards and the implications for both landscape change and the physical infrastructure of communities.

In western Alaska several Inuit communities in low-lying coastal and island areas, including Shishmaref, Kivalina and Little Diomede have been affected by recent climate change and regular extreme weather events and face severe problems due to the erosion and thawing of discontinuous permafrost, as well as the increased frequency and severity of storms. The Iñupiat village of Shishmaref, located on Alaska's Sarichef Island in the Chukchi Sea is home to some 560 people who rely on a subsistence lifestyle of hunting and fishing. Coastal erosion and storm damage have forced the residents to vote in favour of relocating the village to the Alaskan mainland. The residents of Shishmaref and other nearby Alaskan coastal communities may not be the only ones facing such challenges and they may become more widespread in vulnerable coastal regions throughout the global North. Scientists say that permafrost throughout the Arctic will thaw more quickly in spring, but take longer to refreeze in autumn, and permafrost boundaries will gradually move poleward, with most of the ice-rich discontinuous permafrost disappearing by the end of the twenty-first century. The tree line will move further north concurrently. The dark evergreen forests characteristic of the boreal regions south of the tree line are beginning to advance northwards into areas where cedars, firs, pines, spruces, and birches have not been known to grow (ACIA 2005).

Unstable sea ice is beginning to make ice-edge hunting more difficult and dangerous in parts of Greenland, Alaska and northern Canada. I have witnessed how some hunters in northern Greenland become anxious about travelling to the edge of the ice (*sinaaq*) in ways they have never felt before—they say it is "more slippery" (*quasarpoq*) than they have known it, and that they feel more secure travelling by dogsled on the solid ice that is attached to the shore (*qaanngoq*). Changes in snow cover are also causing difficulty in accessing hunting and fishing grounds by dogsled or snowmobile, making local adjustments in winter travel, hunting practices and fishing strategies necessary (Nuttall 2010b). In central west Greenland, hunters and fishers are employing adaptation strategies that combine both reactive and anticipatory interventions undertaken at both the individual and household levels, including travelling to new fishing grounds, seeking alternative sources of income when harvesting activities are not possible, preparing for the unexpected, and an increased reliance on boat transport during winter (Ford and Goldhar 2012). In east Greenland, hunting communities are facing increasing uncertainty and hunters are turning to the growing tourism industry—which in itself is one side-effect of global warming (Buijs 2010). Similar experiences and responses can be found in other parts of the Arctic. Indigenous peoples in Alaska have already reported decreasing snow in autumn and early winter, but substantial snowfall in late winter and early spring. According to local hunters, a lack of snow makes it difficult for polar bears and ringed seals to make dens for giving birth or, in the case of male polar bears, to seek protection from the weather. The lack of ringed seal dens may affect the numbers and condition of polar bears, which hunt ringed seals and often seek out the dens.

People in northern coastal Alaska and in northern Canada are concerned that hungry polar bears may be more likely to approach villages and encounter people, adding a different sense of danger and risk to daily life and the human security of remote communities.

In Nunavut, Inuit hunters have noticed not only the thinning sea ice as an indicator of change, the appearance of birds not usually found in their region (and for which they have no words in Inuktitut) tells them something unusual is happening to the weather. Iñupiat hunters in Alaska, as well as Inuvialuit in the Mackenzie Delta, report that ice cellars dug deep into the permafrost are too warm to keep meat and fish frozen. In Tuktoyaktuk in northwest Canada, for instance, the community's ice cellar was excavated more than 50 years ago to keep seal meat, fish, and whale meat, as well as caribou and other traditional foods frozen. The permafrost is glacially deformed and the ice cellar provides a fascinating glimpse of the ground-ice stratigraphy that was formed during the last glacial-interglacial period. In 2005, I heard hunters in Tuktoyaktuk remark on how much warmer it is in the cellar than usual and watched them prod at meat that was partially defrosted—"It shouldn't be like this," they said, "we worry that our food supplies are beginning to rot away."

CLIMATE CHANGE IN CONTEXT

As the climate changes in the Arctic, and as terrestrial and marine environments are affected and transformed, Inuit and other indigenous peoples of the global North are facing special challenges to the sustainability of livelihoods and communities and, ultimately, to human security. Their abilites to harvest wildlife and food resources are already being tested and while in some cases, such as Berkes and Jolly (2001) report for the Inuvialuit community of Sachs Harbour in the Canadian western Arctic, and Ford and Goldhar (2012) show for Qerqertarsuaq in west Greenland, hunters have come up with reactive and adaptive strategies, for the most part becoming resilient to climate change, and preparing to respond, cope with, and adapt to its impacts, risks and opportunities, thus ensuring human security for northern communities, will require urgent and specific policies and action at both national and international levels. In northern Canada, Fergurson (2011), for instance, discusses how several policy initiatives at the federal and territorial level, as well as local community level responses are targeting the problem of Inuit food insecurity. These approaches are significant, but she calls for a more multi-faceted approach to food insecurity that must emphasize policies that: (1) improve purchasing power by reducing poverty, (2) address the reality of climate change and Inuit adaptation within a warming Arctic environment, and (3) recognize the potential impacts of political interventions and external influences.

It is also important to view whatever changes are happening to the climate of the Arctic within the context of other kinds of changes affecting indigenous societies. Large-scale industrial development, oil exploration and development and mining projects, commercial fishing, and processes of globalization have far-reaching consequences for the Arctic and therefore magnify the likely impacts of weather and climate variation on indigenous peoples and their livelihoods (Nuttall 2010a). Varying scenarios suggest that climate change will have impacts on marine and terrestrial animal populations, affecting their size, structure, reproduction rates and migration routes. But while such scenarios offer storylines for a range of possible futures, Arctic residents, particularly those Inuit communities which depend on living marine resources for their livelihood and cultural survival, say they are feeling these changes already. What these changes do, however, is accentuate the transformations already experienced as a result of broader global processes affecting remote Arctic communities. For Inuit and other circumpolar indigenous peoples, the Arctic has become an environment that contains multiple risks.

The histories of indigenous peoples moving across and living in the Arctic are remarkable, shown not just through the archaeological, historical and ethnographic scholarly record, but by the accounts, narratives, testimonies and present-day lives of indigenous peoples themselves. They have exhibited capacity to adapt social, economic and cultural practices to climate variation and change. Today such flexibility and versatility appears difficult, however. No matter how remote, isolated or small they may be, Arctic communities are tightly tied politically, economically and socially to the national mainstream, as well as being linked to and affected by the global economy, trade barriers, wildlife management regimes, and political, legal and conservation interests. All of this constrains or reduces the ability of indigenous peoples to adapt and be flexible in meeting the challenges posed by climate variability and change. Institutional barriers and legal regimes (such as wildlife management and quota systems) are often perceived by hunters, fishers and herders as a major impediment to adaptation in modern Arctic communities and have reduced the flexibility which has enabled historic adaptation to changing conditions (Anderson and Nuttall 2004; Ford and Goldhar 2012). As Ford and Goldhar (2012) point out, building upon and reinforcing previous work on human-environmental relations and the changing nature of resource management in Greenland (e.g. Nuttall 1992; Dahl 2000), while alternative income sources are increasingly important in light of recent stresses, occupational hunters face restrictions on money-earning from non-hunting and fishing activities, and various hunting and fishing quotas fail to reflect recent alterations in species availability with changing climatic conditions. Hunting regulations in Greenland and the centralized political control of resource management and environmental decision-making, such as the politics of quota allocation, have contributed to the erosion of the moral economy of hunting and fishing and have weakened social networks, increasing vulnerability to projected and anticipated future changes in climate (Ford and Goldhar 2012; Nuttall 1992, 2009).

Climate change is but one of several, often interrelated, problems affecting the livelihoods of northern peoples. Across the Arctic, indigenous communities are exposed to a variety of different drivers of change that increase their vulnerability, affect their resilience, and reduce their capacity to respond effectively. Hunters in Greenland or reindeer herders in Finland, for example, can no longer adapt, relocate or change resource use activities as easily as they may have been able to in the past, because most now live in permanent communities, in greatly circumscribed social and economic situations, where their hunting, herding and fishing activities are determined to a large extent by resource management regimes, conservationist action, political decision-making bodies and global markets which are sometimes far removed from their communities and do not include local knowledge in their formulation and implementation (Heikkinen et al. 2012; Nuttall et al. 2005). And the legacy of the anti-seal hunting campaigns of the 1980s continues to affect Inuit communities in Canada and Greenland in ways in which climate change may never impact them (Lynge 1992; Wenzel 1991).

As the Iñupiat of Shishmaref are discovering, the mobility their ancestors once possessed to respond to shifts in the pattern and state of their resource base is no longer an option. In contemplating the severity of their present situation and the urgency to move to surer and firmer ground that is less exposed to the kinds of extreme weather events that threaten daily life, this Alaskan community literally on the edge is working through the Shishmaref Erosion and Relocation Committee to relocate to the site of Tin Creek under a project jointly coordinated by local, state and federal authorities. But this is a costly venture. Across much of the Inuit world, people have been moved by the state into communities which are dependent on government infrastructure, subsidies and other investment. Inuit in Nunavut, for example, live in communities resulting directly from Canadian government policy, implemented in the 1950s and 1960s, of moving them from nomadic camps into permanent settlements, while the Danish government closed down many small Greenlandic hunting communities in 1960s, relocating people to live in apartment buildings in larger towns on the west coast. In today's social, political and economic climate, seasonal migration to remain in contact with animals and, more broadly, to maintain traditional Inuit hunting livelihoods would seem to be virtually impossible without some form of government assistance or technological innovation. The effect is to limit traditional hunting and fishing activities. And throughout the circumpolar North, wildlife management and conservation strategies, aimed in principle at protecting and conserving wildlife, also often restrict people's rights of access to customary resources (Anderson and Nuttall 2004; Heikkinen et al. 2011). Combined with the Arctic's natural vulnerability and fluctuations in ecosystem processes, this magnifies the potential effects of global climate change on indigenous communities.

HUMAN SECURITY AND THE GOVERNANCE OF ADAPTATION

The indigenous peoples of the Arctic have asserted the importance of understanding the human dimensions of climate change and scholarship has pointed increasingly to the human security aspects. For some Inuit leaders, climate change is a human rights issue and they strive to have their voices heard in international policy-making fora, arguing that Inuit cultural survival is dependent on the continued presence of ice and snow (Nuttall 2009). Yet, policy responses need to be informed by a greater understanding of how potential impacts of climate change are distributed, but also perceived differently across different regions and populations. In Greenland, at the political level, climate change is considered empowering in the context of greater self-government and ambitions for economic— and eventual political—independence from Denmark, and the continued presence of ice and snow is considered by many Greenlandic politicians and business leaders as a hindrance to economic development, such as the potential oil and mining will bring to the country, while such a view is not necessarily shared by members of small communities dependent on hunting and fishing (Nuttall 2009, 2012b, 2012c). I would argue that policy responses should also recognize climate change impacts within the broader context of rapid social and economic change, and in their implementation should underscore the reality that climate change is but one of several problems affecting people and their livelihoods in the Arctic today, which are subject both to the historical development and the contemporary influences of regional and global markets and to the implementation of government policy that either contributes to a redefinition of hunting, herding, and fishing, or threatens to subvert subsistence lifestyles and indigenous ideologies of human–animal relationships.

Applied anthropologist Timothy Finan argues that adaptation, vulnerability, and resilience are social phenomena that convey significant information about local decision-making processes, which are often affected and inhibited by such factors as powerlessness and inequity but which also indicate often remarkable coping and survival strategies. Adaptation is thus a strategy to reduce vulnerability and enhance resilience, but we cannot consider human-environment dynamics without taking into account power, culture, race, class, gender, ethnicity, and so on (Finan 2009). Finan argues for the importance of a livelihoods approach for assessing vulnerability "that formally incorporates natural system change (abrupt or cumulative) into a dynamic human system defined by its multiple asset packages (human, social, political, economic, and physical capitals), the sets of decisions that mobilize and allocate these resources, and the outcomes of these decisions" (Finan 2009: 177). To understand social vulnerability to climate change, Adger and Kelly (1999) discuss the availability of resources and entitlement to those resources by individuals and groups as factors that can influence a society's vulnerability. The governance of vulnerability and how socio-cultural inequality, poverty and powerlessness reduce resilience have also been seen as critical issues for research on resilience in social-ecological systems (e.g. Green 2009; Lazrus 2009).

A vulnerability perspective is thus not necessarily concerned with understanding the outcomes of past responses to change, disruption and disturbance, but focuses attention instead on the potential of anticipatory strategies and future preparedness and responses to change, all of which will be determined by a variety of social, cultural, political and economic factors, influences and practices, environmental, governance and institutional contexts, perceptions of risk, and the severity, nature and duration of exposure to the stress or shock of system change. Resilience may help to explain the capacity of a household or community, or larger society to "absorb" and deal with stress without a change in structure and function.

The extent of vulnerability and resilience to climate change not only depends on cultural aspects and ecosystem diversity, but on the political, legal and institutional rules which govern social-economic systems and social-ecological systems. Parts of the Arctic are unique in terms of the political settlements, land claims and forms of self-government that have been achieved over the last forty years or so. Across the Arctic, indigenous peoples have demanded and fought for the right to self-determination and self-government based on the recognition of historical and cultural rights to the ownership of lands, waters and resources and on long-term occupancy of northern homelands. The governments of some Arctic states have recognized these claims and a number of significant settlements have been negotiated and reached over the past four decades. Notable are the Alaska Native Claims Settlement Act (ANCSA) of 1971, Greenland Home Rule in 1979 and Self-Rule in 2009 and, in Canada, the James Bay and Northern Quebec Agreement (1975), the Inuvialuit Agreement (1984), comprehensive land claims agreements with the Gwich'in and Sahtu Dene in the early 1990s, and the creation of the new territory of Nunavut in 1999, following the Nunavut Land Claims Agreement of 1993. The most complex and unresolved issues relating to the autonomy and self-determination of the Arctic's indigenous peoples are to be found in Russia. Although indigenous minorities of the Russian North were given certain rights and privileges under the Soviets, these rights have not always been recognized and many indigenous groups are seeking forms of self-government and regional autonomy. It seems particularly important that attention is given to the management of resources and to the effectiveness of governance institutions, and critical questions must be asked as to whether they can create additional opportunities to increase resilience, flexibility and the ability to deal with change. How can, for example, new governance mechanisms developed under self-government in Greenland or public government in Nunavut, help (or perhaps they hinder?) people to negotiate and manage the impacts of climate change? In Greenland, Alaska, and northern Canada, are the political and management systems already in place that could assess the impacts of climate change, allowing local and regional authorities to act on policy recommendations to deal with the consequences, and improve the chances for local communities to deal successfully with climate change? How can an assessment and evaluation of past climate change—and the social, economic and political responses (e.g. in the early twentieth century) help in understanding current perspectives and policy responses? The answer to these questions will depend on a range of factors, including the importance of understanding the nature of the relationships between people, communities and institutions if effective policy responses to climate change and human security are to be formulated and implemented.

The Arctic environment and the livelihoods of indigenous peoples are not be changed or influenced by climate change alone and it is crucial that scientific climate change research does not develop scenarios of possible future states of the Arctic environment without putting people in the picture and taking rapid social, cultural, economic and demographic changes into account. To this end, approaches to human security also need to proceed from an understanding of the complexities and nuances of human-environmental relations.

REFERENCES

ACIA (2005), *Arctic Climate Impact Assessment: Scientific Report*. Cambridge, UK: Cambridge University Press.

Adger, W.N. and P.M. Kelly (1999), 'Social vulnerability to climate change and the architecture of entitlements', *Mitigation and Adaptation Strategies for Global Change*, **4**, 253–266.

AMAP (2011), *Snow, Water, Ice and Permafrost in the Arctic (SWIPA): Climate Change and the Cryosphere*. Oslo, Norway: Arctic Monitoring and Assessment Program.

Anderson, David G. and Mark Nuttall (eds) (2004), *Cultivating Arctic Landscapes: Knowing and Managing Animals in the Circumpolar North*. Oxford, UK: Berghahn.

Anisimov, O.A., D.G. Vaughan, T.V. Callaghan, C. Furgal, H. Marchant, T.D. Prowse, H. Vilhjálmsson and J.E. Walsh (2007), 'Polar regions (Arctic and Antarctic)', *Climate Change 2007: Impacts, Adaptation and Vulnerability. Contribution of Working Group II to the Fourth Assessment Report of the Intergovernmental Panel on Climate Change*, M.L. Parry, O.F. Canziani, J.P. Palutikof, P.J. van der Linden and C.E. Hanson (eds), Cambridge, UK: Cambridge University Press, pp.653–685.

Arctic Council (2013), *Arctic Resilience Interim Report 2103* Stockholm: Stockholm Environment Institute.

Barnett, Jon and Neil Adger (2007), 'Climate change, human security and violent conflict', *Political Geography*, **26** (6), 639–655.

Berkes, Fikret and Dyanna Jolly (2001), 'Adapting to climate change: social-ecological resilience in a Canadian western Arctic community', *Conservation Ecology*, **5** (2), 18.

Bird-David, Nurit (1990), 'The Giving Environment: another perspective on the economic system of gatherer-hunters', *Current Anthropology*, **31**, 189–196.

Byers, Michael (2009), *Who Owns the Arctic? Understanding Sovereignty Disputes in the North*. Vamcouver, Canada: Douglas and MacIntyre.

Buijs, Cunera (2010), 'Inuit perceptions of climate change in East Greenland', *Études/Inuit/Studies*, **34** (1), 39–54.

Caulfield, Richard A. (1997), *Greenlanders, Whales and Whaling: Self-Determination and Sustainability in the Arctic*. Hanover, USA: University of New England Press.

Chapin, S, M. Berman, T.V. Callaghan, A. Crepin, K. Danell, B. Forbes, G. Kofinas, D. McGuire, M. Nuttall, O.R. Young and S. Zimov (2005), 'Polar systems', in Millennium Ecosystem Assessment, *Ecosystems and Human Well-Being: Conditions and Trends*, Washington DC, USA: Island Press.

Crate, Susan A. and Mark Nuttall (eds) (2009), *Anthropology and Climate Change: From Encounters to Actions*. Walnut Creek, USA: Left Coast Press.

Dahl, Jens (2000), *Saqqaq: An Inuit Hunting Community in the Modern World*. Toronto, Canada: University of Toronto Press.

Fergurson, Hilary (2011), 'Inuit food (in)security in Canada: assessing the implications and effectiveness of policy', *Queen's Policy Review*, **2** (2), 54–79.

Fienup-Riordan, Ann (1983), *The Nelson Island Eskimo: Social Structure and Ritual Distribution*. Anchorage, USA: Alaska Pacific University Press.

Finan, Timothy (2009), 'Storm warnings: the role of anthropology in adapting to sea-level rise in southwestern Bangladesh', in Susan A. Crate and Mark Nuttall (eds), *Anthropology and Climate Change: From Encounters to Actions*. Walnut Creek, USA: Left Coast Press.

Ford, James D. and Christina Goldhar (2012), 'Climate change vulnerability and adaptation in resource dependent communities: a case study from West Greenland', *Climate Research*, **54**, 181–196.

Green, Donna (2009), 'Opal waters, rising seas: how sociocultural inequality reduces resilience to climate change among indigenous Australians', in Susan A. Crate and Mark Nuttall (eds), *Anthropology and Climate Change: From Encounters to Actions*. Walnut Creek, USA: Left Coast Press.

Grundman, Reiner and Nico Stehr (2010), 'Climate change: what role for sociology?', *Current Sociology*, **58** (6), 897–910.

Guyot, Melissa, Cindy Dickson, Chris Paci, Chris Furgal and Ming Man Chan (2006),

'Local observations of climate change and impacts on traditional food security in two northern Aboriginal communities', *International Journal of Circumpolar Health*, **65** (5), 403–415.

Hastrup, Kirsten (ed.) (2009), *The Question of Resilience: Social Responses to Climate Change*. Copenhagen, Denmark: The Royal Danish Academy of Science and Letters.

Heikkinen, Hannu I., Outi Moilanen, Mark Nuttall and Simo Sarkki (2011), 'Managing predators, managing reindeer: contested conceptions of predator policies in Finland's southeast reindeer herding area', *Polar Record*, **47** (242), 218–230.

Heikkinen, Hannu I., Simo Sarkki and Mark Nuttall (2012), 'Users or producers of ecosystem services? A scenario exercise for integrating conservation and reindeer herding in northeast Finland', *Pastoralism: Research, Policy and Practice*, 2:11, doi:10.1186/2041-7136-2-11.

ICC (2009) *A Circumpolar Inuit Declaration on Resource Development Principles in Inuit Nunaat*, Ottawa: Inuit Circumpolar Council http://inuitcircumpolar.com/files/uploads/icc-files/Declaration_on_Resource_Development_A3_FINAL.pdf

IPCC (2008), *Climate Change and Water: IPCC Technical Paper VI*. Geneva, Switzerland: IPCC.

Kalland Arne and Frank Sejersen (2005), *Marine Mammals and Northern Cultures*. Edmonton, Canada: CCI Press

Krebs, Charles J. and Dominique Berteaux (2006), 'Problems and pitfall in relating climate variability to population dynamics', *Climate Research*, **32** (2), 143–149.

Krupnik, Igor and Dyanna Jolly (eds) (2002), *The Earth is Faster Now: Indigenous Observations of Arctic Environmental Change*. Fairbanks, USA: ARCUS.

Lazrus, Heather (2009), 'The governance of vulnerability: climate change and agency in Tuvalu, South Pacific', in Susan A. Crate and Mark Nuttall (eds), *Anthropology and Climate Change: From Encounters to Actions*. Walnut Creek, USA: Left Coast Press.

Lever-Tracy, Constance (2008), 'Global warming and sociology', *Current Sociology*, **56** (3), 445–466.

Lemke, P., J. Ren, R. Alley, I. Allison, J. Carrasco, G. Flato, Y. Fujii, G. Kaser, P. Mote, R. Thomas and T. Zhang (2007), 'Observations: change in snow, ice and frozen ground', *Climate Change 2007: The Physical Science Basis. Contribution of Working Group I to the Fourth Assessment Report of the Intergovernmental Panel on Climate Change*, S. Solomon, D. Qin, M. Manning, Z. Chen, M. Marquis, K.B. Averyt, M. Tignor and H.L. Miller (eds), Cambridge, UK: Cambridge University Press, pp. 337–384.

Lynge, Finn (1992), *Arctic Wars, Animal Rights, Endangered Peoples*. Hanover, USA: University of New England Press.

Maslowski, Wieslaw, Jaclyn Clement Kinney, Matthew Higgins and Andrew Roberts (2012), 'The future of Arctic sea ice', *Annual Review of Earth and Planetary Sciences*, **40**, 625–665.

Meakin, Stephanie and Tiina Kurvits (2009), *Assessing the Impacts of Climate Change on Food Security in the Canadian Arctic*. Report prepared by GRID-Arendal for Indian and Northern Affairs Canada. Arendal, Norway: GRID-Arendal.

Nuttall, Mark (1992), *Arctic Homeland: Kinship, Community and Development in Northwest Greenland*. Toronto, Canada: University of Toronto Press.

Nuttall, Mark (1998), *Protecting the Arctic: Indigenous Peoples and Cultural Survival*. London, UK: Routledge.

Nuttall, Mark (2009), 'Living in a world of movement: human resilience to environmental instability in Greenland', in Susan A. Crate and Mark Nuttall (eds), *Anthropology and Climate Change: From Encounters to Actions*. Walnut Creek, USA: Left Coast Press.

Nuttall, Mark (2010a), *Pipeline Dreams: People, Environment, and the Arctic Energy Frontier*. Copenhagen, Denmark: IWGIA.

Nuttall, Mark (2010b), 'Anticipation, climate change, and movement in Greenland', *Études/Inuit/Studies*, **34** (1), 21–37.

Nuttall, Mark (2012a), 'Tipping points and the human world: living with change and thinking about the future', *Ambio*, **44** (1), 96–105.

Nuttall, Mark (2012b), 'The Isukasia iron ore mine controversy: extractive industries and public consultation in Greenland', *Nordia Geographical Publications*, **40** (4), 23–34.

Nuttall, Mark (2012c), 'Imagining and governing the Greenlandic resource frontier', *The Polar Journal*, **2** (1), 113–124.

Nuttall, Mark, Fikret Berkes, Bruce Forbes, Gary Kofinas, Tatiana Vlassova and George Wenzel (2005), 'Hunting, herding, fishing and gathering: indigenous peoples and renewable resource use in the Arctic', in ACIA, *The Arctic Climate Impact Assessment*. Cambridge, UK: Cambridge University Press.

Sirina, Anna (2009), 'Oil and gas development in Russia and northern indigenous peoples', in Elana Wilson Rowe (ed.), *Russia and the North*. Ottawa, Canada: University of Ottawa Press.

Stroeve, Julienne C., Vladimir Kattsov, Andrew Barrett, Mark Serreze, Tatiana Pavlova, Marika Holland and Walter N. Meier (2012), 'Trends in Arctic sea ice extent from CMIP5, CPIP3 and other observations', *Geophysical Research Letters*, **39** (16), doi:10.1029/2012GL052676.

Wadhams, Peter (2012), 'Arctic ice cover, ice thickness and tipping points', *Ambio*, **41** (1), 23–33.

Wadhams, Peter and Norman R. Davis (2000), 'Further evidence of ice thinning in the Arctic Ocean', *Geophysical Research Letters*, **27** (24), 3973–3975.

Wang, Muyin and James E. Overland (2009), 'A sea ice free summer Arctic within 30 years?', *Geophysical Research Letters*, **36** (L07502), doi:10.129/2009GL037820.

Wassmann, Paul and Timothy Lenton (2012), 'Arctic tipping points in an Earth systems perspective', *Ambio* **41** (1), 1–9.

Wenzel, George (1991), *Animal Rights, Human Rights*. Toronto, Canada: University of Toronto Press.

12. Climate change and human security in Africa

Sharath Srinivasan and Elizabeth E. Watson

An examination of climate change and human security debates in and about Africa raises questions about representation and the politics of knowledge. Africa, as a continent, holds a special place in debates about climate change and human security: if human security is understood as living a life of dignity, enjoying freedom from want and freedom from fear, then it is in Africa that global climate change is seen as most likely to compromise personal and human security. There, 'unfreedom' from want and 'unfreedom' from fear are commonly seen as connected in a vicious circle of violence and destitution.

In IPCC and other documents, Africa is held up as the continent that is least responsible for climate change, but is likely to feel the worst of its impacts. High levels of poverty, environmental degradation, weak governance and dependence on natural resources in Africa mean that climate change is seen as likely to devastate livelihoods that are already vulnerable (Boko et al. 2007). Africa is also perceived as exhibiting 'weak adaptive capacity' (ibid.), and least able to make the changes necessary to build resilience to shocks that a changing climate may bring. This resonates with reports that single out Africa's unique vulnerability in terms of human security, such as the seminal report of the UN Commission on Human Security (2003). As livelihoods are devastated, conflict and migration are thought likely to result: 'Societies unable to adjust to the new challenges [of climate change] are left with two main options: fight or flee' (Buhaug et al. 2008: 17). Here Africa is a special case as it demonstrates a spectre of what could be if nothing is done: 'a canary in the coal mine, a foretaste of climate-driven political chaos' (Faris, 2007 cited in Hartmann 2010: 236).

In terms of representation and the politics of knowledge, Africa is also a particular case because it has long been the site onto which foreign, even humanity-wide, hopes, desires and fears are projected. Africa, it has been argued, has often played the part of the 'Other', imagined as lacking or absent, incomplete, incapable and uncivilized, in contrast to an ideal 'West' (Mudimbe 1988; Mbembe 2001; Ferguson 2006). Even in the environmental sphere, where the technical matters of soil erosion, forestry and rangeland management might be thought more neutral, understandings have long been influenced by pre-formed, highly valorized models and imported stereotypes. For example, it was often taken for granted and 'received wisdom' that African people did not have the knowledge, foresight, rationality or capability to manage their environments sustainably, and that African individuals and communities were responsible for environmental degradation and desertification (Leach and Mearns 1996). Such 'degradation narratives' (Hartmann 2010) turned complex political processes into technical problems to be fixed, and often justified the management of environments shifting to colonial, state or private hands, at the expense of community access and use of natural resources. Facts and objectivity – knowledge and understanding – have too readily mattered little when well-established narratives guide and frame ways of seeing that are convenient and useful.

Paradoxically, the obstinate tendency to cast Africa as an object apart from the world, or a failed reflection of it, reminds of how perennially important 'Africa-and-the-world' is. This is no less true for the global significance of Africa in debates on climate change and human security. Here, the popular imaginary has already been conjured by the likes of Robert Kaplan (1994, 2001), who depicted an anarchical crisis in Africa of violent competition over the environment as portentous of things to come for the world at large. When, in 2007, UN Secretary-General Ban Ki-moon sought to expose the 'climate culprit' for violence in Darfur, Sudan, he had a captive audience.

The critique of the imaginative construction of Africa and its environment is now well accepted. For thirty years, many environment and development policies have tried to reverse the impact of these ideas with projects that take greater cognisance of local realities and attempt to integrate indigenous knowledge and community participation into their work. Despite initial enthusiasm, such work is not as easy as it might seem; work with communities is messy and goals have been hard won. The history of employing clichés and stereotypes has proved hard to shake off, and is still influential. In addition, as poverty, destitution and environmental degradation *are* a reality evident across much of the continent, the challenge has become to tease out projected fantasy – and all its accompanying baggage – from reality (Ferguson 2006). The challenge is also to develop better understandings of the complexities of local realities, without becoming too idealistic or specific or divorcing these understandings from the wider regional and international picture (ibid.).

Discourses about climate change and human security are relatively new, but they do not exist in a vacuum; they are situated within and interact with this wider discursive history. They also bring their own dynamic: As Jasanoff explains, science (in which climate change discourses are based) operates according to 'abstraction' and that, not uncommonly, 'an impersonal, apolitical, and universal imaginary of climate change, projected and endorsed by science, takes over from the subjective, situated and normative imaginations of human actors engaging directly with nature' (Jasanoff 2010: 235). Global environmental management discourses frequently render local complexities 'illegible' (Adger et al. 2001), and climate change science has prioritized climate change over other drivers of change leading to reductive explanations (Hulme 2011). This chapter examines the ways the problems identified by these writers become magnified when it comes to Africa. There is an inherent danger that, in climate change discourses, the hard-won understandings of the complexities of people-environment relations in Africa may be cast aside and rendered apolitical. The discourses of climate change tend towards monocausal explanations, and prioritize the global scale; again, the local complexities are given insufficient weight in policy matters.

The 'security' optic adds its own urgent dynamic. The human security discourse has also come under considerable criticism for 'securitizing' the field of enquiry, thereby rendering it exceptional. Security crises warrant a kind of 'emergency politics' (Buzan et al. 1998: 24, 29) that unyokes human and social struggles from their embedded context and can justify external intervention from higher scales. The intersection of climate change and human security discourses thus brings to mind earlier critiques of 'environmental security' that arose after the 1987 World Commission on Environment and Development report (see Trombetta 2008). Wolfgang Sachs, writing in 1992, cautioned that,

> the 'survival of the planet' is well on its way to becoming the wholesale justification for a new wave of state interventions in people's lives all over the world . . . while environmentalists have put the spotlight on the numerous vulnerabilities of nature, governments as a result discover a new conflict-ridden area in need of political governance and regulation. (Sachs 1992: 33)

Security discourses readily serve the interests and ideas of actors – from foreign interveners to national states – who enjoy the 'power of naming' and who, in turn, often enact state-centric security logics (Suhrke 1999). When it matters most, freedom from fear dominates over freedom from want, and the fear is a self-referential fear of a world where chaos in peripheries such as Africa destabilizes the global north.

This chapter starts by examining what we know – and what we do not know well enough – about climate change in Africa, and its impact upon freedom from want and freedom from fear. The focus is on how emerging science and local complexity are simplified, reinterpreted and deployed within the new discourse of climate change and human security.

The chapter then shifts to two case studies that exemplify the politics of representation, and of science and politics at different scales. The conflict in Darfur, Sudan, illustrates the ways in which the securitization of climate change activates a particular freedom from fear discourse that gives weight to certain phenomena and causalities over others, directly and indirectly serving selected interests at multiple scales. The case of Marsabit in Kenya, examines the ways in which debates about climate change and freedom from want also tend to overlook certain inconvenient sets of processes, often complex and contingent local ones that are ill-suited to the modalities and objectives of external interveners.

Human security approaches claim to provide a more people-centred approach to understanding climate change and one that also employs 'joined up thinking' (Gasper 2010) across scales and across human-environment domains. There is great potential, therefore, in a human security approach which is 'integrative', 'addresses the wellbeing of individuals from multiple and interrelated perspectives' and 'draws attention to present and emerging vulnerability that is generated through dynamic social, political, economic, institutional, cultural and technological conditions and their historical legacies' (O'Brien et al. 2010: 4–5). The human security approach is also highly normative and ambitious. It frequently calls for system transformation or 're-design' in order to avert crises from climate change (Gasper 2010: 33). In practice, in the rush to meet the challenges before it is too late, the approach risks falling short of the ideal of integrating multiple perspectives on vulnerability, and the same old historic motifs about Africa appear. As before, Africa is seen as most vulnerable and most incapable, and most likely to resort to conflict; the continent serves as a reminder of either how far 'we' have come, or a threat of where 'we' might return. The chapter explores these processes within the human security and climate change discourse, and the ways that, unless vigilance is maintained, this discourse – like many of the environmental discourses in Africa before – could be hijacked to serve certain powerful interests.

THE IMPACT OF GLOBAL CLIMATE CHANGE ON AFRICAN CLIMATES AND THEIR HUMAN SECURITY CONSEQUENCES

Evidence suggests that Africa is warming at a faster rate than the global average (Conway 2009: 7). The high variability and diversity of climates across Africa make conclusions about the impact of global climate change on current climates difficult, and predictions of the future uncertain.

Compounding the heterogeneity is very poor availability of data: the average number of weather stations is reportedly eight times below that recommended by the World Meteorological Organization, and large areas are unmonitored (Washington et al. 2006). In addition, the interconnections between processes that drive African climate – the El-Niño Southern Oscillation (ENSO), tropical convection and the alternation of monsoons – are not well understood (Conway 2009). As a consequence, Africa has proved difficult territory for modellers: 'Africa shows the least agreement between models of all the continents and, apart from relatively small regions, for the majority of Africa the models do not agree on even the sign of change, let alone its magnitude' (Williams and Kniveton 2011: 3).

Conway (2009: 7) summarizes what *is* known: Africa in general is very likely to become warmer and drier, but in some areas it may become colder and wetter, and there is a very high likelihood of 'more extreme weather events' such as droughts and floods. In East Africa, it is thought that rainfall will increase by 10–20 percent, and its distribution will change, with more rain falling from December to January, and less from June to August; the temperature in the region is predicted to increase between 1.5° and 5.8° to 2080 (Toulmin 2009: 24).

Despite the variability and uncertainties, in policy documents, climate change is seen as leading to a 'downturn' in human security in terms of an increased inability to maintain a life of dignity in which basic needs are met, and an increased likelihood of conflict (Alkire 2003). In human security terms the ability to lead lives free from want and free from fear is increasingly compromised. In Africa, climate change has been characterized as a multiplicity of 'stresses' that act as a 'threat multiplier'.[1]

In terms of freedom from want, climate change is thought to lead to food insecurity and increased levels of malnutrition. Agricultural yields are estimated to reduce by as much as 50 percent by 2020 (Conway 2009; Boko et al. 2007). Economies will also be hit by the decline in agriculture, as 'crop net revenues could fall by as much as 90 percent by 2100' (Boko et al. 2007: 435). Water availability, ecosystems, tourism and tourist revenues, are all predicted to be negatively affected. Certain regions on the coasts may be inundated, and there are likely to be increased numbers of deaths from extreme weather events (Conway 2009). Diseases are predicted to increase: as well as the abovementioned malnutrition, malaria is predicted to spread into areas of higher altitude, there may be an increase in cardio-vascular diseases from changes in air quality, and increased levels of diarrhoeal diseases. Evidence that an increase in mean temperature would lead to the devastation of many lives and livelihoods led many African participants at COP15 in Copenhagen to criticize the target proposed by northern countries to limit global temperature increase to only 2°C above pre-industrial levels. Following a speech by Lumumba di-Aping, Sudanese diplomat and Chief Negotiator for the G77 at COP15, many Africans adopted the slogan '2 degrees is suicide'. Di-Aping drew on the IPCC's own calculations to argue that a global temperature increase of 2°C, would mean a 3.5°C increase for Africa, and that '2 degrees Celsius would mean certain death for Africa'.[2]

In terms of freedom from fear, it is a broadly held concern that climate change will lead to large-scale violent conflict in Africa. Emblematic in this logic is the Darfur case, which Ban Ki-moon, UN Secretary-General, described in 2007 as evidence of that war's 'climate culprit':

> Almost invariably, we discuss Darfur in a convenient military and political shorthand – an ethnic conflict pitting Arab militias against black rebels and farmers. Look to its roots, though, and you discover a more complex dynamic. Amid the diverse social and political causes, the Darfur conflict began as an ecological crisis, arising at least in part from climate change. (Ki-moon 2007)

The underlying logic in this and similar claims is that anthropogenic climate change leads to natural resource depletion, scarcity and competition, with disproportionate effects on low-income land and agriculture-dependent households. Wars, of course, need to be fought; the implicitly reductionist and deterministic logic here – inescapable for climate change to remain a meaningful independent variable – is that individuals (especially, idle and disenfranchised young men), political entrepreneurs, communities or even governments, act on the rational calculation that it is worth taking up arms to secure socio-economic position in the struggle over shrinking natural wealth.

The argument is a development from the 'green war' hypothesis (Homer-Dixon and Blitt 1998), which historically has emphasized population growth (Ehrlich and Ehrlich 1970) as one of the major problems and drivers for environmental degradation and natural resources scarcity that leads to conflict and collapse. For one well-known writer, Thomas Homer-Dixon, the population driver underlies an explanation that 'scarcities of critical resources – especially of cropland, freshwater and forests – contribute to violence in many parts of the world . . . stimulat[ing] insurgencies, ethnic clashes and urban unrest' (Homer-Dixon 1999: 12). Hartmann labels this the 'degradation narrative' (Hartmann 2010), that over-population and poverty combine to cause ecological damage, precipitating migration to other ecologically vulnerable areas that in turn fuels political instability.

In recent studies, population growth is still a factor in pressure on resources, but the spectre of rapid resource depletion owing to climate change has come to dominate and, with it, the spectre of devastation has expanded. Again, Homer-Dixon, popularizing his thesis in the *New York Times* in 2007, wrote: 'Climate stress may well represent a challenge to international security just as dangerous – and more intractable – than [the Cold War arms race or the proliferation of nuclear weapons]' (Homer-Dixon 2007). Homer-Dixon's views were echoed to different degrees by a diverse chorus of notables. During the UN Security Council's first debate on climate security, British Foreign Secretary Margaret Beckett was emphatic regarding the connection: 'What makes wars start? Fights over water. Changing patterns of rainfall. Fights over food production, land use' (Reynolds 2007). From the 2006 Stern Review on the Economics of Climate Change to the Norwegian Nobel committee and Nobel Prize winner Al Gore in 2007, looming climate conflict was linked to resource stresses in already over-populated and poor regions of the world, notably Africa.

Taking a closer look, recent research has drawn a strong link between rainfall variability and GDP (Ludwig et al. 2009, cited in Richardson et al. 2011: 109). From 1979 to 2001, dry years had a devastating impact on the GDP of African countries; and years of excessive rainfall did not, as well as years of average rainfall. However, the relationship between reduced interannual rainfall and *conflict* is far from certain. For Hendrix and Glaser (2007: 696), interannual variability in rainfall is 'the most significant climatic variable' driving conflict onset. However the authors did not predict increased variability in sub-Saharan Africa. Ciccone (2011) argues that there is statistically no relationship between conflict and reduced rainfall in sub-Saharan Africa, critiquing previous studies that affirmed this connection for not accounting for positive correlations between conflict and earlier increased rainfall. Hendrix and Salehyan (2012) analyse data for a broader definition of conflict and argue, quite opposite to prevailing wisdom, that abundant rainfall correlates more strongly with violent events.

The debate on the drought/water-shortage and conflict nexus continues to rage. In a detailed study of an African dataset for the years 1960 to 2004 that plots sub-national geo-referenced data on annual precipitation with similarly granular data on the location of civil war onset and the political and socio-economic status of ethnic groups (Theisen et al. 2011), the authors find that despite popular discourse, the drought-conflict connection is not supported. Instead, civil war onset is strongly linked to areas of political marginalization, and 'this statistical regularity is unaffected by abrupt local water shortages' (Theisen et al. 2011: 81–2).

The link between conflict in Africa and climatic warming more generally has also been argued. On assessing a longitudinal national-level dataset,

Burke et al. (2009) claimed that a 1-degree *interannual* increase in temperature directly brought about a 4.5 percent increase in civil conflict in sub-Saharan Africa, ostensibly because of the stress on crop yields caused by warming. In what the authors claimed was the 'the first comprehensive examination of the potential impact of global climate change on armed conflict in sub-Saharan Africa', the study warned of a roughly 54 percent increase in armed conflict incidence by 2030, or an additional 393,000 battle deaths, based on current rates of war-related fatalities. Better governance or economic status did not, they argued, change the causal significance of warming. The study was subject to staunch criticism (see Buhaug (2010) and Richardson et al. (2011) for review), albeit sometimes replacing one causal macro-variable with another to predict violent conflict. Nevertheless, it was easy fodder for the popular press, with pictures of human tragedy in Darfur accompanying a BBC headline, 'Climate "is a major cause" of conflict in Africa'.[3]

For Gleditsch et al. (2007), migration is a key factor in creating conflicts. Yet of course, migration can also be a key adaptation strategy for individuals affected by both conflict and climate change. Again, the causal relation is not at all settled, with Theisen et al. (2011: 85) arguing that although transnational refugee flows might have some effect on the outbreak of armed conflict, 'it is far from obvious that environment-induced migration (to the extent migration can be considered monocausal) will have the same security implications.'

An examination of the African case highlights the ways in which climate change is likely to generate new challenges for people across the continent. The climate change and human security debate illuminates the scale of the problem that is occurring and emphasizes the ways in which issues of personal dignity and well-being – in terms of fear and want – are inter-related. But the African case also illuminates the ways in which the new discourse of climate change and human security can be used in ways that simplify complex situations, and obscure many other important processes taking place. The appropriation and distortion of the academic discourse in policy circles occurs despite its unsettled and formative nature. Climate change is a highly complex set of social and material processes that require greater understanding and yet also urgent policies; this tension has proven difficult to resolve for all concerned. In Africa, it has also become a useful discourse that is deployed by many because it serves certain interests. By further examining some elements of these processes through case studies, the rest of this chapter sheds light on benefits as well as dangers of employing the climate change and human security discourse.

FEAR: DARFUR, AFRICA'S CLIMATE CONFLICT
CAUSE CÉLÈBRE

In Africa, climate change is cast as a 'threat multiplier' amongst multiple, grave threats. Talk of climate change invariably foreshadows upheaval, flight, violence and destruction across the continent. Fears over climate-induced conflict in Africa feed off credible estimations of acute levels of human insecurity where populations are highly reliant on climate-affected livelihoods sectors (notably rain-fed agriculture) and seemingly endemic problems of political instability. They also serve purposes far beyond the continent. Referring to the Darfur case, Brown has lamented that, 'Africans are not really the *intended audience* of the post-Kyoto debate, but they are part of the *evidence* being used to make it' (Brown 2010b: 42, emphasis in original). The 'evidence' is wrapped in a chilling counsel of fear. Yet this fear, abstract and totalizing, is too readily overdetermined, with distorting effects on how specific conflicts and periods of violence are framed, causes interpreted, consequences mitigated and responsibilities attributed. While the climate-induced conflict discourse might be well-meaning in its intentions on one scale (for example, using the spectre of threat to galvanize global governance action), at local scales these distorting effects can interact with conflict dynamics, political contestation and humanitarian and security interventions in not unproblematic ways. Moreover, the scientific debate on causal relationships is far from settled, and this ambiguity allows for a degree of interpretive license in how causal effects are characterized. The risk here is that the very real levels of human insecurity faced by African communities due to violence and upheaval, including in the context of the ecological consequences of climate change, are not more effectively addressed.

We must ground what follows in a basic contextual account of Darfur's history of conflict.[4] Darfur is a large region (the size of France, or Texas) in western Sudan that is relatively under-developed and remote from the capital in Khartoum, and where local inter-group conflicts linked to access to scarce natural resources have intermittently occurred for centuries. The region is home to a diverse and fluid mix of ethnic groups and livelihoods practices that betray simple binary oppositions of 'Arab' *versus* 'African' or 'non-Arab', or 'farmers' *versus* 'pastoralists', and which underlie the complex dynamics of resource governance and conflict. By the 1980s, local conflicts between groups had grown in scale and become more deadly. There is no doubt that ecological crisis – namely, from the early 1970s onwards, a sharp drop in rainfall and desertification in the more arid northern regions and livelihoods stress in semi-fertile and farmed areas – was a structurally relevant condition for this trajectory. However, one critical ingredient that led to large-scale and more politicized conflict was governance failures, especially the Sudanese state's inadequate and oftentimes partisan response to the worsening situation.

In the 1970s, the Sudanese government's partial dismantling and inconsistent use of traditional tribal land governance and dispute resolution systems shifted these functions onto a weak and politically fractured state administration. The authorities' poor response to successive droughts culminated in a deadly famine in the early 1980s (de Waal 2005a). This livelihoods crisis endured thereafter, exacerbated by the retreat of regional authorities and neglect from the central state. Whereas past resource-based conflicts occurred without particular reference to 'Arab' *versus* 'non-Arab' or 'African' distinctions, ethnic polarization had grown during a period when the central government privileged a certain idea of 'Sudanese' identity over pluralism and regional diversity (de Waal 2005b). Because of its strategic location, Darfur was also drawn into the racialized Libya-Chad conflict that lasted until the early 1990s (Burr and Collins 1999). The Sudan government's role in this war, and in manipulating local political structures thereafter, vacillated depending on the tactical alliances of the government of the day. The result was increasingly fractious local politics and vocal anti-government sentiment in a region awash with arms.

The spiral into the civil war that began in 2002–03 also had overwhelmingly political drivers. Divisions within the central ruling party in Khartoum led to the prevailing faction around President Omar el-Bashir conducting a targeted and draconian security operation in Darfur and other regions (Flint and de Waal 2008). Then, amidst the tectonic shifts of peace negotiations to settle the country's long-running 'north-south' civil war, armed rebellion against the government took hold. The conflict escalated rapidly after a major assault on military installations by rebels in April 2003 was answered with a ruthless government counter-insurgency. The situation was evidently one of civil war, not a mere livelihoods conflict. By late 2003, the UN considered the situation in Darfur as the worst humanitarian crisis in the world. By 2006–07, the conflict had forcibly displaced over 2 million civilians and was estimated to have claimed over 400,000 lives (Degomme and Guha-Sapir 2010).

The devastation of violence, death and displacement in Darfur occurred within the context of long-term human insecurity that has roots in livelihoods and ecological crises, but, in line with the argument of Theisen et al. (2011), the degree of fear experienced by Darfur's peoples owes much to the overwhelmingly political factors of marginalization, militarization and manipulation. As the analysis below demonstrates, most of this context, history and political contingency – even the brief account given above – is trivialized with dangerous effects when totalizing frames such as 'climate change', or even 'genocide', take centre stage.

The Darfur conflict had raged for some years before the 'climate change conflict' label rose to prominence in 2006 and 2007. This happened in the context of efforts to put climate change on the agenda of the UN Security Council, for which a foundational link between climate change and 'international peace and security' was essential. When Ban Ki-moon fingered the 'climate culprit' in Darfur in 2007, his was not a lone voice. The United Kingdom Special Representative for Climate Change, John Ashton, also labeled Darfur as the 'first modern climate-change conflict' in April 2007 (see Mazo 2009: 73). In early 2006, the British Defence Secretary, John Reid, had pointed to the 'blunt truth' that Darfur foretold regarding environment, conflict and future dystopias: '[t]he lack of water and agricultural land is a significant contributory factor to the tragic conflict we see unfolding in Darfur. We should see this as a warning sign' (Russell and Morris 2006). Later, in 2008, France's President Nicholas Sarkozy, hosting a so-called Major Economies Meeting on climate change, also deployed this *cause célèbre* to warn of the violent future of climate change: 'In Darfur, we see this explosive mixture from the impact of climate change, which prompts emigration by increasingly impoverished people, which then has consequences in war. If we keep going down this path . . . the Darfur crisis will be only one crisis among dozens of others' (Agence France Press (AFP) 2008).

The economist Jeffrey Sachs perhaps made the claim loudest and longest. As early as November 2004, he told an audience in Oxford University, 'You heard it from me first, Darfur is the world's first climate change war' (Sachs 2004). In 2006, in a piece for *Scientific American*, 'Ecology and political upheaval', Sachs turned again to 'The deadly carnage in Darfur, Sudan . . . which is almost always discussed in political and military terms, has roots in an ecological crisis directly arising from climate shocks' (Sachs 2006).

We must look more closely at the causal explanation relied upon in these general statements. In his book *Common Wealth: Economics for a Crowded Planet* (Sachs 2008), Sachs elaborated his argument regarding Darfur in clear neo-Malthusian terms:

> as the population has soared, the carrying capacity of the land has declined because of long-term diminished rainfall. . . . The striking pattern is the decline of rainfall starting at the end of the 1960s, a pattern that is evident throughout the African Sahel. . . . The results have been predictably disastrous. Competition over land and water has become lethal. (Sachs 2008: 248–9; quoted in Verhoeven 2011: 692)

Ban Ki-moon explained similarly the causal driver:

> Two decades ago, the rains in southern Sudan began to fail. According to UN statistics, average precipitation has declined some 40 percent since the early

1980s. Scientists at first considered this to be an unfortunate quirk of nature. But subsequent investigation found that it coincided with a rise in temperatures of the Indian Ocean, disrupting seasonal monsoons. This suggests that the drying of sub-Saharan Africa derives, to some degree, from man-made global warming. (Ki-moon 2007)

This explanation combines a causal vector of reduced rainfall (oddly, and incorrectly, in 'southern Sudan', quite a distance from Darfur, for which he could have cited credible data) with wider continental weather effects of warming. A similar argument had already been made by Faris in *The Atlantic*, that the 'real fault lines in Darfur' were between 'settled farmers and nomadic herders fighting over failing lands' because of lack of rainfall (Faris 2007).

The epistemic interactions between how Darfur's conflict was instrumentalized and leveraged by commentators and political actors operating at a global scale, and how the research and policy community was analysing and studying the conflict, warrant closer scrutiny. Ki-moon cites 'UN statistics', and a seminal report that gave 'credibility' for the Darfur-as-climate-conflict argument was the UN Environment Programme (UNEP) report, 'Sudan: Post-Conflict Environmental Assessment' (2007). In it, UNEP warned against over-determining the causal role of climate in understanding conflict in Sudan, but nevertheless warned of a 'very strong link between land degradation, desertification and conflict in Darfur', such that the conflict there serves as 'a tragic example of the social breakdown that can result from ecological collapse' (UNEP 2007). UNEP drew directly upon the arguments of Homer-Dixon and others, to make the causal link between overpopulation (people and livestock), climate-related water shortages and environmental crises and conflict in the region. They gave particular focus to long-term desertification (a 50 to 200 km southward shift of the boundary between desert and semi-desert since the 1930s) and the sharp drop in rainfall (notably, of over 30 percent over 50 years in Northern Darfur):

> The scale of historical climate change as recorded in Northern Darfur is almost unprecedented: the reduction in rainfall has turned millions of hectares of already marginal semi-desert grazing land into desert. The impact of climate change is considered to be directly related to the conflict in the region, as desertification has added significantly to the stress on the livelihoods of pastoralist societies, forcing them to move south to find pasture. (United Nations Environment Programme (UNEP) 2007: 60)

There is a sense in the foregoing that climate-induced ecological crisis in Darfur had been gradually and inevitably driving people to a tipping point into collapse and large-scale violence. Yet evidence suggests the rainfall reductions were abrupt rather than incremental. Thus the rainfall effect is less readily understood in terms of interannual variability and reduced rains, which some authors cited earlier have linked causally to violent conflict. Rather, in Darfur as elsewhere in the region, there was a sharp break in long term average rainfall trends around 1970. For Kevane and Gray (2008), the fact that rainfall in Darfur exhibited a flat trend in the thirty years preceding the conflict (1972–2002), albeit with normal interannual variability, underscores a very weak causal relationship between climate change and conflict. A comparative analysis of 38 African countries, 22 of which showed similar structural breaks in rainfall, demonstrated no obvious relationship between such breaks and later conflict.

Besides the rainfall explanation, there remains a more traditional 'neo-Malthusian' argument tying general competition over scarce natural resources to environmental degradation. Even here, the evidence is ambiguous. Brown's examination of Normalized Difference Vegetation Index (NDVI) data to measure 'eco-scarcity' leads him to conclude that the outbreak of conflict was not linked to a proximate worsening of the ecological situation (Brown 2010a). Rather, vegetation growth in the years preceding the outbreak of conflict were better than average over a 25-year period.

Nevertheless, it is rather the attribution of the reduction in rainfall and consequent desertification to the rise in temperature of the Indian Ocean that underpins the 'climate change' explanation. This explanation seems to lay blame squarely at the feet of the historical 'warmers' of the twentieth century, namely the industrialized 'North', and thus is especially valuable to policymakers and activists seeking global climate governance reform, notwithstanding the actual causal link to conflict incidence. Yet the jury is still out on whether the rise in surface temperature of the Indian Ocean that began in the 1950s was anthropogenic in nature. Just as the propensity in the 1970s to lay blame for desertification and drought in the Sahel on 'bad' local land practices and population stress (Charney et al. 1977; Lamprey 1988 on Sudan) has been largely debunked (Swift 1996; Brooks 2004), any blithe attribution of responsibility for the drop in rainfall in the Sahel to anthropogenic climate change risks being discredited as the state of knowledge improves.

Returning to Darfur, the way in which conflict there was framed and the violence named, had already been politically charged when the 'climate conflict' label came to the fore in 2007. During the early stages of the armed rebellion and ensuing ruthless counter-insurgency operation by the state, the Sudanese government had sought to depoliticize the violence and attribute it to 'banditry', 'lawlessness' and 'local' and 'tribal' grievances over resources. The central authorities were already looking to the environment and livelihoods as causes that exculpated their political responsibility well before outsiders activated the 'climate change' argument.

Similarly, it was convenient for Western governments invested in a concurrent peacemaking effort to end the country's long-running 'north-south' civil war to 'localize' the Darfur conflict lest it confuse and disrupt their strategy for securing a negotiated 'peace' (Srinivasan 2012). When, in late 2002, a Darfuri activist had the ear of the UK Special Representative on Sudan and pointed to ominous signs of 'ethnic cleansing', he was reportedly educated on the nature of the conflict being due to stresses upon the 'carrying capacity of the land' (ibid.). When Amnesty International sought to bring global attention to the escalating conflict in 2003, they were rebuffed by British officials for being 'peace spoilers' (Srinivasan 2006). The spectre of 'genocide' did bring Darfur into the global spotlight in 2004, when the 'north-south' peace agreement was all but secure, but it also came at the cost of once again downplaying the complex socio-political history of the conflict and with the effect of further politicizing binary ethno-racial identities (de Waal 2005b).

It is thus of little wonder that researchers on Sudan lamented the risks of Ban Ki-moon's focus on Darfur's 'climate culprit' stemming from the UNEP report. The worry that '[global warming] has become such a trendy issue that everything is being packaged as climate change' (IRIN 2007) was not one of a 'climate sceptic' but of a political analyst aware of the real political dangers inherent in parsimonious pronouncements in high places. Sudan's UN ambassador had the wind in his sails when, addressing a US college gathering in late 2007, he cited Ban Ki-moon's *Washington Post* editorial and explained,

> The major cause of the question of Darfur is the environmental degradation from climate change. Darfur is a classic case of climate change. People have witnessed gradual degradation of the environment and erosion of the resources and desertification and drought that was going on for a long time, since the beginnings of the seventies. (Straw 2007)

Global discourses that evidence African examples to further wider agendas at higher scales contain political judgments that actively shape other discursive spheres and might be leveraged and instrumentalized by local actors in unintended ways. In Africa, the policies, norms and ideas of powerful external actors have arguably long been central to the 'extraversion' strategies of domestic political actors (Bayart 2000).

The securitization of the Darfur conflict in climate change terms undermined more than advanced the human security concerns of the people of the region. Darfur shows us that at the heart of the intersection of climate change and human security lie political ecology contestations at multiple scales. Competition over livelihoods resources in a context of major ecological change is a grave threat multiplier in terms of Africans' freedom from fear across the continent, but a sharp focus on the dynamics of this fear and its causes must fully incorporate, and not elide, the highly sociopolitically contingent nature of large-scale conflict.

WANT: SECURING CLIMATE-THREATENED LIVELIHOODS IN MARSABIT, KENYA

Climate change and human security debates in Africa have historically focused on the natural resources competition-conflict nexus, but, as the Darfur case indicated, underlying this is a broader concern that climate change is leading to a 'downturn' in human security because of its potential devastating impact on livelihoods. The 'multiple stresses' from changes in weather patterns, and more extreme weather events such as drought or flooding, are thought to lead to an inability to meet basic food security, and to create vulnerability to disease and collapse. Images from newspaper and development reports tell a story in which climate change is likely to mean that people's lives are reduced to 'bare life', 'devoid of rights, choices and possibilities' (Elford 2008, from Agamben 1998). It is also thought that living in these conditions means that people 'have nothing to lose' and are more likely to resort to conflict. Policies to secure human dignity are thus made more urgent.

According to the UN, 2011 saw the worst drought in Eastern Africa for sixty years. Commentators have been cautious of directly linking this drought to climate change, but, still, the drought has been held up as an illustration of the extreme vulnerability of people in the region to climate change, and a taste of things to come. As an Oxfam blog comments:

> Attributing the current drought directly to climate change is impossible, but in the words of Sir John Beddington, the UK government's chief scientific adviser . . . 'worldwide, events like this have a higher probability of occurring as a result of climate change'. Moreover, unless something is done, the current suffering offers a grim foretaste of the future – temperatures in east Africa are going to rise and rainfall patterns will change making a bad situation worse.[5]

Even before the 2011 drought, media and development industry discourses were expressing concern about the devastating impact that climate change was already having on the livelihoods of people in the dryland regions of east Africa. For example, after a visit to Moyale on the Ethiopia/Kenya border in 2009, John Vidal, a prominent and well-informed journalist, wrote an article in a UK broadsheet newspaper under the heading 'Climate change is here, it is a reality'. He continued, 'as one devastating drought follows another, the future is bleak for millions in East Africa' (Vidal 2009).

Vidal's observations relate to a dryland area, where rainfall is low and unpredictable. In Marsabit County on the Kenyan side of the border, the people are pastoralists who have historically depended on camel or cattle herding, combined with sheep and goats. Mobility was important to these livelihoods as it allowed herders to move to areas where grass had grown following rain, and to move away from grazed areas to allow grass to recover. In the media and development industry discourses, these pastoralist livelihoods are now seen as particularly precarious and under threat from climate change. As climate change begins to bite, the suggestion is that this livelihood is no longer sufficient. For example, as Vidal continued, quoting a 'climate advisor' to a development NGO:

> [Climate change is] not in the imagination or a vision of the future. [And] climate change adds to the existing problems. It makes everything more complex. It's here now and *we have to change*. (ibid.; italics ours)

These ideas are also propounded by development organizations working in the region. One of the largest NGOs in the area claimed in a report that:

> As a result of climate change, *the ecosystem within Marsabit is no longer favourable to pastoralism*. The communities need to diversify their livelihood in order to reduce risk to natural disaster like drought. (NGO Report 2010; italics ours)

These views of the situation in Marsabit are not unique: they are just one example of a wider mainstream perspective on climate change, human security and pastoralism. Catley and Aklilu, for example, found that the view that 'pastoralism is in crisis and non-viable, as evident from increasing levels of pastoral destitution' (2013: 85) was a prominent narrative among humanitarian and development actors in the Horn of Africa in 2010, and that it gained strength from 'an emerging sub-narrative around climate change' (ibid.: 85). Returning to Marsabit, we find that little evidence is presented to support claims that the 'ecosystem is no longer favourable to pastoralism'. As Catley and Aklilu (2013) suggested for the Horn of Africa more generally, the existence of large numbers of people in the area who have lost animals and depend on food aid is considered evidence enough, despite the fact that their pathways into destitution are likely to have been various and not necessarily climate-related.[6]

In Marsabit, two further lines of argument that relate to climate change and human security are marshalled as evidence for the 'pastoralism is non-viable' view. The first is the argument that conflict has already resulted from competition for resources, and that therefore, unless something is done, further conflict is likely to result.[7] The simplifications of, and problems with, this logic have been discussed above. In the second line of argument, references are made to scientific analyses of livestock's impact on the environment – at a global scale – in order to make the case that, even if pastoralism were 'working' at the local level, it is now understood to be too environmentally damaging *at a global scale* to be an option. For example, alongside a discussion of the problems facing pastoralist livelihoods in Marsabit, a report of the NGO discussed above cites a prominent Food and Agriculture Organization (FAO) study by Steinfeld et al. (2006) called *Livestock's Long Shadow*. A UN press release that accompanied the launch of this FAO report captures its main message:

> What causes more greenhouse gas emissions, rearing cattle or driving cars? Surprise! According to a new report published by the United Nations Food and Agriculture Organization, the livestock sector generates more greenhouse gas emissions as measured in CO_2 equivalent – 18 percent – than transport. It is also a major source of land and water degradation.[8]

By referring to the Steinfeld et al. report, the NGO implies that livestock-based systems are problematic from a wider global climate perspective, because they are high greenhouse gas emitters and sources of environmental degradation. Yet the relevance of these claims to Marsabit can be questioned (see below). First, however, more general questions can be raised about the 'pastoralism is non-viable' narrative and the extent to which climate change is leading to such a downturn in human security. These narratives and ideas are often unquestioned because they fit with the common sense expectations of observers that pastoralists, who live in 'harsh' and 'marginal' environments and who are so closely dependent on nature, should easily be pushed into destitution by climate change. They appear as the archetypal people living on the edge, for whom the smallest nudge will push them over, with sometimes violent consequences for many others near and far. Here, fears about climate change have been used to breathe new life into older narratives in which:

> for many decades, governments regarded pastoralism as 'backward', economically inefficient and environmentally destructive, leading to policies that have served to marginalise and undermine pastoral systems. (Nassef et al. 2009: ii)

The main premises of this historic narrative have been critiqued before (Niamir-Fuller and Turner 1999). But the premises that relate to climate change are new and merit further investigation.

First, it may seem obvious that 'marginal' and dryland environments may be the first and worst hit by climate change, but in practice no easy conclusions can be made (Ayantunde et al. 2011). It is extremely difficult to make accurate or certain statements about climate change's impact on the environment on which pastoralists depend (Thornton et al. 2009). The impact of climate change on grasslands and livestock systems depends on complex interactions between temperature, rainfall and CO_2 emissions. Thornton et al.'s (2009) review of scientific research in this area concluded, unsurprisingly, that under rising temperatures 'average biomass generally increased for warm-season grasses and decreased for cool-season forbs and legumes'. . . but also that 'there are likely to be smaller impacts on livestock yields per se . . . because of the ability of livestock to adjust consumption' (2009: 116). A reduction in rainfall is likely to have a stronger negative effect on pasture and livestock productivity. But, predictions suggest that, if anything, rainfall in East Africa may increase and, despite anecdotes and fears, as yet no overall drying trend in the region has been measured (Catley and Aklilu 2013). Thornton et al. (2009: 120) also make the point that 'the tropics and sub-tropics contain a wealth of animal genetic resources that could be utilized in relation to heat-stress-related issues' as local livestock breeds are well-adapted to heat-stress. In summary, the science relating to the impact of climate change on dryland pasture is not yet comprehensive or certain. If the ecosystem in Marsabit is 'no longer favourable to pastoralism', it has not yet been conclusively demonstrated that this is as a consequence of climate change (as claimed), or that such an outcome is inevitable.

Second, questions can be raised about the extent to which Steinfeld et al.'s (2006) study is helpful for understanding the local processes taking place in, for example, northern Kenya. Steinfeld et al.'s (2006) study focuses on the global impact of livestock and is an enormously complex subject. It involves combining data on the environmental impact of different types of livestock (pigs, chickens, cows, small ruminants, etc.), reared under different conditions (extensive, intensive; small-scale, large-scale; commercial, subsistence; high-tech, low-tech; global north, global south), and it explores the impact of livestock on direct pollution through emitting nutrients and organic matter, pathogens and drug residues, gases such as methane (directly and from waste). It also explores the impact of livestock on water use, biodiversity and on the degradation of land used for grazing and feedstocks. The study also calculates that there is an opportunity cost to grazing lands – if trees were grown there instead, for example, then a more positive (instead of negative) impact could be made on the global environment.

For Steinfeld et al. (2006: xx), pastoralists like those of Marsabit contribute to land degradation ('extensive grazing still occupies and degrades vast areas of land'), and they have a negative impact on biodiversity through habitat destruction and conflict with wildlife. The report points to the enormous significance to global greenhouse gas levels of negative externalities from livestock systems which are intensive, vertically integrated, geographically concentrated and in which production has been up-scaled through various technological innovations. And this suggests that, in terms of greenhouse gas emissions, extensive grazing systems may do better. But at the same time, it criticizes extensive pastoralism systems because of their use of wide areas of land which could be put to other more productive purposes, and concludes that

> intensification – in terms of increased productivity both in livestock production and in feed crop agriculture – can reduce greenhouse gas emissions from defor-estation and pasture degradation (Steinfeld et al. 2006: xxi–xxii)

A careful and systematic analysis of this influential report is still awaited, but reservations have already been expressed about the application of this kind of analysis to areas like northern Kenya (NRI 2010; Oba 2011), where it serves to cement the view that pastoralism is not part of the planet's sustainable future. The low concentration of livestock in regions like northern Kenya – and the types of livestock reared – raise further questions about whether its share of the 18 percent of global greenhouse gases calculated as produced by the livestock industry is large enough to warrant concern. Caution might also be exercised in relation to the broad and generalized claims that pastoralists are environmentally degrading. As discussed already, the environmentally degrading pastoralist is a 'received wisdom' that has often proved wrong in the past (Swift 1996; Sandford 1983; Scoones 1994). The stereotype was present in Hardin's 'tragedy of the commons' (1968) thesis, in which pastoralists graze in competition with each other, and with no concern for the future. An enormous amount of scholarly work has demonstrated how pastoralists cooperate together to regulate access to and use of grazing, and employ impressive reserves of knowledge in their use of grasslands and their grazing practices (Ostrom 1990). It has also shown how dryland environments are non-equilibrium environments, and how grazing patterns are flexible in order to maintain a livelihood under conditions of uncertainty (Scoones 1994):

> Unlike ecologists, herders can distinguish between landscapes that are vulnerable to heavy grazing and degrade rapidly, and those that resist degradation. Where there is greater risk of degradation, landscapes are grazed for brief periods during the wet season . . .. Using the trajectories of vegetation change, herders are able to alter their herd composition and modify grazing movements. (Oba 2013: 32)

Where pastoralists do overgraze and degrade their lands, all too often it is because their patterns of movement have been disrupted, and key resources have been lost (Oba 2013; Feyissa and Schlee 2009).

The discourse that climate change is driving human insecurity in terms of livelihood devastation occludes many of these positive dimensions of pastoral livelihoods. As with the discourse about violent conflict, many people and organizations 'buy in' to the discourse for multiple reasons. As well as fears about the scale of the impact of climate change (which, as it is difficult to precisely determine, is in some ways a fear of the unknown), for NGOs and other organizations, the climate change and human security discourse gives new impetus to their activities and combats compassion fatigue.

The discourse that 'pastoralism is non-viable given the challenge of climate change' is totalizing, but it is not uncontested: in some research and development domains a strong counter-argument has been made that pastoralism developed in the first instance as an adaptation to high climate variability, and – particularly because of its use of mobility – is very well placed to cope with challenges posed by further climate variability and unpredictability (Grahn 2008; Davies and Nori 2007). Research has suggested that pastoralists already make significant and unrecognized contributions to local, regional and national economies (Hesse and MacGregor 2006; Nassef et al. 2009), and are engaged in entrepreneurial and long-distance trade (Catley et al. 2013; Pavanello 2009). In this literature, pastoralist areas are not without problems, but they are considered one of the most efficient, resilient and productive uses of dryland environments: 'the weight of evidence suggests that "modern", commercialized forms of livestock-keeping and irrigated farming are not as productive as customary forms of pastoralism' (Catley et al. 2013).

There are signs that the 'pastoralism is resilient' discourse has begun to influence development policies and practices, particularly with the publication in 2010 of a forward-looking African Union policy document (see Catley et al. 2013; Schlee and Shongolo 2012). But in many cases the older dominant 'pastoralism is non-viable' narrative has prevailed, in which older motifs of pastoralists as anachronistic and environmentally degrading have tended to be reproduced. Here global-scale science about climate change and downturns in human security triumphs over understandings of complex local situations. Pastoralists are cast as global polluters as well as degrading their own environments. This framing of people-environment relations matters because it misrepresents pastoralists and legitimizes a shift to support for quite different forms of livelihood or uses of land and resources and major external interventions. Steinfeld et al. (2006) for example, advocate soil carbon sequestration programmes for rangelands, which they envisage would generate revenues when combined with mechanisms such as the Clean Development Mechanism. They acknowledge, however, that such sequestration programmes are still in their infancy and that ensuring the participation of local people in these programmes is difficult. More generally, carbon finance projects rely on a win-win philosophy that the problems of a global climate can be addressed while also improving the lives of the poor; unfortunately experience so far suggests the latter part of this equation has so far often failed to materialize (Fairhead et al. 2012; Sandbrook et al. 2010).

Two other kinds of alternative developments underway ought also to be mentioned: the first is the Kenyan Government's ambitious national 'blueprint' for development, 'Vision 2030'. This 'Vision' emphasizes infrastructure development such as road, railway and pipeline construction across northern Kenya in order to facilitate trade and oil-related development. Other avenues for development emphasized in Vision 2030 include tourist development (such as the construction of two 'Sun City'-type 'resort cities' in northern Kenya), irrigation development and livestock marketing. Many of these development goals appear to represent a return to a large-scale top-down high-tech modernizing style of development which relies on a 'trickle down' of economic growth that historically often failed to occur. But the poor are not forgotten. A Social Protection Programme is also envisaged, and as part of this, the UK Department for International Development is funding the Hunger Safety Net Programme which provides cash transfers to the most destitute in northern Kenya. In this programme, beneficiaries and other members in the communities are being registered 'biometrically' for smartcards which ostensibly enable more efficient targeting, delivery and monitoring of the cash transfers. Under a 'security optic' one might wonder if the Kenyan state, which has long found the mobile pastoralist populations difficult to manage, finds this biometric registering of its population a fortuitous by-product of this pro-poor initiative. As November 2012 has seen 'the most deadly attack on Kenya's police since independence',[9] when 42 security personnel were killed in an ambush by Turkana cattle rustlers, the very real and pressing need for this security is underlined.

The second area of development that must also be mentioned is the wider context of Africa in general as the site of foreign investment – 'the last investment frontier'[10] – and that lands and water are increasingly sought for development by new investors. Where countries (like Qatar) are investing in lands for agricultural production (as they are reported to be in Kenya), these investors aim to protect themselves from perceived potential global and national food insecurities, but their actions are likely to have impacts on the food securities of others.

It is impossible to do justice here to the range of developments taking place in the region, but even this cursory overview suggests a pattern in the processes is evident. Policy discourses about climate change draw on global scale science and calculations to rethink policies that are appropriate for regions like Marsabit. The human security approach claims to be integrative, to examine issues across scales and across the human-environment divide, but, all too often, the more detailed local level investigations are brushed aside because of the complexities of such endeavours, and the uncertainties inherent to them. Larger scale, generalized narratives dominate and stereotypes of African pastoralists as incapable, environmentally degrading, quick to fight and fatalistic are re-deployed (Galaty 2013; Mortimore 2010). These narratives help to make the case for 'system re-design' (Gasper 2010), and for more radical and top-down forms of development. While this global scale, universal, impersonal and apolitical imaginary continues to dominate, it seems unlikely that the pastoralists themselves will be able to influence the nature of developments, or that their own human security will be improved.

CONCLUSION

Africa has featured prominently in debates about climate change and human security for both compelling and troublesome reasons. A strong narrative has emerged in which climate change is understood as driving a downturn in freedom from fear and freedom from want, and that it is in Africa, that the two 'unfreedoms' are linked in particularly pernicious and dangerous ways. Here Africa appears as a cautionary tale that raises awareness of the urgency of the problem and the need to galvanize action to tackle climate change and its impacts. A benign interpretation would be that the climate change and human security discourse has the capacity to galvanize action across the usual divides (north, south, rich, poor) because of the global nature of climate change, or even that it draws attention to the need for richer countries to take responsibility for the problems they have caused. The evidence presented here suggests a less generous interpretation is in order: that the climate change/human security narrative relies heavily on older motifs of Africa as more likely to resort to violence, more unstable, more environmentally degrading, and less able to meet the challenges that life (or the climate) may bring. Paradoxically, in Africa 'the future is now' for climate change and human security precisely because of discursive frames that trap contemporary African societies, economies and polities in distant places and times. Such elements to the narrative risk placing conceptual distance between writers, their intended audiences and the subjects of such analyses, perhaps comfortably, suggesting that the challenges of climate change and human security are elsewhere.

The nature of ideas about Africa influence and constitute relations between actors in the international effort to address climate change and to define who has the power to decide and to act. It also has an impact on what happens in Africa, and the policies that are considered legitimate and desirable. In the climate change and human security narrative, climate change tends to dominate as the main driver of, and explanatory variable for, physical and livelihood security at the expense of other factors upon which its effects are highly contingent. Such recourse to monocausal explanations has happened, and has been critiqued, before (for example with regard to population growth). But here the focus on climate change and human security brings with it very particular dynamics which impact on both local politics and international policy.

First, because climate change is global in nature it gives impetus to the idea that the 'problem' is beyond the control and capacity of local communities, and inadvertently risks helping to exculpate other local actors from responsibility. Secondly, and relatedly, specific geographies and histories are occluded. For example, the eliding of specific geographies was clearly seen in the Darfur case, where Darfur and South Sudan were equated. Where exactly the problem is, and what exactly is happening, does not matter much, given the scale of the problem. In addition, as seen in the Marsabit case, the ways in which local environmental changes and management practices intersect with longer-term climate dynamics are poorly understood and integrated into policies. Histories are similarly elided. Many of the vulnerabilities of local communities are a result of their relationship to the state, or to conflicts, or to forms of development (such as the expansion of commercial agriculture or conservation areas), and yet – if discussed at all – they are relegated to secondary factors when compared to the scale of the climate change and human security challenge. By giving little recognition to the geographical and historical specificities, human insecurities (political or livelihood; fear or want) are depoliticized.

Thirdly, the climate change and human security narrative with regard to Africa has a very particular emotional register: that of fear. Climate change invokes fear because it is not known what its impacts will be, or how it might be controlled. The ease with which older motifs of Africa as a space of chaos and anarchy (Ferguson 2006) are reproduced in amplifying this fear is tremendously worrying, given how much work has been done.

These dynamics of the climate change and human security narrative matter because they legitimize the idea that Africa and Africans need to be 'saved' by new and powerful technologies, forms of transnational governance, waves of humanitarian relief and modes of investment. Where these initiatives misunderstand local realities and pay little heed to the primary importance of working with and for local populations and from their historical and social worldview, they are unlikely to improve the human security of communities in Africa. The 'survival of the planet', as Sachs presciently put it, again risks being 'the wholesale justification for a new wave of state interventions in people's lives all over the world'. Now, the domestic state is not the only actor at play – foreign states, NGOs and investors (foreign and domestic) are increasingly influencing what takes place in Africa, frequently in the name of addressing climate change.

Securing Africans' freedom from fear and want is enmeshed with others' fear of climate conflicts, of the burdens of widespread destitution and of climate change more generally. We know that human security discourse owes its central 'freedom from want and fear' motif to US President Franklin D. Roosevelt's 1941 State of the Union address. Yet Roosevelt famously made another reference to fear, at his inaugural address in 1933, in which he urged Americans that 'the only thing we have to fear, is fear itself,' a 'nameless, unreasoning, unjustified terror which paralyzes needed efforts to convert retreat into advance.' Especially when a focus on human security is weakly integrative and locally grounded, the securitization of climate change risks fomenting a somewhat unbounded and totalizing fear that, while useful at higher scales and for various intervening actors, might perniciously distract away from the real work needed to address the more proximate drivers of unfreedom and insecurity experienced by African populations.

NOTES

1. http://www.fco.gov.uk/en/global-issues/climate-change/priorities/global-security/ Accessed 16 August 2012. See also Council of the European Union (2008): 'Climate change is a threat multiplier which threatens to overburden states and regions which are already fragile and conflict prone.'
2. Lumumba Di-Aping, 2009, http://www.youtube.com/watch?v=aAcp0uHDBBU.
3. http://news.bbc.co.uk/2/hi/science/nature/8375949.stm.
4. For overviews of the conflict that pay close attention to history, politics and context, see: Flint and de Waal (2008), Daly (2010) and Prunier (2008). The cursory summary here draws on these and other more detailed analyses.
5. Green (2011), 'Famine and climate change – what's the link?' http://www.oxfamblogs.org/fp2p/?p=6440 Accessed 18 July 2012.
6. See Fratkin and Roth (2005) for further information on reasons for destitution.
7. For example, a 2006 Christian Aid report uses the case of the 2005 'Turbi massacre' near Marsabit, in which 53 people were killed, as evidence of climate-related conflict (http://www.christianaid.org.uk/Images/climate-of-poverty.pdf). Mwangi's (2006) more detailed investigation of the conflict finds its causes in regional political struggles and also points to the involvement of foreign-based militias. He concludes that any

relationship between conflict and competition for scarce resources is 'debatable' (2006: 89).

8. http://www.un.org/apps/news/story.asp?newsID=20772&CR1=warning Accessed 19 July 2012.
9. http://www.bbc.co.uk/news/world-africa-20392510.
10. http://www.forbes.com/sites/moneybuilder/2012/08/08/africa-the-last-investment-fron tier.

REFERENCES

Adger, W.N., Benjaminsen, T.A., Brown, K. and Svarstad, H. (2001), 'Advancing a political ecology of global environmental discourses', *Development and Change*, **32**, 681–715.
Agamben, G. (1998), *Homo Sacer: Sovereign Power and Bare Life*. Stanford: Stanford University Press.
Agence France Press (AFP) (2008), *Climate Change: Progress at Polluters' Talks, But Obstacles Ahead* [cited 1 September 2012]. Available from http://afp.google.com/article/ALeqM5i5F-QlDqPpaTzK4YavRCOgrWCgtw.
Alkire, Sabina (2003), 'A conceptual framework for human security', Working Paper No. 2, Centre for Research on Inequality, Human Security and Ethnicity, University of Oxford, UK.
Ayantunde, A.A., Leeuw, J. de, Turner, M.D. and Said, M. (2011), 'Challenges of assessing the sustainability of (agro)-pastoral systems', *Livestock Science*, **139** (1–2), 30–43.
Bayart, Jean-François (2000), 'Africa in the world: a history of extraversion', *African Affairs*, **99** (395), 217–267.
Boko, M. et al. (2007), 'Africa. Climate change 2007: Impacts, adaptation and vulnerability. Contribution of Working Group II', in M.L. Parry, O.F. Canziani, J.P. Palutikof, P.J. van der Linden and C.E. Hanson (eds.), *Fourth Assessment Report of the Intergovernmental Panel on Climate Change*. Cambridge, UK: Cambridge University Press, pp. 433–467.
Brooks, N. (2004), 'Drought in the African Sahel: long term perspectives and future prospects', Tyndall Centre Working Paper 61.
Brown, I. A. (2010a), 'Assessing eco-scarcity as a cause of the outbreak of conflict in Darfur: a remote sensing approach', *International Journal of Remote Sensing*, **31** (10), 2513–2520.
Brown, Oli (2010b), 'Campaigning rhetoric or bleak reality? Just how serious a security challenge is climate change for Africa?', in Heinrich-Böll-Stiftung (ed.), *Securing Africa in an Uncertain Climate*. Cape Town, South Africa: Heinrich Böll Foundation Southern Africa.
Buhaug, H. (2010), 'Climate not to blame for African civil wars', *Proceedings of the National Academy of Sciences*, **107** (38), 16477–16482.
Buhaug, H., Gleditsch, N.P. and Ole Magnus Theisen, O.M. (2008), 'Implications of climate change for armed conflict', Paper presented at the Social Dimensions of Climate Change workshop, http://siteresources.worldbank.org/INTRANETSOCIALDEVELOPMENT/Resources/SDCCWorkingPaper_Conflict.pdf.
Burke, Marshall B., Edward Miguel, Shanker Satyanath, John A. Dykema, and David B. Lobell (2009), 'Warming increases the risk of civil war in Africa', *Proceedings of the National Academy of Sciences*, **106** (49), 20670–20674.
Burr, Millard and Robert O. Collins (1999), *Africa's Thirty Years War: Libya, Chad, and the Sudan, 1963–1993*. Boulder, USA: Westview Press.
Buzan, B., O. Wæver, and J. Wilde (1998), *Security: A New Framework for Analysis*. Boulder, USA: Lynne Rienner.
Catley, A. and Aklilu, Y. (2013), 'Moving up or moving out? Commercialization, growth and destitution in pastoralist areas', in A. Catley, J. Lind and I. Scoones (eds), *Pastoralism and Development in Africa: Dynamic Change at the Margins*. London, UK: Earthscan for Routledge, pp. 85–97.

Catley, A., Lind, J. and Scoones, I. (eds) (2013), *Pastoralism and Development in Africa: Dynamic Change at the Margins*. London, UK: Earthscan for Routledge.

Charney, J., W.J. Quirk, S.H. Chow, and J. Kornfield (1977), 'A comparative study of the effects of albedo change on drought in semi-arid regions', *Journal of the Atmospheric Science*, **187**, 434–435.

Ciccone, Antonio (2011), 'Economic shocks and civil conflict: a comment', *American Economic Journal: Applied Economics*, **3** (4), 215–227.

Conway, Gordon (2009), 'The science of climate change in Africa: impacts and adaptation', *Grantham Institute for Climate Change, Discussion Paper*, 1, 46.

Council of the European Union (2008), 'Climate change and international security', Report from the Commission and the Secretary-General/High Representative to the European Council, 3 March. Brussels, Belgium: CEU.

Daly, M.W. (2010), *Darfur's Sorrow: The Forgotten History of a Humanitarian Disaster*. 2nd edn. Cambridge, UK: Cambridge University Press.

Davies, J. and Nori, M. (2007), 'Change of wind or wind of change? Climate change, adaptation and pastoralism', summary of an online conference prepared for the World Initiative for Sustainable Pastoralism, Nairobi, Kenya: IUCN.

de Waal, Alex (2005a), *Famine That Kills: Darfur, Sudan*. Rev. edn, Oxford Studies in African Affairs. New York, Oxford: Oxford University Press.

de Waal, Alex (2005b), 'Who are the Darfurians? Arab and African identities, violence and external engagement', *African Affairs*, **104** (415), 181–205.

Degomme, Olivier and Debarati Guha-Sapir (2010), 'Patterns of mortality rates in Darfur conflict', *The Lancet*, **375**, 294–300.

Ehrlich, Paul R. and Anne H. Ehrlich (1970), *Population, Resources, Environment: Issues in Human Ecology*, series of biology books. San Francisco, USA: Freeman.

Elford, L. (2008), 'Human rights and refugees: building a social geography of bare life in Africa', *African Geographical Review*, **27** (1), 65–79

Fairhead, J., Leach, M. and Scoones, I. (2012), 'Green grabbing: a new appropriation of nature?', *Journal of Peasant Studies*, **39** (2), 237–261.

Faris, S. (2007), *The Real Roots of Darfur* (April) [cited 1 September 2012]. Available from http://www.theatlantic.com/magazine/archive/2007/04/the-real-roots-of-darfur/3057 01/?single_page=true.

Ferguson, J. (2006), *Global Shadows: Africa in the Neoliberal World Order*. Durham, USA: Duke University Press.

Feyissa, D. and Schlee, G. (2009), 'Mbororo (Fulbe) migrations from Sudan into Ethiopia', in G. Schlee and E.E. Watson (eds), *Changing Identifications and Alliances in North-East Africa: Volume II: Sudan, Uganda, and the Ethiopia-Sudan Borderlands*. Oxford, UK: Berghahn, pp 157–180.

Flint, Julie, and Alex de Waal (2008), *Darfur: A New History of a Long War*. Revised and updated edn. London, UK: Zed Books.

Fratkin, E. and E. A. Roth (eds) (2005), *As Pastoralists Settle: Social, Health and Economic Consequences of Pastoral Sedentarization in Marsabit District*. Kenya. New York, USA: Kluwer Academic Publishers.

Galaty, J. (2013), 'Land grabbing in the Eastern African Rangelands', in A. Catley, J. Lind and I. Scoones (eds), *Pastoralism and Development in Africa: Dynamic Change at the Margins*. London, UK: Earthscan for Routledge, pp. 143–153.

Gasper, D. (2010), 'The idea of human security', in K. O'Brien, A.L. St Clair and B. Kristoffersen (eds), *Climate Change, Ethics and Human Security*. Cambridge, UK: Cambridge University Press, pp. 23–46.

Gleditsch, N.P., Nordas, R. and Salehyan, I. (2007), *Climate Change and Conflict: The Migration Link: Coping With Crisis*. International Peace Academy.

Grahn, R. (2008), *The Paradox of Pastoral Vulnerability*. Oxford, UK: Oxfam.

Hardin, G. (1968), 'The tragedy of the commons', *Science*, **162**, 1243–1248

Hartmann, B. (2010), 'Rethinking climate refugees and climate conflict: rhetoric, reality and the politics of policy discourse', *Journal of International Development*, **22**, 233–246.

Hendrix, Cullen S, and Idean Salehyan (2012), 'Climate change, rainfall, and social conflict in Africa', *Journal of Peace Research*, **49**, 35-50.

Hendrix, Cullen S., and Sarah M. Glaser (2007), 'Trends and triggers: climate, climate change and civil conflict in Sub-Saharan Africa', *Political Geography*, **26**, 695–715.

Hesse, C. and MacGregor, J. (2006), 'Pastoralism: drylands' invisible asset', Issue Paper 142, IIED.

Homer-Dixon, T. F. (1999), *Environment, Scarcity, and Violence*. Princeton, USA: Princeton University Press.

Homer-Dixon, T. F. (2007), 'Terror in the weather forecast', *The New York Times*, 24 April.

Homer-Dixon, T. F. and Jessica Blitt (1998), *Ecoviolence: Links Among Environment, Population and Security*. Oxford, UK: Rowman & Littlefield.

Hulme, M. (2011), 'Reducing the future to climate: a story of climate determinism and reductionism', *Osiris*, **26** (1), 245–266.

IRIN (2007), *Sudan: Climate Change – Only One Cause Among Many for Darfur Conflict*, 28 June [cited 1 September 2012]. Available from http://www.irinnews.org/Report/72985/SUDAN-Climate-change-only-one-cause-among-many-for-Darfur-conflict.

Jasanoff, S. (2010), 'A new climate for society', *Theory Culture Society*, **27**, 233–253.

Kaplan, R. D. (1994), 'The coming anarchy: how scarcity, crime, overpopulation, tribalism, and disease are rapidly destroying the social fabric of our planet', *Atlantic Monthly*, **February**, 44–76.

Kaplan, R. D. (2001), *The Coming Anarchy: Shattering the Dreams of the Post Cold War*. New York, USA: Vintage Books.

Kevane, Michael, and Leslie Gray (2008), 'Darfur: rainfall and conflict', *Environmental Research Letters*, **3** (3).

Ki-Moon, B. (2007), *A Climate Culprit in Darfur*, 16 June [cited 1 September 2012]. Available from http://www.washingtonpost.com/wp-dyn/content/article/2007/06/15/AR2007061501857_pf.html.

Lamprey, H.F. (1988), 'Report on desert encroachment reconnaissance in northern Sudan: 21 October to 10 November, 1975', *Desertification Control Bulletin*, **17**, 1–7.

Leach, M. and Mearns, R. (eds) (1996), *The Lie of the Land: Challenging Received Wisdom on the African Environment*. Oxford, UK: James Currey.

Ludwig, F., P. Kabat, S. Hagemann and M. Dorlandt (2009), 'Impacts of climate variability and change on development and water security in sub-Saharan Africa', *IOP Conference Series: Earth and Environmental Science*, **6**, 292002.

Mazo, J. (2009), 'Chapter Three: Darfur: The First Modern Climate-Change Conflict', *The Adelphi Papers*, **49** (409), 73–86.

Mbembe, A. (2001), *On the Postcolony*. Berkeley, USA: University of California Press.

Mortimore, M. (2010), 'Adapting to drought in the Sahel: lessons for climate change', *WIREs Climate Change*, **1**, 134–143.

Mudimbe, V.Y. (1988), *The Invention of Africa: Gnosis, Philosophy, and the Order of Knowledge*. Bloomington, USA: Indiana University Press.

Mwangi, G.O. (2006), 'Kenya: conflict in the "Badlands": the Turbi massacre in Marsabit District', *Review of African Political Economy*, **33** (107), 81–91.

Nassef, M., Anderson, S. and Hesse, C. (2009), *Pastoralism and Climate Change: Enabling Adaptive Capacity*. London, UK: ODI.

Niamir-Fuller, M. and Turner, M.D. (1999), 'A review of recent literature on pastoralism and transhumance in Africa', in M. Niamir-Fuller (ed.), *Managing Mobility in African Rangelands: The Legitimization of Transhumance*. London, UK: Intermediate Technology Publications, pp. 18–46.

NRI (2010), 'Pastoralism Information Note 5: Pastoralism and climate change', http://www.new-ag.info/assets/pdf/pastoralism-and-climate-change.pdf.

O'Brien, K., St Clair, A.L. and Kristoffersen, B. (2010), 'The framing of climate change: why it matters', in K. O'Brien, A.L. St Clair and B. Kristoffersen (eds), *Climate Change, Ethics and Human Security*. Cambridge, UK: Cambridge University Press, pp. 3–22.

Oba, G. (2011), 'Book review: Steinfield, H., Mooney, H.A., Schneider, F. and Neville, L. E., *Livestock in a Changing Landscape: Drivers, Consequences, and Responses* (Volume 1) and Gerber, P., Mooney, H.A., Dijkman, J., Tarawali, S. and de Haan, C., *Experiences and Regional Perspectives* (Volume 2)', *Pastoralism: Research, Policy and Practice*, **1**, 10.
Oba, G. (2013), 'The sustainability of pastoral production in Africa', in A. Catley, J. Lind and I. Scoones (eds), *Pastoralism and Development in Africa: Dynamic Change at the Margins*. London, UK: Earthscan for Routledge, pp. 29–36.
Ostrom, E. (1990), *Governing the Commons: The Evolution of Institutions for Collective Action*. Cambridge: Cambridge University Press.
Pavanello, S. (2009), *Pastoralists' Vulnerability in the Horn of Africa: Exploring Political Marginalisation, Donors' Policies and Cross-border Issues – Literature Review*. London, UK: ODI HPG.
Prunier, G. (2008), *Darfur: A 21st Century Genocide*, 3rd edn. Ithaca, USA: Cornell University Press.
Reynolds, Paul (2007), *Security Council Takes on Global Warming* [cited 1 September 2012]. Available from http://news.bbc.co.uk/2/hi/6559211.stm.
Richardson, K., Steffan, W. and Liverman, D. (2011), *Climate Change: Global Risks, Challenges and Decisions*. Cambridge, UK: Cambridge University Press.
Russell, K. and N. Morris (2006), 'Armed forces are put on standby to tackle threat of wars over water' (28 February) [cited 1 September 2012]. Available from http://www.independent.co.uk/environment/armed-forces-are-put-on-standby-to-tackle-threat-of-wars-over-water-467974.html.
Sachs, Jeffrey (2004), '2004 OXONIA Inaugural Lecture: The end of poverty', 13 October, Said Business School: University of Oxford. Digital audio recording, on file with author.
Sachs, J. (2006), *Ecology and Political Upheaval* (July) [cited 1 September 2012]. Available from https://www.scientificamerican.com/article.cfm?id=ecology-and-political-uph.
Sachs, J. (2008), *Common Wealth: Economics for a Crowded Planet*. London, UK: Allen Lane.
Sachs, W. (1992), *The Development Dictionary: A Guide to Knowledge as Power*. London, UK: Zed.
Sandbrook, C., Nelson, F., Adams, W. and Agrawal, A. (2010), 'Carbon, forests and the REDD paradox', *Oryx*, **44** (3), 330–334.
Sandford, S. (1983), *Management of Pastoral Development in the Third World*. New York, USA: John Wiley and Sons.
Schlee, G. and Shongolo, A.A. (2012), *Pastoralism and Politics in Northern Kenya and Southern Ethiopia*. Oxford, UK: James Currey.
Scoones, I. (ed.) (1994), *Living with Uncertainty: New Directions for Pastoral Development in Africa*. London, UK: Intermediate Technology.
Srinivasan, Sharath (2006), *Minority Rights, Early Warning and Conflict Prevention: Lessons from Darfur*. London, UK: Minority Rights Group International.
Srinivasan, S. (2012), 'The politics of negotiating peace in Sudan', in D. Curtis and G.A. Dzinesa (eds), *Peacebuilding, Power and Politics in Africa*. Athens: Ohio University Press, pp. 195–211.
Steinfeld, H., P. Gerber, T. Wassenaar, V. Castel, M. Rosales, and C. de Haan (2006), *Livestock's Long Shadow: Environmental Issues and Options*. Rome, Italy: FAO.
Straw, Becky (2007), 'Sudanese ambassador: "Darfur is a classic case of climate change."', 16 November [cited 1 September 2012]. Available from http://www4.lehigh.edu/news/newsarticle.aspx?Channel=%2FChannels%2FNews%3A+2007&WorkflowItemID=539838e6-7971-4a07-ab19-a2ea3e5125ce.
Suhrke, Astri (1999), 'Human security and the interests of states', *Security Dialogue*, **30** (3), 265–276.
Swift, J. (1996), 'Desertification: narratives, winners and losers', in M. Leach and R. Mearns (eds), *The Lie of the Land: Challenging Received Wisdom on the African Environment*. Oxford, UK: James Currey, pp. 73–90.

Theisen, Ole Magnus, Helge Holtermann and Halvard Buhaug (2011), 'Climate wars? Assessing the claim that drought breeds conflict', *International Security*, **36** (3), 79–106.

Thornton, P.K., van de Steeg, J., Notenbaert, A. and Herrero, M. (2009), 'The impacts of climate change on livestock and livestock systems in developing countries: a review of what we know and what we need to know', *Agricultural Systems*, **101**, 113–127.

Toulmin, C. (2009), *Climate Change in Africa*. London, UK: Zed books.

Trombetta, Maria Julia (2008), 'Environmental security and climate change: analysing the discourse', *Cambridge Review of International Affairs*, **21** (4), 585–602.

United Nations Commission on Human Security (2003), *Human Security Now: Protecting and Empowering People*. New York, USA: United Nations Commission on Human Security.

United Nations Environment Programme (UNEP) (2007), *Sudan Post Conflict Environment Assessment*. Nairobi: United Nations Environment Programme.

Verhoeven, Harry (2011), 'Climate Change, Conflict and Development in Sudan: Global Neo-Malthusian Narratives and Local Power Struggles', *Development and Change*, **42**, 679-707.

Vidal, J. (2009), 'Climate change is here, it is a reality', *The Guardian*, 3 September http://www.guardian.co.uk/environment/2009/sep/03/climate-change-kenya-10-10 accessed 12 March 2013.

Washington R., Harrison, M., Conway, D., Black, E., Challinor, A., Grimes, D., Jones, R. Morse, A., Kay, G. and Todd, M. (2006), 'African climate change: taking the shorter route', *American Meteorological Society*, **87**, 1355–1366.

Williams, C.J.R. and Kniveton, D.R. (eds) (2011), *African Climate and Climate Change: Physical, Social and Political Perspectives*. London, UK: Springer.

World Commission on Environment and Development (1987), *Our Common Future*. Oxford, UK: Oxford University Press.

PART IV

RESPONSES TO THE THREATS POSED BY CLIMATE CHANGE TO HUMAN SECURITY

13. Climate change and human security: the individual and community response
C. Michael Hall

INTRODUCTION

In a general sense security is 'the condition of being protected from or not exposed to danger' (Barnett, 2001a, p.1). Environmental, social and economic issues, as well as the system of multi-level governance by which such concerns are governed, now clearly lie within contemporary understandings of security (Boulding, 1991; Lowi and Shaw, 2000). The broadening of ideas of security beyond national defence is reflective of the changed notion of politics in the post cold-war period. Indeed, at the international level the shift towards notions of common and collective security, although not universally accepted and even actively opposed by some administrations, has been integral in making the environment, and hence climate change, a security issue (Page and Redclift, 2002), as well as linking sustainable development to security concerns (Hall et al., 2003; O'Brien et al., 2010a; IPCC, 2012). According to the IPCC (2012, p.293) 'The impacts of climate extremes and weather events may threaten human security at the local level (*high agreement, medium evidence*)'. Indeed, O'Brien et al. (2010b, p.4) argue that climate change is 'a problem that can *only* be resolved by focusing on climate change as an issue of human security, which includes a thorough investigation of what it means for humans to be "secure"' (this author's emphasis).

The relationship between climate change, security and sustainable development also frames a new research agenda, and the potential development of a broad range of answers to the questions of 'security from what and secure to do what?' (Hall et al., 2003, pp.8–9). Especially as, despite concerns over various resources, the security impacts of climate change often take less direct and 'more multifarious routes' (Barnett, 2001a, p.1) than direct international conflict.

This chapter explores some of the issues relating to the emerging human security and climate change agenda at the community and individual scale. It argues that the challenge is as much conceptual with respect to paradigms of behaviour and governance as it is to the realities of climate change. Indeed, the intellectual capacity to consider other ways of doing

337

and decision-making are potentially the most important of all. The chapter is divided into three main sections. The first discusses the ways in which issues of human security and climate change response are framed. Such a discussion is regarded as fundamental to considering capacities to change and interventions to assist and enable such change. The chapter then goes on to examine the multi-level nature of responses to human security and climate change and the way that communities and individuals are embedded within broader social and economic structures. The critical role of trust and values in communities is noted but it is also stressed that communities should not be romanticized and it is vital to recognize that they may be ridden with conflict that makes appropriate sustainable solutions extremely hard to achieve. Similarly, the lack of capacity or willingness of individuals to respond to climate change, even if from an outsider's perspective this is the rational thing to do, is noted, and the challenges that this provides for communication of risk and security is highlighted. Finally, the chapter returns to the vital role of different paradigms of behaviour and governance and the implications that this has for the nature of interventions and capacity to change behaviour given the potential of the 'lock in' of communities and individuals within certain socio-technical systems.

FRAMING HUMAN SECURITY

The short and long-term economic, eco-system, health and welfare risks that climate change poses to human well-being are now acknowledged in both scientific and, increasingly, policy circles. However, like climate change itself the impacts of which are not evenly distributed in space and time, so policy and institutional responses are also uneven. Yet, this makes local approaches even more important with vulnerability at the local level attributable 'to social, political, and economic conditions and drivers including localized environmental degradation and climate change' (IPCC, 2012, p. 293). Indeed, Barnett (2001b) argues that environmental insecurity in the case of the less developed countries is the double vulnerability of people that arises when underdevelopment and impoverishment are compounded by anthropogenic environmental change. Such a perspective clearly grounds climate change in the environmental, economic and, vitally, social dimensions of sustainable development that the United Nations Development Program (UNDP) frames as an issue of human security, which is:

> . . . concerned with how people live and breathe in a society, how freely they exercise their many choices, how much access they have to market and social opportunities – and whether they live in conflict or peace.

Human security is not a concern with weapons – it is a concern with human life and dignity. (UNDP, 1994, pp. 22–23)

Following Alkire (2003), Barnett and Adger defined human security as 'the condition where people and communities have the capacity to manage stresses to their needs, rights, and values' (2007, p. 640). Their approach also reflects the notion that human security is an accentuated discourse on vulnerability, which is itself a major discourse in the climate change literature with respect to the potential adverse effects of climate change on ecosystems, infrastructure, economic sectors, social groups, communities and regions (Füssel and Klein, 2006; Füssel, 2007; IPCC, 2007, 2012; O'Brien et al., 2007).

As with notions of vulnerability, the assessment of human security requires considering levels of risk, susceptibility to loss, and capacity to recover (Barnett and Adger, 2007). 'Addressing disaster risk and climate extremes at the local level requires attention to much wider issues relating to sustainable development' (IPCC, 2012, p. 293). However, like ideas of vulnerability and risk, human security tends to be 'more socially constructed than objectively determined', with the consequence that human security 'tends to be attached to the most important of vulnerable entities – for example the nation (national security), basic needs (human security), income (financial security) and property (home security)' (Barnett, 2001a, p. 2). Yet recognition that notions of security, vulnerability and risk are socially constructed has clear political and scientific implications for defining particular risks and framing the problem of climate change at various scales, particularly as much social science runs counter to the construction of global change research in narrowly technical and reductionist scientific terms by the IPCC and other national and international bodies (Demeritt, 2001a, 2006; Hall, 2013a). Examining human security therefore means engaging in the broader issues of integrating social and biophysical perspectives in climate change research (Füssel and Klein, 2006), along with accompanying ontological shifts (Turnpenny et al., 2010). However, this also requires that particular attention be given to the ways in which both security threats and responses are framed and how different assumptions about individual and community behaviours are interrelated to different modes of climate change governance (Hall, 2013b).

Knowledge is produced locally and globally in different forms (Slocum, 2010), including with respect to the social dimensions of the communities that study climate change and the roles of social science in climate-change models and projections (Demeritt, 2001a, 2001b, 2006; Yearley, 2009). Demeritt's (2001a, b) examination of the construction of climate change science and the politics of science is particularly informative (see also the response of Schneider, 2001). Demeritt was not denying the existence of climate change but rather focused on the way in which climate change science is constructed and how this leads into political issues (see also Wainwright, 2010). 'For the most part, climate change model projections have been driven by highly simplistic business-as-usual scenarios of human population growth, resource consumption, and GHG emissions at highly aggregated geographic scales' (Demeritt, 2001a, p. 312) that operate at a global scale, rather than framing the problem in terms of alternative, and no less relevant forms, such as the structural imperatives of the capitalist economy that drives emissions; the north-south gap in terms of emissions; or regionalized, community or individual conceptions that focus on issues of poverty and deprivation. 'By treating the objective physical properties of [greenhouse gases] in isolation from the surrounding social relations serves to conceal, normalize, and thereby reproduce those unequal social relations' (Demeritt, 2001a, p. 316).

The appeals of formal quantitative evaluation methods are social and political as much as technical and scientific; particularly they are also more credible from a public perspective of natural science (Hall, 2013a): 'insofar as adherence to rigidly uniform and impersonal and in that sense "procedurally objective". . . rules limits the scope for individual bias or discretion and thereby guarantees the vigorous (self-)denial of personal perspective necessary to make knowledge seem universal, trustworthy and true' (Demeritt 2001a, p. 324).

Issues surrounding how climate change research and its results are constructed scientifically are clearly significant for understanding the relationship between human security at the individual and community level to climate change and the debate surrounding that relationship for a number of reasons (Hall, 2013a). First, it helps explain why anthropogenic climate change has primarily been defined as an environmental rather than a political or economic problem, or one that requires framing in terms of the imperatives of the capitalist economic system and its alternatives. Second, even though there has been a call for greater social science information to be brought into the climate change assessment process this has primarily been assessed in terms of neo-classical economic contributions (e.g. Stern, 2007), with the need for action to minimize the effects of climate change also dominated by formal modelling (Dietz and Stern, 2008). Yet there are numerous major weaknesses in such models with respect to predicting behavioural responses to climate change (Gossling and Hall, 2006). Furthermore, the neoclassical economic utilitarian view of value has been subjected to a number of different philosophical and moral critiques as to the way people engage with their physical and social environment and

Table 13.1   Ontologies of ecology

Methodological approach	Ontology	Epistemology
Reductionism	Properties of wholes are always found among the properties of their parts	Knowledge of the parts is both necessary and sufficient to understand the whole
Mechanism	Properties of wholes are of the same kind or type of those parts	Knowledge of the kind or type of the cause suffices to understand the type of kind of the effect
Emergentism	There is at least one property of some wholes not possessed by any of their parts. Parts can exist independently of the whole, and novel properties of wholes can be lost via submergence when a system is reduced to its parts	Knowledge of the parts and their relations is a necessary but not a sufficient condition to understand the whole
Organicism	Recognize the existence of emergent properties of wholes. Once a whole has appeared, its parts cannot exist or be understood independently of a whole	Knowledge of the whole is a necessary condition to understand the parts and vice-versa
Holism	The emergent novel properties of the whole can be understood without further consideration of the parts and their relationships. The basic unit the whole – wholes are independent of parts	Knowledge of the parts is neither necessary nor sufficient to understand the whole

Sources: After Blitz, 1992: 175–178; Keller and Golley, 2000; Hall, 2013a.

actually behave (Demeritt and Rothman, 1999) and which will be returned to in more detail below.

Also of significance is the way within which climate change is widely understood through strategic cyclical scaling (Root and Schneider, 1995), with small-scale, detailed case studies informing and being informed by large-scale, comprehensive statistical and simulation studies. However, such an approach has an implicit ontological position with respect to the relationship between the parts and the whole that is only one of several potential epistemic and ontological positions with respect to knowledge of ecological entities (Table 13.1) (Blitz, 1992; Hall, 2013a; Keller and Golley, 2000). This is a point of critical importance in trying to understand how in human ecological terms, individuals interact with larger socio-political groupings such as communities, regions, and nations in responding to climate change, and the assumptions upon which behavioural interventions rest. While the focus of human security is the individual (Barnett and Adger, 2007), the way in which individuals are regarded as being affected by different larger scale processes and therefore respond and adapt to them will depend on the way in which such multi-scalar relationships are framed (Cohen, 2007).

Recognition of the relationship between disaster risk reduction and climate change adaptation (O'Brien et al., 2008; IPCC, 2012) has been profound for the framing of the human security and climate change issue (O'Brien et al., 2010b). As opposed to research and policy formulation on adaptation that focuses on biophysical impacts, disaster risk research has tended to recognize the broader socio-economic causes and consequences of what are otherwise presented as 'natural disasters', including poverty, inequality, market failures, and policy failures (Barnett and Adger, 2007). As a result the relationships between climate change and sustainable development were strengthened (Beg et al., 2002; Swart et al., 2003; Halsnæs et al., 2008; Adger et al., 2009b; IPCC, 2012), particularly with respect to poverty reduction initiatives and global initiatives such as the Millennium Development Goals, and ameliorating the human consequences of climate change (DFID, 2006; UNDP, 2008). As O'Brien et al. (2008, p. 5) observe, 'Enhancing human security in the twenty-first century is about responding to climate change and disaster risks in ways that not only reduce vulnerability and conflict, but also create a more equitable, resilient and sustainable future'. Indeed, the Intergovernmental Panel on Climate Change (IPCC) (2012) define human security in terms of two main aspects, 'It means, first, safety from such chronic threats as hunger, disease, and repression. And second, it means protection from sudden and hurtful disruptions in the patterns of daily life – whether in homes, in jobs, or in communities. Such threats can exist at all levels of national income and development' (IPCC, 2012, p. 572). Given such a definition it is then not surprising that the IPCC (2012, p. 20) concluded, 'A prerequisite for sustainability in the context of climate change is addressing the underlying causes of vulnerability, including the structural inequalities that create and sustain poverty and constrain access to resources'.

THE MULTI-LEVEL NATURE OF HUMAN SECURITY RESPONSES TO CLIMATE CHANGE

The relationship between sustainable development and vulnerability – the propensity or predisposition to be adversely affected (IPCC, 2012) –

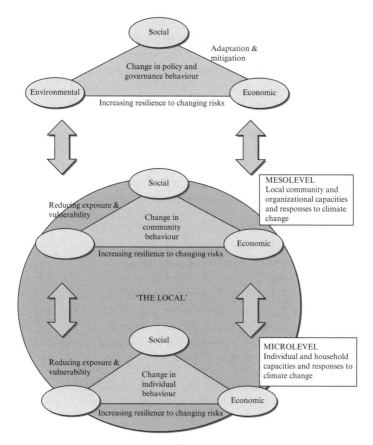

Figure 13.1 Multi-level aspects of climate change, human security, and sustainable development

provides a natural springboard into examining responses to climate change and human security. However, difficulties in progressing and operationalizing the concept of sustainable development also point to some of the human security challenges faced in responding to climate change.

Figure 13.1 illustrates the multi-level nature of climate change, human security and sustainable development. It suggests that climate change as an issue for sustainable development and security requires attention at all scales from global and national to the local and the individual; with interactions among the various scales being an important aspect of human response (Wilbanks, 2003, 2007; O'Brien et al., 2004; Yohe et al., 2007). However, given that the effects of climate change are geographically variable and the capacity to manage risks and adapt to change is also unevenly distributed within and across nations, regions, communities, and households (Barnett and Adger, 2007), 'climate change-related stresses upon and strategies for sustainable development are often most usefully considered in a place-based context, where an integrated view of complex interrelationships is more tractable and strategies for action can be made more tangible' (Wilbanks, 2003, p.S147). This is especially the case given the difficulties in achieving clear global and national strategies to combat climate change. Similarly, O'Brien et al. (2008, p.19) highlight, 'Local-level experiences can be considered the front-line of impacts from hazards and extreme events, thus they can provide important insights on the most urgent challenges associated with extreme weather events in a changing climate'.

THE CHALLENGE OF THE LOCAL

Although 'the local' is clearly of great significance in responding to climate change it is vital not to romanticize it. As research in the fields of resource management and planning indicates, communities are not the embodiment of innocence (Millar, 1996; Young, 2001; Suryanata and Umemoto, 2003, 2005; Joyce and Satterfield, 2010),

> on the contrary, they are complex and self-serving entities, as much driven by grievances, prejudices, inequalities, and struggles for power as they are united by kinship, reciprocity, and interdependence. Decision-making at the local level can be extraordinarily vicious, personal, and not always bound by legal constraints. (Millar and Aiken, 1995, p.629)

Such observations reinforce O'Brien et al.'s (2008, p.18) observation, 'that risk reduction and adaptation strategies must be carefully tailored to individual, household and community needs. Approaches that treat communities as homogeneous (i.e., able to adapt or reduce risks as a group) are prone to failure.'

Change is a normal part of the human experience. However, climate change, especially in the context of increased frequency of high-magnitude weather events and 'natural disasters' (IPCC, 2012), may serve to hasten rates of change above those that are 'comfortable' for many people. Rapid or sudden place change can serve to dramatically alter the web of relations that residents have with place (Hay, 1998; Connor et al., 2004; Albrecht et al., 2007; Hanna et al., 2011; Biggs et al., 2012), and substantially affect personal, household and community well-being, thus directly impacting senses of security. As Millar and Aiken (1995, p.620) commented,

Conflict is a normal consequence of human interaction in periods of change, the product of a situation where the gain or a new use by one party is felt to involve a sacrifice or changes by others. It can be an opportunity for creative problem solving, but if it is not managed properly conflict can divide a community and throw it into turmoil.

There is a substantial body of literature on appropriate community-level approaches and strategies to engage communities in seeking to improve their resilience and well-being and lessen their vulnerabilities and risks (e.g. Tobin, 1999; Paton and Johnston, 2001; David et al., 2003; Hall, 2008; Ensor and Berger, 2009; Comfort et al., 2010; Krishnaswamy et al., 2012). Such approaches have also been long identified as significant for sustainability practices (Kelly, 2009). For example, Ostrom (1990) noted the following interrelated factors of sustainable development at the community level:

- clearly defined boundaries;
- harmony between appropriation and provision rules and local conditions;
- participation by all interested parties in changes that may affect them;
- accountable monitoring;
- graduated sanctions administered by an accountable authority
- low cost and readily accessible mechanisms for conflict resolution
- recognition by governments of the rights to organize;
- appropriate licensing provisions, monitoring, enforcement, conflict resolution and organizational arrangements, for those regimes which are part of large systems of governance.

Yet despite awareness of appropriate community practices in academic literature and reports (O'Brien et al., 2008; IPCC, 2012), knowledge is not necessarily translated into action (White et al., 2001). Indeed, the four possible explanations provided by White et al. (2001) as to why more is lost while more is known are arguably more pertinent than ever in examining individual and community response to climate change: (1) knowledge continues to be flawed by areas of ignorance; (2) knowledge is available but not used effectively; (3) knowledge is used effectively but takes a long time to have effect; and (4) knowledge is used effectively in some respects but is overwhelmed by increases in vulnerability and in population, wealth, and poverty.

A key issue in both acceptance and effective use of knowledge in communities is the issue of trust. Where trust is absent, cooperative or voluntary collective action is impossible, particularly in 'commons' situations that rely on the 'curbing of opportunistic impulses toward individual exploitation' (Millar, 1996, p. 207). Trust therefore provides for a sufficient number of reciprocal and cooperative actions to occur such that there will be a greater return to all stakeholders than would be forthcoming through individual exploitation (Brann and Foddy, 1987). Trust requires a sufficiently common set of values between community members in order to operate. Therefore, attention in many community planning and climate change response exercises needs to be given to the social and political context within which human security occurs and value conflict arises. As Millar and Aitken (1995, pp. 623–624) noted: 'we must be more concerned with how the local society and resource base are organized . . . communities exist within a web of kinship, physical interdependency, and social obligation, and . . . cannot be separated from the social issues of property and morality'.

Trust in relation to climate change is arguably even more significant than many other planning issues given the high profile of climate change sceptics in the media and politics. Although there is widespread agreement within the science community on the urgency of action to deal with climate change, such agreement does not necessarily exist at the community level (Gössling et al., 2012a). Such a situation reflects the importance of recognizing that relevant information is individually and socially contextualized and actioned. 'Concepts and tools do not necessarily motivate behaviour change where individuals are not motivated to change or perceive barriers to doing so' (Whitmarsh et al., 2011, p. 58). This is something that may have significant repercussions in responding to the human security dimensions of climate change.

For the majority of people in an increasingly urbanized world, the understanding of environmental issues is often limited to abstract or vague concepts. Climate change poses major challenges to communicators and educators as, 'It is a risk "buried" in familiar natural processes such as temperature change and weather fluctuations . . . and has low salience as a risk issue because it cannot be directly experienced' (Whitmarsh et al., 2011, p. 57; see also Grothmann and Patt, 2005; Weber, 2006). There is some evidence to support this observation given that research suggests that the attitudes change when people experience weather events that they believe are associated with climate change (Hall, 2006; Tervo-Kankare et al., 2013). For example, in December 2012 following the impacts of Hurricane Sandy, the results of a US poll by the Associated Press-GfK found rising concern about climate change among Americans, with 80 percent citing it as a serious problem for the US, up from 73 percent in 2009. According to Goldenberg (2012), 'Some of the doubters said in follow-up interviews that they were persuaded by personal experience: such as record temperatures, flooding of New York City subway tunnels, and news of sea ice melt in the Arctic and extreme drought in the mid-west'. Although changes in attitudes may be encouraging, if communities wait for an event to "make them believe" then the prospects for engendering more sustainable and secure communities are not good. Communications are therefore rightly highlighted as extremely important, 'which suggests an emphasis on presenting knowledge in a community's own language, through innovative media, and in understandable non-scientific terms' (O'Brien et al., 2008, p. 20).

While public participation is seen as a standard mechanism to deal with community planning issues, it should be noted that simply hosting a public meeting, a common consultation strategy, will not by itself make it more likely that conflicts will be resolved. Indeed, they may well lead to even greater conflict between parties and serve to reinforce rather than change positions and draw people closer to agreement. 'Public meetings may help to identify conflicts, but they cannot resolve or manage them. While it is true they allow everyone to have his or her say, the root causes ... are often neglected' (Millar and Aiken, 1995, p. 627). The problem has often been a focus on the technique – the public meeting – rather than the process and the creation of social relations and what the hoped-for outcome of the process actually is (Umemoto and Suryanata, 2006). Too often, processes have been interest based rather than values based. However, if long-term agreement and common ground between stake-holders is sought then attention must be given to the values of those who are affected (Millar and Yoon, 2000; Ryan and Wayuparb, 2004; Evans and Garvin, 2009). Therefore, community strategies require appropriate communication strategies to be integrated with community engagement in a manner that is responsive to local situations (Hall, 2008; van Aalst et al., 2008). O'Brien et al. (2008, p. 20), for example, stress the importance of 'working with trusted local intermediaries who have a firm understanding of community circumstances and dynamics, basing new activities, technologies or practices on existing coping practices.'

There appears considerable opportunity for reducing vulnerability at the local level via community-based adaptation. Bottom-up approaches promote locally-appropriate measures and empowerment, and encourage greater ownership of risk reduction and adaptation actions and strategies to increase resilience and security (O'Brien et al., 2008; Ensor and Berger, 2009; McEvoy et al., 2010). Such approaches also fit in with the clear need identified by the IPCC for adaptation assessments to

consider not only the technical feasibility of certain adaptations but also the availability of required resources, the costs and side effects of adaptation, the knowledge about those adaptations, their timeliness, the incentives for the adaptation actors to actually implement them, and their compatibility with individual or cultural preferences . . . (Schneider et al., 2007, p. 796)

Despite enthusiasm for bottom-up approaches there is also recognition that there are non-technical limits to adaptation capacities (Adger et al., 2007, 2009a). For example, Ensor and Berger (2009, p. 227), noted 'the limits that culture places on the freedom of individuals and communities to embrace change', while Adger et al. (2007) identified a number of different types of non-technical and non-ecological limits to adaptation including informational and cognitive limits, social and cultural limits, institutional political limits, and financial limits. Therefore, the state remains an important means to influence the capacity of communities and individuals to respond to climate change.

CARBON CAPABILITY AND APPROACHES TO INDIVIDUAL AND COMMUNITY CHANGE

Carbon capability is 'the ability to make informed judgments and to take effective decisions regarding the use and management of carbon, through both individual behavior change and collective action' (Whitmarsh et al., 2009, p. 2). In seeking to understand the different modes of policy intervention in changing behaviours with respect to climate change, there is a need to understand the situated meanings associated with climate change, sustainability, and human security (Hall, 2013b). That is, how individuals translate and apply knowledge in their daily lives and decision-making (Whitmarsh et al., 2011). Carbon capability has three core dimensions:

- decision-making/cognitive/evaluative (technical, material and social aspects of knowledge, skills, motivations, understandings and judgements);
- individual behaviour or 'practices' (e.g., energy conservation, reducing risk); and
- broader engagement with systems of provision and governance (e.g., voting, protesting, creating alternative infrastructures of provision).

Under a carbon literacy approach the assumption is that as individuals gain knowledge, perhaps as a result of educational campaigns, they change individual behaviours and practices. In contrast, a carbon capability approach argues that much behaviour is inconspicuous, habitual and routine, rather than the result of conscious decision-making. This means that individual cognitive decisions about behaviour and risk (Marx et al., 2006), and accompanying practices are constrained by social systems of provision and the rules and resources of macro-level structures and institutions (Whitmarsh et al., 2009). Carbon capability does not imply that education and communication to try and encourage behaviour change is worthless. Rather it highlights that the capacity for individual and community behavioural change needs to be understood in a wider context (Hall, 2013b, 2014).

The notion of carbon capability inherently means looking beyond an understanding of behavioural change that suggests that information provision is sufficient for consumers to make 'rational' and 'appropriate' choices. However, it also raises the need to understand the underlying assumptions of different approaches to behaviour change. Three major contemporary approaches can be recognized in approaching issues of decision-making, behaviour and sustainable consumption: utilitarian, social/psychological and systems of provision/institutional (Seyfang, 2011) (Table 13.2).

The utilitarian approach to behavioural change aims to appeal to rational actors who maximize their utility with information to overcome an 'information deficit' and encourage 'rational behaviour' and send appropriate signals to the market (Hall, 2013b, 2014). For example, in examining elderly people's perceptions of heat wave risks, Wolf et al. (2009) showed that because individuals did not perceive their own vulnerability, they did little to adapt. However, individual responses to climate change may not be as rational as many assessments of adaptive behaviour assume (Grothmann and Patt, 2005). Information overload can lead to difficulties in decision-making (Seyfang, 2011), while also of importance is the role of social norms and routines that are not subject to rational cost-benefit calculations, including notions of community and fairness in economic outcomes. Therefore, decision-making is not a unidirectional and sequential process and is instead incremental and at times multi-directional (O'Brien et al., 2008, p. 22).

The focus on satisficing in decision and policy-making has assumed renewed significance with respect to climate change solutions (Thynne, 2008), including transition management (Gössling et al., 2012b). It is an underlying dimension of public policy interest in 'nudging' (Thaler and Sunstein, 2008). The approach suggests that the goal of public policy-making should be to steer citizens towards making positive decisions as individuals and for communities while preserving individual choice. Acting as 'choice architects' policy-makers organize the context, process and environment in which individuals make decisions and, in so doing, they exploit 'cognitive biases' to manipulate people's choices (Alemanno,

Table 13.2 *Approaches to understanding decision-making and behavioural change*

Approach	Scale	Understanding of decision-making	Tools to achieve behavioural change	Dominant forms of governance
Utilitarian (green economics)	Individual	Cognitive information processing on basis of rational utility-maximization	Green labelling, tax incentives, pricing, education	Markets (marketization and privatization of state instruments)
Social & psychological (Behavioural economics; green consumerism; ABC model)	Individual	Response to psychological needs, behaviour and social contexts Dominant paradigm of 'ABC': attitude, behaviour, and choice	Nudging – making better choices through manipulating the behavioural environment Social marketing in order to encourage behavioural change	Markets (marketization and privatization of state instruments) Networks (public-private partnerships)
Systems of provision/ Institutions (degrowth; community-based approaches)	Community, society, network	Constrained/shaped by socio-technical infrastructure and institutions	Short-supply chains, local food security; developing local competencies and capacities	Hierarchies (nation state and supranational institutions) Communities (public-private partnerships, communities)

Sources: Adger et al., 2009b; Hall, 2009, 2011a, 2011b, 2013b; Seyfang, 2011; Burgess, 2012; Gössling and Hall, 2013.

2012). According to Bovens (2009) and Alemanno (2012) a policy instrument qualifies as a nudge when it satisfies the following:

- the intervention must not restrict individual choice;
- it must be in the interest of the person being nudged;
- it should involve a change in the architecture or environment of choice;
- it implies the strategic use of cognitive biases;
- the action it targets does not stem from a fully autonomous choice (e.g. lack of full knowledge about the context in which the choice is made).

In the UK the influence of the concept on policy initiatives, including emissions reduction, is seen in David Cameron reportedly making *Nudge* (Thaler and Sunstein, 2008) obligatory reading for his colleagues before election (Burgess, 2012), and the MINDSPACE report (Dolan et al., 2010), which suggests 'approaches based on "changing contexts" – the environment within which we make decisions and respond to cues – have the potential to bring about significant changes in behaviour at relatively low cost' (Dolan et al., 2010, p. 8). As the MINDSPACE discussion paper reported

> ... most policymakers are focused on this bigger picture – often known as culture change ... There may be occasions when the power of argument alone can eventually [bring] such culture change, such as gender equality or race relations. But generally the broader sweep of policy history suggests that such change is driven by a mix of both broad social argument and small policy steps. Smoking is perhaps the most familiar example. Over several decades the behavioural equilibrium has shifted from widespread smoking to today's status as an increasingly minority activity. Better information; powerful advertising (and the prohibition of pro-smoking advertising); expanding bans; and changing social norms have formed a mutually reinforcing thread of influence to change the behavioural equilibrium. There is every reason to think that is a pattern that we will see repeated in many other areas of behaviour too, from sexual behaviour to carbon emissions. (Dolan et al., 2010, p. 77)

Both nudging and the closely related concept of social marketing recognize that decision-making and consumption are multilayered phenomena that are full of meaning including roles as signifiers of identity, cultural and social affiliations, and relationships (Seyfang, 2011). The potential of social marketing to influence sustainable behaviours in relation to climate change has become increasingly argued (Downing and Ballantyne, 2007; Peattie et al., 2009). However, although social/psychological approaches to behaviour change recognize the need to make major changes, and although social norms are often cited as driving factors, there is little scope for wondering about how needs and aspirations come to be as they are, i.e. they do not fundamentally question structures and paradigms (Shove, 2010), which is not the case with the systems of provision approach. Furthermore, the social/psychological approach may not be effective in situations where individuals deny risks, feel powerless to act, or have little adaptive capacity (O'Brien et al., 2008). Indeed, Burgess (2012) suggests that application of behavioural solutions assumes a fundamentally pliant and passive population that attaches limited value to individual liberty and autonomy in modern 'risk society', something that would appear to be at odds with goals of empowerment in community-based approaches to human security and climate change (O'Brien et al, 2008).

The systems of provision approach focuses on the contextual collective societal institutions, norms, rules, structures and infrastructure that constrain individual decision-making, consumption and lifestyle practices. The significance of the approach is that it highlights that particular socio-technical systems constrain choice to that available within the system of provision, and can therefore 'lock in' individuals and communities to particular ways of behaving (Lorenzoni et al., 2007, 2011; Maréchal, 2010; Seyfang, 2011). The systems of provision approach has focused on the development of alternative systems such as food networks for the enhancement of local food security and local economies, e.g. farmers' markets, and short supply chains, e.g. Fair Trade (Gössling and Hall, 2013). Such initiatives are important as they also point the way to the development of new forms of post-carbon communities and localism that may be essential in responding to climate and other forms of environmental change (Monaghan, 2012).

The utilitarian (green labeling, tax incentives, pricing, education) and social/psychological (nudging, social marketing) approaches are grounded in an 'ABC model' (Hall, 2013b, 2014). Social change is thought to depend upon values and attitudes (A), which are believed to drive the kinds of behaviour (B) that individuals choose (C) to adopt. 'The combination of A and B and C generates a very clear agenda for effective policy, the conceptual and practical task of which is to identify and affect the determinants of pro-environmental behavior' (Shove, 2010, p. 1275; see also Hinkel, 2011). As Shove (2010) suggests, present ABC strategies of state intervention frame environmental damage as a consequence of individual actions and assume that given better information or incentives individuals will choose to act and behave differently. But in so doing they do not fundamentally question structures/paradigms and their role in influencing individual behaviours and understandings of risk and vulnerability (Hall, 2013b).

... policy makers are highly selective in the models of change on which they draw, and that their tastes in social theory are anything but random. An emphasis on individual choice has significant political advantages and in this context, to probe further, to ask how options are structured, or to inquire into the ways in which governments maintain infrastructures and economic institutions, is perhaps too challenging to be useful . . . (Shove, 2010, p. 1283)

CONCLUSION: INDIVIDUAL AND COMMUNITY RESPONSE TO HUMAN SECURITY AND CLIMATE CHANGE – BEYOND ABC?

The chapter has suggested that our understandings of the ways in which individuals and communities respond to climate change and human security issues are related to different paradigms of behaviour. Nudging and social marketing are of some value in encouraging individual and community change (Burgess, 2012). However, without also facing up to the implications of structure and institutions then the likelihood of being 'locked-in' to socio-technical systems of provision is greatly increased. Furthermore, different concepts of behavioural change are also linked to different modes of governance (Hall, 2011b, 2013b, 2014).

Each of the conceptualizations of governance is related to the use of particular sets of policy instruments (Hall, 2011b). Given the artificiality of any policy–action divide, these instruments and modes of governance can also be connected to different conceptual approaches to policy implementation and state intervention (Hall, 2009). In something of a chicken and the egg situation, it is not really possible to say what came first – the mode of governance or the intervention mechanism for behavioural change that implements public policies. But what is important is that the mode of governance and the manner of intervention become mutually reinforcing. One cannot be adequately understood without the other (Hall, 2011a). Yet mutual reinforcement between modes of governance and intervention creates a path dependency in which solutions to increased human security in response to climate change are primarily identified within 'green growth' arguments for greater efficiency and market-based solutions, and an ideology that frames the problem of sustainable development and risk reduction in terms of individual decision-making and responsibility (Bailey and Wilson, 2009; Shove, 2010; Whitmarsh et al., 2011).

One of the appeals of behavioural economics and social marketing approaches for policy-making, such as nudging, is the promise of cost effectiveness; achieving 'more for less', particularly in public sector expenditure. The practical emphasis is on 'smart' solutions that supposedly do not involve more resources; an imperative in recessionary times with many governments committed to reducing spending (Burgess, 2012), and have clear indicators. However, the cause-and-effect relations of even seemingly straightforward nudge proposals are not nearly as predictable or measurable as expected (Hinkel, 2011) and the approach may be incapable of 'solving' 'wicked problems' such as sustainable consumption and climate change 'in which multiple systems, operating with different norms and at different scales, interact with one another to produce emergent behavior that can be exceptionally difficult, if not impossible, to predict' (Selinger and Powys White, 2012, p.29), such as rebound and backfire effects (Hall et al., 2013). As Selinger and Powys White (2012, p.29) go on to note, 'nudge proponents do not even attempt to discuss at all the complex interventions that are needed to shift diverse citizens in diverse parts of the world away from a CO_2 intensive society or adapt to the changes that might accompany the perpetuation of CO_2 intensive industries, infrastructures, and lifestyles'.

A further significant and complicating issue is that 'Given that the ABC is the dominant paradigm in contemporary environmental policy, the scope of *relevant* social science is typically restricted to that which is theoretically consistent with it' (Shove, 2010, p.1280) (this author's emphasis). As Shove acknowledges, the ABC paradigm is not just a theory of behavioural change:

> it is also a template for intervention which locates citizens as consumers and decision makers and which positions governments and other institutions as enablers whose role is to induce people to make pro-environmental decisions for themselves and deter them from opting for other, less desired, courses of action . . . (Shove, 2010, p.1280)

This is therefore a critical issue in seeking to respond to issues of human security and climate change. Given the inter-relationship between how human behaviour is understood, and modes of governance and intervention, there is the likelihood that how individuals and communities are enabled to respond is being constrained in many cases.

'Lock-in' to particular socio-technical systems of provision is a major constraint to avoidance of dangerous climate change (Bailey and Wilson, 2009; Maréchal, 2010). Such lock-ins occur at multiple scales of governance. In such a situation policy learning is extremely difficult (Hall, 2011a), with changes often occurring only at the margin that do not challenge the basic growth paradigm and/or ways of doing. As Shove suggests

> . . . paradigms and approaches which lie beyond the pale of the ABC are doomed to be forever marginal no matter how interactive or how policy-engaged their advocates might be. To break through this log jam it would be

necessary to reopen a set of basic questions about the role of the state, the allocation of responsibility, and in very practical terms the meaning of manageability, within climate-change policy . . . (Shove, 2010, p. 1283)

This situation creates a very real problem for communities seeking to do 'other' and highlights the difficulties of mainstreaming adaptation into sustainable development, even though it will enhance human security (O'Brien et al., 2008; Adger et al., 2009b; IPCC, 2012). If enhancing human security in the twenty-first century is, as O'Brien et al. (2008) suggest, about responding to climate change in ways that not only reduce vulnerability and conflict, but also create a more equitable, resilient and sustainable future, then it is also about finding ways to respond politically and deal with issues of power. Highlighting issues such as 'equity' (O'Brien et al., 2008), 'repression' (IPCC, 2012, p. 572), and 'structural inequality' (IPCC, 2012, p. 20) is a statement about the nature of current systems of provision and their inadequacy. These are also clear statements that require strong community political leadership to be tackled and to continue to be highlighted as constraints on human security. Just as importantly they are significant challenges to the academic community that works on human security and climate change to not only continue to speak out on such issues in policy documents and reports but also to the essential value of critical social science within the scientific community of climate change as well.

REFERENCES

Adger, W.N., S. Agrawala, M.M.Q. Mirza, C. Conde, K. O'Brien, J. Pulhin, R. Pulwarty, B. Smit and K. Takahashi (2007), 'Assessment of adaptation practices, options, constraints and capacity', in M. Parry, O. Canziani, J. Palutikof, P. van der Linden and C. Hanson (eds), *Climate Change 2007: Impacts, Adaptation and Vulnerability. Contribution of Working Group II to the Fourth Assessment Report of the Intergovernmental Panel on Climate Change*, Cambridge, UK: Cambridge University Press, pp. 717–743.

Adger, W.N., S. Dessai, M. Goulden, M. Hulme, I. Lorenzoni, D. Nelson, L. Naess, J. Wolf and A. Wreford (2009a), 'Are there social limits to adaptation to climate change?', *Climatic Change*, **93**, 335–354.

Adger, W.N., I. Lorenzoni and K. O'Brien (eds) (2009b), *Adapting to Climate Change: Thresholds, Values, Governance*, Cambridge, UK: Cambridge University Press.

Albrecht, G., G.-M. Sartore, L. Connor, N. Higginbotham, S. Freeman, B. Kelly, H. Stain, A. Tonna, and G. Pollard (2007), 'Solastalgia: the distress caused by environmental change', *Australasian Psychiatry*, **15** (S1), 95–98.

Alemanno, A. (2012), 'Nudging smokers – the behavioural turn of tobacco risk regulation', *European Journal of Risk Regulation*, **1/2012**, 32–42.

Alkire, S. (2003), *A Conceptual Framework for Human Security*, CRISE Working Paper 2, Oxford: Queen Elizabeth House.

Bailey, I. and G. Wilson (2009), 'Theorising transitional pathways in response to climate change: technocentrism, ecocentrism, and the carbon economy', *Environment and Planning A*, **41**, 2324–2341.

Barnett, J. (2001a), *Security and Climate Change*, Tyndall Centre Working Paper No. 7, Norwich, UK: Tyndall Centre for Climate Change Research, University of East Anglia.

Barnett, J. (2001b), *The Meaning of Environmental Security: Ecological Politics and Policy in the New Security Era*, London, UK: Zed Books.

Barnett, J. and W.N. Adger (2007), 'Climate change, human security and violent conflict', *Political Geography*, **26**, 639–655.

Beg, N., J. Corfee Morlot, O. Davidson, Y. Afrane-Okesse, L. Tyani, F. Denton, Y. Sokona, J.P. Thomas, E.L. La Rovere, J.K. Parikh, K. Parikh and A.A. Rahman (2002), 'Linkages between climate change and sustainable development', *Climate Policy*, **2**, 129–144.

Biggs, D., C.M. Hall and N. Stoeckl (2012), 'The resilience of formal and informal tourism enterprises to disasters – reef tourism in Phuket', *Journal of Sustainable Tourism*, **20**, 645–665.

Blitz, D. (1992), *Emergent Evolution: Qualitative Novelty and the Levels Of Reality*, Boston, USA: Kluwer.

Boulding, E. (1991), 'States, boundaries and environmental security in global and regional conflicts', *Interdisciplinary Peace Research*, **3** (2), 78–93.

Bovens, L. (2009), 'The ethics of nudge', in T. Grüne-Yanoff and S. Ove Hansson (eds), *Preference Change: Approaches from Philosophy, Economics and Psychology*, Berlin, Germany: Springer, pp. 207–220.

Brann, P. and M. Foddy (1987), 'Trust and the consumption of a deteriorating common resource', *Journal of Conflict Resolution*, **31**, 615–630.

Burgess, A. (2012), '"Nudging" healthy lifestyles: the UK experiments with the behavioural alternative to regulation and the market', *European Journal of Risk Regulation*, **1/2012**, 3–16.

Cohen, M.J. (2007), *Food Security: Vulnerability Despite Abundance*, Coping With Crisis Working Paper Series, New York, USA: International Peace Academy.

Comfort, L., T. Birkland, B. Cigler and E. Nance (2010), 'Retrospectives and prospectives on Hurricane Katrina: five years and counting', *Public Administration Review*, **70**, 669–678.

Connor, L., G. Albrecht, N. Higginbotham, S. Freeman and W. Smith (2004), 'Environmental change and human health in Upper Hunter communities of New South Wales, Australia', *EcoHealth*, **1** (2, Supplement), SU47–SU58.

David, R.G., S. Brody and R. Burby (2003), 'Public participation in natural hazard mitigation policy formation: challenges for comprehensive planning', *Journal of Environmental Planning and Management*, **46**, 733–754.

Demeritt, D. (2001a), 'The construction of global warming and the politics of science', *Annals of the Association of American Geographers*, **91**, 307–337.

Demeritt, D. (2001b), 'Science and the understanding of science: a reply to Schneider', *Annals of the Association of American Geographers*, **91**, 345–348.

Demeritt, D. (2006), 'Science studies, climate change and the prospects for constructivist critique', *Economy and Society*, **35**, 453–479.

Demeritt, D. and D. Rothman (1999), 'Figuring the costs of climate change: an assessment and critique', *Environment and Planning A*, **31**, 389–408.

Department for International Development (DFID) (2006), *Eliminating World Poverty: Making Governance Work for the Poor, A White Paper on International Development*, London, UK: Department for International Development.

Dietz, S. and N. Stern (2008), 'Why economic analysis supports strong action on climate change: a response to the Stern Review's critics', *Review of Environmental Economics and Policy*, **2**, 94–113.

Dolan, P., M. Hallsworth, D. Halpern, D. King and I. Vlaev (2010), *MINDSPACE: Influencing Behaviour Through Public Policy*, London, UK: Cabinet Office and Institute for Government.

Downing, P. and J. Ballantyne, J. (2007), *Tipping Point or Turning Point? Social Marketing & Climate Change*, London, UK: Ipsos MORI Social Research Institute.

Ensor, J. and Berger, R. (2009), 'Community-based adaptation and culture in theory and practice', in W.N. Adger, I. Lorenzoni and K. O'Brien (eds), *Adapting to Climate*

Change: Thresholds, Values, Governance, Cambridge, UK: Cambridge University Press, pp. 227–239.

Evans, J. and T. Garvin (2009), '"You're in oil country": moral tales of citizen action against petroleum development in Alberta, Canada', *Ethics Place and Environment*, **12**, 49–68.

Füssel, H.-M. (2007), 'Vulnerability: a generally applicable conceptual framework for climate change research', *Global Environmental Change*, **17** (2), 155–167.

Füssel, H.-M. and R.J.T. Klein (2006), 'Climate change vulnerability assessments: an evolution of conceptual thinking', *Climatic Change*, **75**, 301–329.

Goldenberg, S. (2012), 'Extreme weather more persuasive on climate change than scientists', *The Guardian*, 14 December.

Gössling, S. and C.M. Hall (2006), 'Uncertainties in predicting tourist flows under scenarios of climate change', *Climatic Change*, **79** (3–4), 163–73.

Gössling, S. and C.M. Hall (2013), 'Sustainable culinary systems: an introduction', in C.M. Hall and S. Gössling (eds), *Sustainable Culinary Systems: Local Foods, Innovation, and Tourism & Hospitality*, London, UK: Routledge, pp. 3–44.

Gössling, S., D. Scott, C.M. Hall, J. Ceron and G. Dubois (2012a), 'Consumer behaviour and demand response of tourists to climate change', *Annals of Tourism Research*, **39**, 36–58.

Gössling, S., C.M. Hall, F. Ekström, A. Brudvik Engeset and C. Aall (2012b), 'Transition management: a tool for implementing sustainable tourism scenarios?', *Journal of Sustainable Tourism*, **20**, 899–916.

Grothmann, T. and A. Patt (2005), 'Adaptive capacity and human cognition: the process of individual adaptation to climate change', *Global Environmental Change, Part A*, **15**, 199–213.

Hall, C.M. (2006), 'New Zealand tourism entrepreneur attitudes and behaviours with respect to climate change adaption and mitigation', *International Journal of Innovation and Sustainable Development*, **1**, 229–237.

Hall, C.M. (2008), *Tourism Planning*, 2nd edn, Harlow, UK: Pearson Prentice-Hall.

Hall, C.M. (2009), 'Archetypal approaches to implementation and their implications for tourism policy', *Tourism Recreation Research*, **34**, 235–245.

Hall, C.M. (2011a), 'Policy learning and policy failure in sustainable tourism governance: from first and second to third order change?', *Journal of Sustainable Tourism*, **19**, 649–671.

Hall, C.M. (2011b), 'A typology of governance and its implications for tourism policy analysis', *Journal of Sustainable Tourism*, **19**, 437–457.

Hall, C.M. (2013a), 'The natural science ontology of environment', in A. Holden and D. Fennell (eds), *The Routledge Handbook of Tourism and the Environment*, Abingdon, UK: Routledge, pp. 6–18.

Hall, C.M. (2013b), 'Framing behavioural approaches to understanding and governing sustainable tourism consumption: beyond neoliberalism, "nudging" and "green growth"?', *Journal of Sustainable Tourism*, DOI:10.1080/09669582.2013.815764.

Hall, C.M. (2014), *Tourism and Social Marketing*, Abingdon, UK: Routledge.

Hall, C.M., D.J. Timothy and D. Duval (2003), 'Security and tourism: towards a new understanding', *Journal of Travel & Tourism Marketing*, **15**, 1–18.

Hall, C.M., D. Scott and S. Gössling (2013), 'The primacy of climate change for sustainable international tourism', *Sustainable Development*, **21** (2), 112–121.

Halsnæs, K., P. Shukla, and A. Garg (2008), 'Sustainable development and climate change: lessons from country studies', *Climate Policy*, **8**, 202–219.

Hanna, E.G., E. Bell, D. King, and R. Woodruff (2011), 'Climate change and Australian agriculture: a review of the threats facing rural communities and the health policy landscape', *Asia-Pacific Journal of Public Health*, **23** (2 suppl.), 105S–118S.

Hay, R. (1998), 'A rooted sense of place in cross-cultural perspective', *Canadian Geographer*, **42**, 245–266.

Hinkel, J. (2011), 'Indicators of vulnerability and adaptive capacity: towards a clarification of the science–policy interface', *Global Environmental Change*, **21**, 198–208.

IPCC (2007), *Climate Change 2007: Impacts, Adaptation, and Vulnerability. Contribution* of Working Group II to the Fourth Assessment Report of the Intergovernmental Panel on Climate Change*, M.L. Parry, O.F. Canziani, J.P. Palutikof, P.J. van der Linden and C.E. Hanson (eds), Cambridge, UK: Cambridge University Press.

IPCC (2012), *Managing the Risks of Extreme Events and Disasters to Advance Climate Change Adaptation. A Special Report of Working Groups I and II of the Intergovernmental Panel on Climate Change*, C.B. Field, V. Barros, T.F. Stocker, D. Qin, D.J. Dokken, K.L. Ebi, M.D. Mastrandrea, K.J. Mach, G.-K. Plattner, S.K. Allen, M. Tignor and P.M. Midgley (eds), Cambridge, UK, and New York, USA: Cambridge University Press.

Joyce, A.L. and T.A. Satterfield (2010), 'Shellfish aquaculture and First Nations' sovereignty: the quest for sustainable development in contested sea space', *Natural Resources Forum*, **34**, 106–123.

Keller, D.R. and F.B. Golley (eds) (2000), *The Philosophy of Ecology. From Science to Synthesis*, Athens, USA: University of Georgia Press.

Kelly, E.D. (2009), *Community Planning: An Introduction to the Comprehensive Plan*, 2nd edn, Washington DC, USA: Island Press.

Krishnaswamy, A., E. Simmons and L. Joseph (2012), 'Increasing the resilience of British Columbia's rural communities to natural disturbances and climate change', *BC Journal of Ecosystems and Management*, **13** (1). Available at: http://www.jem.forrex.org/forrex/index. php/jem/article/view/168, accessed 20 January 2013.

Lorenzoni, I., S. Nicholson-Cole and L. Whitmarsh (2007), 'Barriers perceived to engaging with climate change among the UK public and their policy implications', *Global Environmental Change*, **17**, 445–459.

Lorenzoni, I., G. Seyfang and M. Nye (2011), 'Carbon budgets and carbon capability: lessons from personal carbon trading', in I. Whitmarsh, S. O'Neill and I. Lorenzoni (eds), *Engaging the Public with Climate Change: Behaviour Change and Communication*, London, UK: Earthscan, pp. 31–46.

Lowi, M. and B. Shaw (eds) (2000), *Environment and Security: Discourses and Practices*, New York, USA: St. Martins Press.

Maréchal, K. (2010), 'Not irrational but habitual: the importance of "behavioural lock-in" in energy consumption', *Ecological Economics*, **69**, 1104–1114.

Marx, S., E. Weber, B. Orlove, A. Leiserowitz, D. Krantz, C. Roncoli and J. Philips (2006), 'Communication and mental processes: experiential and analytic processing of uncertain climate information', *Global Environmental Change*, **17**, 47–58.

McEvoy, D., P. Matczak, I. Banaszak and A. Chorynski (2010), 'Framing adaptation to climate-related extreme events', *Mitigation and Adaptation Strategies for Global Change*, **15**, 779–795.

Millar, C. (1996), 'The Shetland way: morality in a resource regime', *Coastal Management*, **24**, 195–216.

Millar, C. and D. Aiken (1995), 'Conflict resolution in aquaculture: a matter of trust', in A. Boghen (ed.), *Coldwater Aquaculture in Atlantic Canada*, 2nd edn, Moncton, Canada: Canadian Institute for Research on Regional Development, pp. 617–645.

Millar, C. and H. Yoon (2000), 'Morality, goodness and love: a rhetoric for resource management', *Ethics, Place & Environment*, **3**, 155–172.

Monaghan, P. (2012), *How Local Resilience Creates Sustainable Societies: Hard to Make, Hard to Break*, London, UK: Routledge.

O'Brien, K., L. Sygna and J.E. Haugen (2004), 'Vulnerable or resilient? A multi-scale assessment of climate impacts and vulnerability in Norway', *Climatic Change*, **64**, 193–225.

O'Brien, K., S. Eriksen, L.P. Nygaard and A. Schjolden (2007), 'Why different interpretations of vulnerability matter in climate change discourses', *Climate Policy*, **7**, 73–88.

O'Brien, K., L. Sygna, R. Leichenko, W.N. Adger, J. Barnett, T. Mitchell, L. Schipper, T. Tanner, C. Vogel and C. Mortreux (2008), *Disaster Risk Reduction, Climate Change Adaptation and Human Security*, Report prepared for the Royal Norwegian Ministry of Foreign Affairs by the Global Environmental Change and Human Security (GECHS) Project, GECHS Report 2008:3, Oslo, Norway: Ministry of Foreign Affairs.

O'Brien, K.L., A.L. St Clair and B. Kristoffersen (eds) (2010a), *Climate Change, Ethics, and Human Security*, Cambridge, UK: Cambridge University Press.

O'Brien, K.L., A.L. St Clair and B. Kristoffersen (2010b), 'Towards a new science on climate change', in K. O'Brien, A.L. St Clair and B. Kristoffersen (eds), *Climate Change, Ethics, and Human Security*, Cambridge, UK: Cambridge University Press, pp. 215–227.

Ostrom, E. (1990), *Governing the Commons: The Evolution of Institutions for Collective Action, The Political Economy of Institutions and Decisions*, Cambridge, UK: Cambridge University Press.

Page, E. and M. Redclift (eds) (2002), *Human Security and the Environment: International Comparisons*, Cheltenham, UK and Northampton, MA, USA: Edward Elgar.

Paton, D. and D. Johnston (2001), 'Disasters and communities: vulnerability, resilience and preparedness', *Disaster Prevention and Management*, **10**, 270–277.

Peattie, K., S. Peattie and C. Ponting (2009), 'Climate change: a social and commercial marketing communications challenge', *EuroMed Journal of Business*, **4**, 270–286.

Root, T.L. and S.H. Schneider (1995), 'Ecology and climate: research strategies and implications', *Science*, **26** (5222), 334–341.

Ryan, P. and N. Wayuparb (2004), 'Green space sustainability in Thailand', *Sustainable Development*, **12**, 223–237.

Schneider, S.H. (2001), 'A constructive deconstruction of deconstructionists: a response to Demeritt', *Annals of the Association of American Geographers*, **91**, 338–344.

Schneider, S.H., S. Semenov, A. Patwardhan, I. Burton, C.H.D. Magadza, M. Oppenheimer, A.B. Pittock, A. Rahman, J.B. Smith, A. Suarez and F. Yamin (2007), 'Assessing key vulnerabilities and the risk from climate change', in M. Parry, O. Canziani, J. Palutikof, P. van der Linden and C. Hanson (eds), *Climate Change 2007: Impacts, Adaptation and Vulnerability. Contribution of Working Group II to the Fourth Assessment Report of the Intergovernmental Panel on Climate Change*, Cambridge, UK: Cambridge University Press, pp. 779–810.

Selinger, E. and Powys White, K. (2012), 'Nudging cannot solve complex policy problems', *European Journal of Risk Regulation*, **1/2012**, 26–31.

Seyfang, G. (2011), *The New Economics of Sustainable Consumption: Seeds of Change*, Basingstoke, UK: Palgrave Macmillan.

Shove, E. (2010), 'Beyond the ABC: climate change policy and theories of social change', *Environment and Planning A*, **42**, 1273–1285.

Slocum, R. (2010), 'The sociology of climate change: research priorities', in J. Hagel, T. Dietz and J. Broadbent (eds), *Workshop on Sociological Perspectives on Global Climate Change*, Arlington, USA: National Science Foundation, pp. 133–137.

Stern, N. (2007), *The Economics of Climate Change: The Stern Review*, Cambridge, UK: Cambridge University Press.

Suryanata, K. and K.N. Umemoto (2003), 'Tension at the nexus of the global and local: culture, property, and marine aquaculture in Hawaii', *Environment and Planning A*, **35**, 199–214.

Suryanata, K. and K. Umemoto (2005), 'Beyond environmental impact: articulating the "intangibles" in a resource conflict', *Geoforum*, **36**, 750–760.

Swart, R., J. Robinson and S. Cohen (2003), 'Climate change and sustainable development: expanding the options', *Climate Policy*, **3** (Suppl. 1), S19–S40.

Tervo-Kankare, K., C.M. Hall and J. Saarinen (2013), 'Christmas tourists' perceptions of climate change in Rovaniemi, Finnish Lapland', *Tourism Geographies*, 15, 292–317 DOI: 10.1080/14616688.2012.726265.

Thaler, R.H. and Sunstein, C.R. (2008), *Nudge: Improving Decisions about Health, Wealth and Happiness*, London, UK: Yale University Press.

Thynne, I. (2008), 'Symposium introduction: climate change, governance and environmental services: institutional perspectives, issues and challenges', *Public Administration and Development*, **28**, 327–339.

Tobin, G.A. (1999), 'Sustainability and community resilience: the holy grail of hazards planning?', *Global Environmental Change Part B: Environmental Hazards*, **1**, 13–25.

Turnpenny, J., M. Jones and I. Lorenzoni (2010), 'Where now for post-normal science? A critical review of its development, definitions, and uses', *Science Technology Human Values*, doi: 10.1177/0162243910385789.

Umemoto, K. and K. Suryanata (2006), 'Technology, culture, and environmental uncertainty considering social contracts in adaptive management', *Journal of Planning Education and Research*, **25**, 264–274.

UNDP (United Nations Development Program) (1994), *Human Development Report 1994*, New York, USA: Oxford University Press for UNDP.

UNDP (2008), *Fighting Climate Change: Human Solidarity in a Divided World, 2007/2008 Human Development Report*, New York, USA: Oxford University Press for UNDP.

van Aalst, M.K., Cannon, T. and Burton, I. (2008), 'Community level adaptation to climate change: the potential role of participatory community risk assessment', *Global Environmental Change*, **18** (1), 165–179.

Wainwright, J. (2010), 'Climate change, capitalism, and the challenge of transdisciplinarity', *Annals of the Association of American Geographers*, **100**, 983–991.

Weber, E.U. (2006), 'Experience-based and description-based perceptions of long-term risk: why global warming does not scare us (yet)', *Climatic Change*, **77**, 103–120.

White, G.F., R.W. Kates and I. Burton (2001), 'Knowing better and losing even more: the use of knowledge in hazards management', *Global Environmental Change Part B: Environmental Hazards*, **3**, 81–92.

Whitmarsh, L., S. O'Neill, G. Seyfang and I. Lorenzoni (2009), 'Carbon capability: what does it mean, how prevalent is it, and how can we promote it?', Tyndall Working Paper 132. Norwich, UK: Tyndall Centre for Climate Change Research, University of East Anglia.

Whitmarsh, L., G. Seyfang and S. O'Neill (2011), 'Public engagement with carbon and climate change: to what extent is the public "carbon capable"?', *Global Environmental Change*, **21**, 56–65.

Wilbanks, T.J. (2003), 'Integrating climate change and sustainable development in a place-based context', *Climate Policy*, **3** (Suppl. 1), S147–S154.

Wilbanks, T.J. (2007), 'Scale and sustainability', *Climate Policy*, **7** (4), 278–287.

Wolf, J., I. Lorenzoni, R., Few, V. Abrahmson and R. Raine (2009), 'Conceptual and practical barriers to adaptation: an interdisciplinary analysis of vulnerability and response to heat waves in the UK', in W.N. Adger, I. Lorenzoni and K. O'Brien (eds), *Adapting to Climate Change: Thresholds, Values, Governance*, Cambridge, UK: Cambridge University Press, pp. 181–196.

Yearley, S. (2009), 'Sociology and climate change after Kyoto: what roles for social science in understanding climate change?', *Current Sociology*, **57**, 389–405.

Yohe, G.W., R.D. Lasco, Q.K. Ahmad, N.W. Arnell, S.J. Cohen, C. Hope, A.C. Janetos and R.T. Perez (2007), 'Perspectives on climate change and sustainability', in M. Parry, O. Canziani, J. Palutikof, P. van der Linden and C. Hanson (eds), *Climate Change 2007: Impacts, Adaptation and Vulnerability. Contribution of Working Group II to the Fourth Assessment Report of the Intergovernmental Panel on Climate Change*, Cambridge, UK: Cambridge University Press, pp. 811–841.

Young, E. (2001), 'State intervention and abuse of the commons: fisheries development in Baja California Sur, Mexico', *Annals of the Association of American Geographers*, **91**, 283–306.

14. Climate change, human security and the built environment
Karen Bickerstaff and Emma Hinton

INTRODUCTION

In the UK, the consumption of energy within domestic housing accounts for roughly one third of the country's annual carbon dioxide emissions (DECC, 2011). Homes have been framed by a range of stakeholders, from energy companies to governments and campaigning groups, as an important site for reducing energy use and, more particularly, meeting national climate change goals. Improving the efficiency of existing homes (and commercial buildings) is critical to such efforts. The UK housing stock is renewed at a rate of only about 1 percent a year and, as such, most of the homes we will occupy – up to at least 2050 – already exist (Beddington, 2008: 4299). Here, we also focus on energy used to provide heating and sustain comfort – recognizing that space heating and cooling account for the lion's share of domestic energy use in most western societies (Shove, 2003: 396), Comfort, and an adequate standard of warmth, is also a critical aspect of human security. A recent report on fuel poverty (Hills, 2012) claims a 'profoundly disappointing' 3 million households will be fuel-poor[1] by 2016, despite the introduction of government measures intended to tackle the problem.

Government policy to improve the energy performance of the existing (domestic) housing stock has, from the 1990s onwards, pursued two dominant (if, at times, connecting) models of transformation. The first prioritizes the behaviour (and choices) of individual energy consumers. Efforts to engage householders in changing their behaviour in order to reduce domestic emissions have traditionally been underpinned by notions of responsible and rational individuals, free in their 'choices' (but also implicitly morally governed to make the correct choice). The route to change, seen in Government-led campaigns such as 'Helping the Earth Begins at Home' (Hinchliffe, 1996; Blake, 1999), 'Are You Doing Your Bit?' and 'Act on CO2' (also DEFRA, 2008), and the Energy Saving Trust's annual campaigns, was the provision of information to fill a presumed public 'knowledge' deficit (Owens, 2000) and inculcate more (eco)rational attitudes, beliefs and values.

At the same time we can also identify a more systemic, and technocratic, approach to delivering change, reflected in a suite of intervention-based programmes designed by national government, local governments, energy companies and civil society actors, aimed at materially and technologically re-engineering the domestic environment to ensure that reductions in GHG emissions are achieved and to make participation 'easy' (Marres, 2008). This structural approach to change can be seen in a string of recent government 'demand management' initiatives: the Framework for Pro-Environmental Behaviours (Defra, 2008) sought to influence behaviour through the installation of particular technologies; CERT (the Carbon Emissions Reduction Target), and CESP (the Community Energy Saving Programme) required energy utilities to promote a range of technologies to customers – often at a discount; and, more recently, the Green Deal will see efficiency technologies distributed via the private sector in tandem with novel financing mechanisms. The UK Government is also committed to the mass roll out of smart meters – providing consumers with real-time information on their energy consumption – in theory 'enabling them to better control and manage their energy use, and reduce emissions' (DECC, 2012a). There is also considerable interest in the potential of intelligent heating controls to ensure 'that interactions between heat supply and storage technologies and occupant's heating preferences are optimised' (DECC, 2012b: 35).

Yet, in practice, both paradigms have fallen short of expectations in terms of delivering transformation in patterns of energy consumption. Our aim in this chapter is to engage with these dominant models of social transformation, and their critiques – particularly in relation to issues of materiality and agency. From this we argue for a socio-technical approach to securing decarbonization of the built environment – one that recognizes the social and material constitution of everyday domestic practices.

TRANSFORMING THE ENERGY INTENSITY OF DOMESTIC LIFE

There are two dominant models of agency that underpin policy rhetoric and practice aimed at transforming the energy (and carbon) intensity of domestic life: the individualist paradigm puts human conscious agency at the centre of analysis and the second, systemic, model centres attention on the structural conditions for action, with a focus on technological innovation (Benton and Redclift, 1994; Spaargaren, 2011). The UK government's policies for the promotion of more sustainable consumption behaviours have tended to embrace the individualist model of behaviour change (Jackson, 2005; Defra, 2008). Behaviour change strategies are heavily rooted in theories of planned behaviour (Ajzen, 1991), a strand of psychological literature which assumes that people behave rationally, that behaviours are driven by beliefs and values and that lifestyles and tastes are expressions of personal choice (Shove et al., 2012: 3). The assumption is that economically rational actors, equipped with the necessary technical and economic information, will consistently act more responsibly through their personal consumption choices (Guy, 2006). Shove (2010) refers to this as the Attitude-Behaviour-Choice (ABC) model of behaviour change. It is an approach that has been criticized for failing to recognize the complexities involved in addressing global environmental problems like climate change and for focusing on individuals abstracted from their structural (social, cultural or sociotechnical) contexts. Crucially, authors have pointed out that the actions of individuals (as consumers) cannot simply be read as resulting from free choices, but are inevitably shaped by the practical constraints imposed by the social, infrastructural and material conditions that people have to negotiate in organizing domestic life (Spaargaren, 2011; Shove, 2003, 2010; Hobson, 2006; Guy, 2006; Hinchliffe, 1996; Benton and Redclift, 1994). A slew of behaviour change campaigns seeking to educate and empower consumers as citizens – and premised on a link between attitudes, behaviours and autonomous choice – have been disappointing in their effects. Whilst levels of general environmental awareness have increased, this awareness has proven a very weak predictor for actually performed environmental behaviours (Spaargaren, 2011; Hobson, 2003; Jackson, 2005; Kollmuss and Agyeman, 2002; Owens, 2000).

Partly in response to the limited achievements of political strategies seeking information-induced behavioural change, we can identify a policy turn towards a more structural paradigm of transformation – with an emphasis on literally and universally engineering participation in domestic energy efficiency. Change, from this perspective, depends upon external forces – social and political structures and technology itself – bearing down on the detail of daily life (Shove, 2010). Institutional actors like companies, organizations, and local and central government will deliver transformation in consumption through the development and distribution of efficient technologies, devices and products (Spaargaren, 2011). It is an approach which constructs consumers as passive recipients of approved technologies: people will optimally utilize these objects in the fashion intended by their creators (Hinchliffe, 1995: 94).

Of course such programmes, and the associated devices, do not, and cannot, neatly script public participation in climate politics through more efficient domestic practices (Marres, 2011). There is, for instance, an established and growing body of evidence of a performance gap between building design (for efficiency) and delivery – in terms of energy use – which has been linked to a lack of understanding of how energy is actually used in a home (NHBC, 2012). Similar issues have been raised in relation to specific 'efficiency' technologies: programmed heating controls that don't deliver shorter heating periods (Shipworth et al., 2010), and smart meters that have been disappointing in terms of occupant engagement and changing behaviours (Darby, 2006; Hargreaves et al., 2010; Strengers, 2011). There remains little understanding of the processes by which energy monitoring devices are able to facilitate even modest reductions in energy consumption (Hargreaves et al., 2010). In fact, it has been suggested that devolving greater control of energy services to technology may work against efficiency goals – and entrench and legitimize 'non-negotiable' high demand practices, disengaging people from any need to consider and question them (Darby, 2010; Strengers, 2011). These findings highlight the need for alternative ways of thinking about energy consumption and improving household efficiencies.

RETHINKING AGENCY: MATERIALIZING EFFICIENCY PRACTICES

A movement in the social sciences has begun to re-evaluate the materiality of social life; attempting to understand how things relate to human thought or action, but without returning to 'deterministic realisms' (Thrift, 2005: 134). In fields as diverse as science and technology studies, sociology, geography, anthropology and political theory the question of how things constrain, inform and constitute our political, ethical and everyday lives has provoked increasing attention (see, for instance, Barry, 2001; Bennett, 2004; Marres and Lezaun, 2011; Barad, 2003). The Actor Network tradition has been critical in (re)conceptualizing human agency as always inextricably tied to the specific socio-material arrangements of which we are part, as already distributed to 'our' infrastructures, instruments and artefacts. From this perspective, human agency only emerges as agentic through our relations to 'foreign' materialities we are all too eager to label as mere objects (Bennett in Khan, 2009: 93).

In work on energy and buildings, conventional materialist accounts of thermal comfort have focused on mechanical and physiological understandings, which construct comfort as 'the provision and maintenance of a fixed set of thermal, luminous and acoustic conditions' (Cole et al., 2008: 324). From this, it follows that a standardized measure of comfort can be determined and, through the application of technology, universally delivered. Crucially, recent authors (within a socio-technical tradition) have articulated a far more differentiated and complex understanding of comfort – recognizing the social and physical relations that limit and enable energy consumption for comfort, cooking, cleanliness etc. (Guy, 2006; Shove et al., 2012). Socio-technical accounts of stability and change in energy consumption have, in particular, placed emphasis on social practices: seeking to understand the context of particular routines and habits, and the ways in which ideas about the right way of doing something circulate through the social body and relate to implicitly assimilated standards and conventions, and to the material objects involved (Shove, 2003, 2010; Crosbie and Guy, 2008; Gram-Hanssen, 2010, 2011). Shove and Pantzar (2007) stress the interplay of images (cultural representations), stuff (tangible physical entities, and the materials of which they are made) and skill (the learned, often tacit, bodily ability to master specific material and cultural affordances) in the performance of everyday life – arguing that practices emerge, persist, shift and disappear when connections between these elements are made.

Social practice theories prioritize neither human agency and choice, nor external structures (Shove and Walker, 2007; Spaargaren, 2011). Rather, it is practices that enlist and recruit people willing and able to carry them (Shove and Pantzer, 2007). From this position agency does not pre-exist or determine activities but is a product of the recursive relationship between practices and their performance. Whilst offering a compelling account of the reproduction of everyday life, proponents of practice theory differ in how they conceptualize the relations between materials, people and practices, and consequently in their reading of agency. As other authors have noted, the theorization of the material elements of social practices has been only partially developed (Gram-Hanssen, 2011; Shove et al., 2012; Strengers and Maller, 2012). Here then, we seek to better understand the play of things themselves – that is, how objects are implicated in transforming not only the organization of everyday life, but also political subjectivities – and the implications of this for securing meaningful change in domestic energy efficiency.

Kersty Hobson's (2006) research into the Sydney-based sustainable living programme 'Green Home' is suggestive here, revealing the ways in which 'eco-efficient' domestic technologies such as recycling bins and shower timers facilitate specific forms of ethical environmental practices – 'moralizing' their users (cf. Jelsma, 2003; Hobson, 2006). Here, 'morality is as much in the things we use as in the minds of people' (Jelsma, 2003: 103). Cole et al. (2008) similarly argue that buildings may be constructed and managed to inculcate particular forms of energy consumption practices in their users, which may provide space for building users to rethink individual and social norms of comfort and how they are practised by choosing to accept a greater degree of seasonal variation in their indoor environments. Noortje Marres' work on the politics of (energy efficiency) objects, has similarly explored the capacities of carbon calculation technologies (carbon calculators, smart meters etc.) to 'materialize' political participation with the environment. Critically for Marres (2011), and in contrast to Hobson's analysis, these interventions make possible a form of effortless public engagement – devices that in effect absolve domestic subjects from certain moral-political responsibilities.

In the discussion below we present a review of empirical research, focusing on our own study which followed the effects of specific efficiency interventions on routine social practices. Our purpose in doing so is to explore how things matter not only in transforming 'normal' (energy consuming) practices but also in enabling new forms of political engagement with the environment.

CASE STUDY RESEARCH

The account which follows draws, in particular, from our own research looking at how the everyday energy practices of householders changed as a result of the introduction of efficiency devices and energy system changes. An overview of the 15 case study households is provided in Table 14.1 (and a fuller discussion of the various projects can be found in Hinton et al., 2012).

Whilst this sample is by no means representative, there is considerable heterogeneity in terms of geographical location, housing type, tenure and household composition (for full details see Hinton et al., 2012). Householders 1–9 were interviewed prior to and following interventions, and householders 10–15 were interviewed post interventions only.[2] The primary devices studied can be categorized as follows:

1. Heating system transformations: installation of gas central heating (from electric storage heaters) and installation of solar water heating or PV (electricity) systems.
2. Feedback devices: feedback lights that linked a 'warm' orange glow to a desirable temperature (18–21C); a radiator feedback device that provided cues on energy use through a series of lights – if a nearby window was open whilst the radiator was warm or hot, the light would change to red (both developed by Loughborough Design School); a number of electronic consumption meters displaying one or more measures (kilowatts, money and/or CO_2 emissions).

Table 14.1 Participant profiles

Interviewee	Location	Housing type	Household (age)	Intervention(s)
1. Jenny	Merthyr Tydfil	Terraced house	50–59	Solar water heating Thermal feedback light
2. Hannah	Merthyr Tydfil	Semi-detached house	60–69 (1); 30–39 (1); 20–29 (1); 9 or under (1)	Radiator/window device
3. Peter	Merthyr Tydfil	Flat in sheltered housing complex	60–69	From electric storage heaters in sheltered housing to 2 bed semi-detached bungalow with GCH
4. Louise and Steve	Merthyr Tydfil	Semi-detached house	50–59 (x2)	Feedback light
5. Gemma and Jason	Merthyr Tydfil	Semi-detached house	50–59 (x2)	Wattbox (for 1 week); radiator feedback
6. Julie	Merthyr Tydfil	Terraced house	50–59 (x2)	Wattbox
7. Amy	Merthyr Tydfil	Semi-detached house	20–29; 9 or below (x2)	No intervention
8. Kath	Torfaen	Flat in tower block	60–69	Retrofit (solid wall insulation, replacement windows, replacement of electric storage heaters with GCH)
9. John	Torfaen		60–69	
10. Sarah	Ham and Petersham [Low Carbon Zone and British Gas Green Streets]	Terraced house	50–59 (x2)	Green Streets: Solar PV, energy monitor LCZ: Home energy survey and eco-starter kit (e.g., eco-kettle, low energy lightbulbs)

Table 14.1 (continued)

Interviewee	Location	Housing type	Household (age)	Intervention(s)
11. Fiona and Paul	Ham and Petersham	End of terrace house	60–69 (x2)	
12. Vicky	Ham and Petersham	Terraced house	30–39 (1) 10–19 (x2)	Green Streets: Solar PV, loft insulation
13. Linda	Ham and Petersham	Semi-detached house	40–49 (x2) 10–19 (x2)	Green Streets: Energy monitor LCZ: Home energy survey and eco-starter kit (e.g. eco-kettle, low energy lightbulbs)
14. Cathy	Hyde Farm Green Streets	Terraced maisonette	50–59	Home energy survey and eco-starter kit: (e.g. energy saving sockets, reflective panels for behind radiators; access to an energy monitor)
15. Patricia	Hyde Farm Green Streets	Terraced house	50–59	As above – but also took up the option of home draft-proofing

3. Automation of thermal comfort: the Wattbox heating control system (developed by IESD, De Montfort University) is an 'intelligent' control technology that provides an alternative to the traditional time clock and thermostat. The system controls both central and water heating, and is designed to reduce energy consumption by automatically monitoring and learning occupant behavioural patterns and temperature preferences, and through this more effectively (and efficiently) managing the heating system. Householders can use the device to view their energy consumption and room temperature, and can use the interface to modulate their comfort by increasing or decreasing space or water heating.

INTERVENING IN PRACTICES: DISRUPTION AND CHANGE

In this section we look at how particular devices and technologies, designed to promote energy efficiency in the home, have effected the reorganization of domestic practices, particularly with regard to heating and comfort.

In our own research (Hinton et al., 2012) six participants had, prior to or during the study period, a change to their heating systems (from electric to gas central heating (GCH) or the addition of solar water heating/solar PV systems). Three households had moved to GCH (away from electric storage heaters), which had forcefully challenged routines and enabled new competences around the direct control of the heating system to evolve. A couple of these participants had only heated, and in one case only occupied, a relatively small area of their home during the winter period, and all three would use doors as a means to improve their thermal comfort, 'topping up' with, or completely relying on, a portable electric heater. The rhythms of charging and discharging heat, associated with the storage heaters, were inflexible and participants talked about feeling trapped and unable to 'risk' turning the heaters off from autumn to spring. The immediate difference brought about by the new GCH system was articulated very much in terms of a practical and embodied experience of comfort. New, relatively stable, patterns of managing comfort emerged quite rapidly – a product of new competences (in using the GCH system), knowledge of the financial costs, and corporeal experience. One participant, John,[3] talked about wearing 'normal' clothes rather than using multiple layers; and he no longer wore hats, coats and gloves indoors. Nor was he closing off (and not heating) the living room in wintertime. John had increased the background temperature of his flat from 18 to 21.5 degrees C – a temperature he had struggled to achieve before the retrofit – and thus anticipated few financial savings. Another participant, Kath (who had been part of a wider retrofit programme, including insulation) talked about a discernible improvement in her feeling of comfort following the system change but had, over the course of the year, felt little need to turn the heating on. For her the GCH offered responsiveness and security – it allowed her to turn off the heating without fearing the consequences of a decline in external temperature. As a result her bills had fallen sharply. Critically, the components of the GCH system (thermostat, boiler controls, and thermostatic radiator valves (TRVs)) enabled householders to play a more active role in the maintenance of comfort, as they differentially experienced it over time and across the space of their homes. All three delighted in their newly acquired thermal management opportunities, skills and ultimately greater control over the experience of comfort: indeed, not one of the householders showed any interest in the heating programmer function (and a means of automation) offered by the GCH systems. All elected to control the boiler manually.

Where participants had taken up opportunities to have solar water heating or solar PV systems fitted,[4] all spoke about new ways of interacting with their heating systems – linking corporeal experience, weather observations and an array of domestic appliances. For several participants the new *in situ* power generation had prompted marked shifts in the spatial and temporal organization of particular domestic activities. Participants talked about putting the dishwasher, washing machine or other devices on during the day if the weather was sunny to make use of the PV electricity, and drying the clothes on the line instead of using a tumble dryer. Jenny (and others) talked about a process of accommodation following installation (in her case relating to a solar hot water system): she was happy to delegate control to the degree that she recognized the technology to be enabling (financially). As such, she had re-evaluated her expectation of continuous hot water: 'you learn to, [. . .] know when you've got water. But this time of year I don't expect [continuous] hot water'. Although Jenny was willing to accept some diminution of control for economy, she was also quick to assert her own agency within this new socio-material assemblage, pointing out that the move to solar water heating had not significantly disrupted longstanding social practices: 'the system would run to suit me, not me change to suit the system'.

Our research also followed a number of technologies that provided feedback on energy use and indoor temperature. Passive devices, such as thermometer cards and feedback lights, provided visual cues on temperature (ideal, too hot, too cold). A number of participants spoke about the cards and kept them in highly visible spaces (e.g. the mantelpiece).

Critically, and from use, they recognized, and had internalized, the idea of comfort as being 18 degrees C – as somehow the 'right' temperature. This material representation of 'the ideal' internal temperature had quite evidently impinged on their cognitive and embodied understanding of comfort, and from this, how they managed their heating systems:

> Kath: [I try and keep it at] 18 and that's . . . you know, and then I can adjust the radiators to what I want. If it's bitter cold then it will go up to 22; there's no need to put it on any more than that.
> Interviewer: So, why 18 and why 22?
> Kath: Because that thing they give . . . [. . .] Off E.ON, yeah, it says on there: 18 comfortable, 22 if it's really cold – something like that on it. Then there was one down the bottom, danger, hypothermia and all, isn't it. It's a guideline for people to go by. And yeah, it's been a good guideline.

Through her participation in a low carbon scheme in London, Linda had been provided with an energy monitor (as part of an eco-starter kit). The device had been very popular with the entire family and had been used quite actively and deliberately to make more visible the energy demands of a range of home appliances – in other words how 'greedy' they were (Hargreaves et al., 2010). Linda invested considerable effort in calibrating the monitor and selecting the appropriate measure: in her family's case pounds sterling.

> Linda: I'd set it to money because that worked really well with [names friends]. [. . .] I sort of went through this [booklet] and worked out what I wanted to do and then I decided that the only thing that really drove people in this house was looking at pounds, shillings and pence [. . .]. And, because, you know, it doesn't actually mean a lot. If I say to somebody like Simon or Rebecca, oh, you know, the washing machine's using so many kilowatts, they say, so what, you know, what does that mean? They can't relate to it.

This expanded understanding of the financial significance of a range of domestic practices had led to a number of long term changes in household 'doings' – around hot drinks, dishwashing, washing and drying clothes (cf. Hargreaves et al., 2010).

Energy and temperature feedback devices did not always lead to positive (or even stable) changes to household routines. In some cases the thermal feedback devices, in signalling a shift in temperature (e.g. when the radiator was cooling down), would prompt a modification of comfort practices in a way that actually increased thermal output:

> Gemma: And then . . . now, sometimes I notice, we'll be sitting here and I think: oh, the white light is on there. And for some reason the radiator was going off. So, it was good because otherwise perhaps you wouldn't notice. And then you

> think: oh, I don't feel so warm now and the heating's on. [. . .] I'd turn up the thermostat for a bit. And when I see it come back then onto the orange then I'd know it's okay now.
> Interviewer: So, it actually made you increase your heating use?
> Gemma: If I was feeling cold.

For the most part shifts in practices connected with thermal feedback tools were limited, fragile and prone to reversal. So whilst people did talk about new capacities and skills afforded by signals and new information, this impinged only peripherally on their management of comfort. As Jason put it, the efficiency measures simply became 'invisible once they'd been put in' and use fell off (cf. Hargreaves et al., 2010).

Finally, our research involved the trialling, in two households, of an intelligent heating control system – the Wattbox (described above). Of note, members of both households articulated, quite explicitly, a sense that the technology had wrested control of comfort away from them, demanding some radical shifts in their normal routines – changes which they found unsettling and difficult to accept. They also felt that the 'smart' technology had resulted in the heating being on more than it would if manually controlled and based on the bodily sensation of comfort. Julie talked about the technology as being difficult to adapt to – noting that it was switching the heating off when she and her partner were trying to increase output. 'I have tried to use it [. . .] but it just stays on for about five minutes and still knocks off'. For both households the inflexibility and inability to (re)negotiate the distribution of control, precipitated fairly immediate disruption. In each case, the Wattbox appeared to demand such a dramatic change in the organization of, and responsibility for, comfort that its operation was interpreted as disruptive and even dysfunctional. Julie, for instance, refers to persistent 'breakdowns' and of constantly being 'locked out' of the heating system. It's an observation that ties in with other work (e.g. Cooper, 1998: 184) demonstrating that efforts to promote the most technically rational design to heat management – requiring sealed buildings and passive use – were resisted by what engineers saw as 'irrational users' who preferred systems which, although less efficient, provided greater flexibility and active control.

Interestingly, in the case of Julie's household, a manual override box was fitted to the device, as a precaution, following a problem with the boiler interface, to prevent a loss of heating. This enabled Julie to increase her manual control of the heating system and, initially at least, she (or her partner) frequently forced the temperature of the house up manually. By the time of the final interview Julie was still using the manual override facility – though she appeared to let the Wattbox take control in certain circumstances – 'it was easier' – and to recognize some benefits to automation, to the degree that she was not only keen to retain the Wattbox but also to better understand its capabilities.

THE POLITICS OF THINGS: TRANSFORMING ENERGY RELATIONS

In this section we want to move on to argue that things do not merely mediate practices – as an 'ingredient' in the transformation of routines and habits – but that efficiency objects themselves possess certain powers of engagement, to open up and transform moral-political engagements with energy.

Many of our participants talked about how living with various efficiency objects had made them think about the production of energy and comfort in spatially and temporally more complex ways. The capacity of devices to make energy ontologically something different, to make it present and somehow tangible, transformed not only understandings but also the stuff of everyday practices. Linda talked about how, over the last 3–4 months, she had set up an energy monitor to provide daily household targets (in terms of money spent on electricity) – and that the use of these targets had been instrumental in literally materializing the energy intensity (and costs) of household practices. There was a sense of pride in what had been achieved and the process of monitoring and revising practices had seemingly become normalized or habitual – at least in terms of the time period of use:

> Interviewer: So, does that ever get old? Do you think, I know how much this is going to use now, or do you still have a little bit of a look anyway?
> Linda: I still have a little bit of a look. I think we all do. I mean it may wear off but I think it's kind of, I think now that the behaviour's changed you want to know. [. . .] I think it has definitely changed our behaviour, definitely.

Fiona and Paul drew a similar connection with the use of a smart meter and their extremely low level background use of electricity. Paul approximated that in sunny conditions the solar panels would generate sufficient electricity to cover these background items. They also drew on their experience with water metering and how this had had a far reaching impact on their practices – leading to a rapid reduction in the amount of water used. The meter had enabled them to directly link water supply with their use and, in Paul's words, to 'really come to terms with the potential saving that they could make'. Fiona also talked about her experience with an eco-kettle she had been given as part of an eco-pack under the Low Carbon Zone scheme. She found the appliance difficult to use and as a result reverted to using her old kettle. However, as a result of interacting with the device and engaging with its 'politics of efficiency' she now 'always . . . not measure exactly, but I'm always careful about how much water I put in an ordinary kettle'.

For Gemma, her engagements with the radiator feedback device were implicated in a more gradual and subtle politicization around energy. Although she was unable to identify a direct change in any aspects of her routines she did, later on in the interview, point to a more fundamental shift in how she thought about energy and practised comfort:

> Well, I do stop and think; before I put the heating on, you know, I will do other things, and probably more so than I would have. I do stop and think; [. . .] I think it is a waste when you're using that if you're in a t-shirt and if you've got no socks on, and then you go and put the heating on; it does seem a waste to me. [. . ..] So, I stop and think: right, do I really need this heating on now because in about half an hour I'm going to be making food, you know. So, I'd rather go and get a cardigan, go and get some socks; whereas perhaps before I would have been in a t-shirt, you know, with the heating on – which is a bit ridiculous.

Sarah spoke of a transformative shift in the way her husband thought about energy – a consequence of (his) new capacities to monitor the energy generated by PV panels, fitted three months earlier under the Green Streets project. For her husband the solar energy represented 'a harvest' that he was 'always checking, to see how much electricity [the panels] have produced'. So in this case the panels worked to materialize energy, sustaining a desire for, and practices of conservation. Sarah drew the analogy with farming – and the way a farmer would relate to crops: 'you don't waste what you produce'. In a similar way, Strengers and Maller (2012) found that energy and water systems that are materially present, and exhibit traits of scarcity, may engage householders as co-managers of their everyday practices – within which 'energies' become an active rather than passive element of everyday routines (cf. Strengers and Maller, 2012). In contrast, the participants who had moved to a new GCH system, although talking about the technology as empowering and life-changing, showed no real changes in terms of their conception and relations to energy – it remained a passive and invisible resource.

A number of other participants spoke, effectively, of the interpersonal recruitment of new practitioners – and how the technologies themselves played a critical role in this transmission. Hargreaves et al. (2010) note a similar effect, though stressing this was the 'least common behavioural effect' in their study. However, their analysis focused on interpersonal prompts – that is, interviewees encouraging other people to reduce consumption, rather than the political effects of the devices themselves. Linda, for instance, commented on how 'useful' another member of her family had found the monitor – to the extent that they had subsequently bought their own monitor 'they made that a priority'.

In discussing her experiences with the energy monitor, Cathy was clear that the device itself hadn't been particularly useful or interesting to her – she had found it difficult to find time to read the instructions and set the monitor up. Far more powerful was a project meeting at which another resident spoke about the capacities of the monitor and how the device had transformed some of his routines. It was a (second-hand) engagement that made certain routine inefficiencies visible, and from this (some) practices of energy conservation became possible, and crucially desirable, for Cathy. Vicky described a similar process of material participation by proxy. Several participants talked about the impacts the presence of solar panels had had on neighbours:

> Paul: So most of them [neighbours] know that I have got the solar panels, because they're fairly obvious on the roof. So a couple of people have come up to me and said, oh I've . . . I saw them installing it. The chap opposite said, oh, it's really exciting, isn't it? So he . . . so they're interested and quite keen. And we're having a solar panel workshop in the library.

Indeed, Vicky's experience of efficiency interventions, provided through an earlier scheme, had inspired her to volunteer as a Street Champion in the Low Carbon Zone.[5] As she noted 'seeing the thermal image, getting the insulation, actually seeing the benefits myself, made me think gosh, this is fantastic; other people should be doing this.'

There is also some suggestion that devices were implicated in revealing sources of household differences, precipitating 'disputes' and even competition – such that rhythms of energy-using practices (occasionally) became an active topic of household conversation (see also Hargreaves et al.'s (2010) discussion of monitors as a form of surveillance between household members, and a source of household level disputes).

The temperature feedback devices provided a tool with which to (more effectively) negotiate between family members whether the heating should be turned up or down. Linda talked about how her teenage son had used the energy monitor to locate (financially) the costs of a whole range of household practices – and that this engagement had been instrumental in transforming not only some of his own routines but also a broader relationship to energy conservation.

> Linda: Initially they weren't really interested until Sam saw the machine working in the kitchen and he got very excited, particularly when it proved that his Xbox was quite low energy use. And then he spent the next week running up

and down telling me every time it had gone up when dad was using the shower, Katie had left her television on or something. So, he was trying to score bonus points all the time, and it's worked with him. He's actually quite good, and then he put in the EON plug himself and set it, I didn't have to set it up, he set it up in his bedroom. So, that's the standby plug. [. . .] [It] was just something that got him hooked.

CONCLUSIONS

In this chapter we have explored efforts to move society to a low carbon future through effecting transformations in domestic energy behaviours. We have reviewed dominant models of change, which emphasize either the individual or structural barriers to efficiency behaviours, and demonstrated how these are wanting. Instead, and drawing on findings from recent qualitative research tracing the effects of physical interventions in the home, we have sought to extend thinking on the place of technology and materiality in reconfiguring everyday energy practices – particularly, though not exclusively, around comfort.

We followed a number of different interventions – from energy system changes, to heat management devices, through to feedback devices – which had more or less direct effects on extant (energy-consuming) routines. The more successful technologies sustained an ongoing interaction between people and things – particularly apparent when the device was seen to be enabling (financially or in terms of managing comfort). So, for instance, a move from electric to gas central heating was universally discussed as freeing, and a means of reasserting (and redistributing) agency between social and technological actors. In some instances, energy monitors supported the interpretation of consumption across financial, environmental or physical registers, and as such enabled the household(ers) to tailor information to their needs and interests. These dynamic socio-technical interactions promoted not only new 'efficiency' practices but their repetition and ultimately their embedding in normal life. Other passive feedback devices proved less (directly) effective in prompting change – they supported only limited householder interaction and in many cases were viewed as an irrelevance, and, after some initial interest, were soon forgotten. New practices were not sustained and they lacked the capacity to engage on an ongoing basis. The intelligent heat control system, which was foremost about automating heating management and thus a redistribution of control, was interpreted in critical terms, and as failing to deliver efficiency changes. During the trial, the diminution of personal control led to a reading of the technology as dysfunctional and ultimately to its rejection. What is clear, then, across these diverse technologies is that issues of control – its expression through interactions of social and material actors, and capacities to renegotiate this agency – are critical in understanding not only why new practices emerged but also how they were sustained and reproduced.

Our account went on to address the political and moral effects of efficiency devices themselves, pointing to the ways in which particular technologies – in materializing dynamic energy flows across the home, and between production and consumption – made energy something ontologically different. It wasn't so much that they made participation easy (Marres, 2008) or that they had very direct moralizing effects (Hobson, 2003). Rather they helped to make explicit what Giddens (1984) has termed the practical consciousness – the tacit and routine character of daily life – and in so doing made novel engagements with the environment possible. The presence of PV panels, and (some) feedback devices, revealed contentious (and energy intensive) practices within households and between householders, prompting (often some time after use or installation) moral-political reflections on the structure or need for particular practices, led to conversations about energy use and flows and played a critical role in the recruitment of other practitioners.

For policy, this sociotechnical analysis of (changing) practices underlines the limits of a political focus on removing universal barriers to energy-efficient behaviours (apathy, ignorance, lack of financial interest), and suggests, by attending to the materiality of innovation, a number of ways we might promote more efficient energy use.

Recent UK energy policy, notably the 'Green Deal', seeks to effect large scale transformation in consumption through the diffusion of efficiency objects and materials (insulation, smart meters, intelligent controls) and by making the financing of new measures more attractive. There have been many criticisms levelled at the Green Deal itself, not least the emphasis on particular interventions, at the expense of others, the limited benefits for low-income families, the degree to which financial benefits for householders will be realized and the efficacy of communications to promote take-up (Harrabin, 2012). Our aim is not to suggest that structural factors (such as costs and access to technology) are irrelevant to promoting change in practices,[6] but that this approach tends to discount the ways these technologies, once in place, have effects – which may well run counter to expert and policy expectations. Here we make two suggestions. Firstly, there is a need to better understand how households manage heating and other energy services, and linked to this the distribution of agency between householders, technologies, the fabric of the home and wider socio-technical systems (to include social landlords, energy companies etc.). From here we might be better positioned to align interventions to households and develop contexts that make certain (energy efficient) practices more likely to hold. Some suggestive findings are presented here by parallel work trialling the heating control system, and which demonstrated very real efficiency gains. Crucially the device was trailed in households with regular heating patterns (already using programmers) and in social housing tenants that had recently switched from GCH (for thermal and water heating) to ground-source heat pumps[7] (Stafford, 2012; Boait and Rylatt, 2010) – a structural change that demanded a delegation of responsibility for heating to technology. In both circumstances the Wattbox technology fitted in with these (newly) established relations of agency.

Secondly, we would stress the need for policy and innovation to work with the political capacities of efficiency objects and the wider energy systems they form a part of – specifically in making energy relations, and the practices they support, more transparent and open to change. As Strengers and Maller (2012: 757) argue, resourcefulness comes 'not only or necessarily through education or from financial rewards and penalties', but through the physical experience of 'making, sorting, treating, coordinating and using energies'.

ACKNOWLEDGEMENTS

The Engineering and Physical Sciences Research Council (EPSRC) and E.ON UK for providing the financial support for research under the Carbon, Control & Comfort project (EP/G000395/1). Particular thanks for contributions along the way go to Harriet Bulkeley, Peter Boait (IESD, De Montfort University), and colleagues at the Loughborough Design School. The usual disclaimers apply.

NOTES

1. Meaning the money they have left, after paying for the energy required to heat a home to an adequate standard of warmth, will leave them below the official poverty line.
2. The case study households were drawn from two separate projects (though linked to the same research programme). As such there were some differences in the frequency of data collection – although the broad content of the interviews remained the same.
3. In all cases pseudonyms have been used.
4. Through either a social housing landlord or a local low carbon/energy efficiency scheme.
5. Under the Ham and Petersham LCZ Vicky was a Street Champion, volunteering to engage residents living in the area to take pro-environmental measures.
6. Indeed, there is a very real need for research that attends to the structuring of opportunity

and access to technology – and the ways in which power and inequality shape the evolution and diffusion of efficiency practices (cf. Shove and Pantzar, 2007).

7. Research has demonstrated the ways in which the heat-pump system impacts on the ability of householders to dynamically manage their thermal comfort (Lilley et al., 2010).

REFERENCES

Ajzen, I. (1991), 'The theory of planned behavior', *Organizational Behavior and Human Decision Processes*, **50**, 179–211.

Barad, K. (2003), 'Posthumanist performativity: toward an understanding of how matter comes to matter', *Signs*, **28**, 801–831.

Barry, A. (2001), *Political Machines: Governing a Technological Society*, London, UK: The Athlone Press.

Beddington, J. (2008), 'Managing energy in the built environment: rethinking the system', *Energy Policy*, **36**, 4299–4300.

Bennett, J. (2004), 'The force of things: steps toward an ecology of matter', *Political Theory*, **32**, 347–372.

Benton, Ted, and Redclift, Michael (1994) 'Introduction', in Michael Redclift and Ted Benton (eds), *Social Theory and the Global Environment*, 1–13. London, UK and New York, US: Routledge.

Blake, J. (1999), 'Overcoming the "value-action gap" in environmental policy: tensions between national policy and local experience', *Local Environment*, **4**, 257–278.

Boait, P.J., and Rylatt, R.M. (2010), 'A method for fully automatic operation of domestic heating', *Energy and Buildings*, **42**, 11–16.

Cole, R.J., Robinson, J., Brown, Z., and O'Shea, M. (2008), 'Re-contextualizing the notion of comfort', *Building Research and Information*, **36**, 323–336.

Cooper, Gail (1998), *Air Conditioning America: Engineers and the Controlled Environment, 1900–1960*, Baltimore, USA: Johns Hopkins University Press.

Crosbie, T., and Baker, K. (2010), 'Energy-efficiency interventions in housing: learning from the inhabitants', *Building Research & Information*, **38**, 70–79.

Crosbie, T., and Guy, S. (2008), 'Enlightening energy use: the co-evolution of household lighting practices', *Int. J. Environmental Technology and Management*, **9**, 220–235.

Darby, S. (2006), *The Effectiveness of Feedback on Energy Consumption: A Review for DEFRA of the Literature on Metering, Billing and Direct Displays*, Environmental Change Institute: University of Oxford, UK. http://www.eci.ox.ac.uk/research/energy/downloads/smart-metering-report.pdf.

Darby, S. (2010), 'Smart metering: what potential for householder engagement?', *Building Research and Information*, **38**, 442–457.

Department of Energy and Climate Change (DECC) (2011), *Great Britain's Housing Energy Fact File* http://www.decc.gov.uk/assets/decc/11/stats/climate-change/3224-great-britains-housing-energy-fact-file-2011.pdf.

Department of Energy and Climate Change (DECC) (2012a), *Smart Meters* http://www.decc.gov.uk/en/content/cms/tackling/smart_meters/smart_meters.aspx.

DECC (2012b) *The Future of Heating: A Strategic Framework for Low Carbon Heat in the UK* http://www.decc.gov.uk/assets/decc/11/meeting-energy-demand/heat/4805-future-heating-strategic-framework.pdf.

Department for Environment, Food and Rural Affairs (DEFRA) (2008), *A Framework for Pro-environmental Behaviours* http://www.defra.gov.uk/publications/files/pb13574-behaviours-report-080110.pdf.

Giddens, Anthony (1984), *The Constitution of Society*, Cambridge, UK: Polity Press.

Gram-Hanssen, K. (2010), 'Standby consumption in households analyzed with a practice theory approach', *Journal of Industrial Ecology*, **14**, 150–165.

Gram-Hanssen, K. (2011), 'Understanding change and continuity in residential energy consumption', *Journal of Consumer Culture*, **11**, 61–78.

Guy, S. (2006), 'Designing urban knowledge: competing perspectives on energy and buildings', *Environment and Planning C: Government and Policy*, **24**, 645–659.

Hargreaves, T., Nye, M., and Burgess, J. (2010), 'Making energy visible: a qualitative field study of how householders interact with feedback from smart energy monitors', *Energy Policy*, **38**, 6111–6119.

Harrabin, Roger (2012), 'Cameron hears Green Deal concerns', BBC Online, 16 May, http://www.bbc.co.uk/news/science-environment-18074650.

Hills, J. (2012), 'Getting the measure of fuel poverty: the final report of the fuel poverty review', http://www.decc.gov.uk/en/content/cms/funding/Fuel_poverty/Hills_Review/Hills_Review.aspx.

Hinchliffe, S. (1996), 'Helping the earth begins at home: the social construction of socio-environmental responsibilities', *Global Environmental Change: Human and Policy Dimensions*, **6**, 53–62.

Hinchliffe, S. (1995), 'Missing culture: energy efficiency and lost causes', *Energy Policy*, **23**, 93–95.

Hinton, E., Bickerstaff, K., and Bulkeley, H. (2012), *Deliverable 25: Understanding and Changing Comfort Practices* (copy available from the authors on request).

Hobson, K. (2003), 'Thinking habits into action: the role of knowledge and process in questioning household consumption practices', *Local Environment*, **8**, 95–112.

Hobson, K. (2006), 'Bins, bulbs, and shower timers: on the "techno-ethics" of sustainable living', *Ethics, Place and Environment*, **9**, 317–336.

Jackson, T. (2005), 'Motivating sustainable consumption: a review of evidence on consumer behaviour and behavioural change', http://www.c2p2online.com/documents/MotivatingSC.pdf.

Jelsma, J. (2003), 'Innovating for sustainability: involving users, politics and technology', *Innovation: The European Journal of Social Science Research*, **16**, 103–116.

Khan, Gulshan (2009), 'Agency, nature and emergent properties: an interview with Jane Bennett', *Contemporary Political Theory*, **8**, 90–105.

Kollmuss, A., and Agyeman, J. (2002), 'Mind the gap: why do people act environmentally and what are the barriers to pro-environmental behavior?', *Environmental Education Research*, **8**, 239–260.

Lilley, D., and Moore, N. (2010), *Intensive Sample Social Fieldwork Thematic Analysis: Harrogate*, Copy available from the author(s) on request.

Marres, N. (2008), 'The making of climate publics: eco-homes as material devices of publicity', *Dirtinition*, **9**, 17–45.

Marres, N. (2011), 'The costs of public involvement: everyday devices of carbon accounting and the materialization of participation', *Economy and Society*, **40**, 510–533.

Marres, N., and Lezaun, J. (2011), 'Materials and devices of the public: an introduction', *Economy and Society*, **40**, 489–509.

National House-Building Council (NHBC) (2012), *Low and Zero Carbon Homes: Understanding the Performance Gap*, Milton Keynes, UK: NHBC Foundation.

Owens, S. (2000), '"Engaging the public": information and deliberation in environmental policy', *Environment and Planning A*, **32**, 1141–1148.

Owens, S., and Driffill, L. (2008), 'How to change attitudes and behaviours in the context of energy', *Energy Policy*, **36**, 4412–4418.

Shipworth, M., Firth, S.K., Gentry, M.I., Wright, A.J., Shipworth, D.T., and Lomas, K.J. (2010), 'Central heating thermostat settings and timing: building demographics', *Building Research & Information*, **38**, 50–69.

Shove, Elizabeth (2003), *Comfort, Cleanliness and Convenience: The Social Organisation of Normality*, Oxford, UK: Berg.

Shove, E. (2010), 'Beyond the ABC: climate change policy and theories of social change', *Environment and Planning A*, **42**, 1273–1285.

Shove, E., and Pantzar, M. (2007), 'Recruitment and reproduction: the carriers of digital photography and floorball', *Human Affairs*, **17**, 154–167.

Shove, E., and Walker, G. (2007), 'CAUTION! Transitions ahead: politics, practice, and sustainable transition management', *Environment and Planning A*, **39**, 763–770.

Shove, E., Pantzar, M., and Watson, M. (2012), *The Dynamics of Social Practice: Everyday Life and How it Changes*, London, UK: Sage.

Spaargaren, G. (2011), 'Theories of practices: agency, technology, and culture: exploring the relevance of practice theories for the governance of sustainable consumption practices in the new world-order', *Global Environmental Change*, **21**, 813–822.

Stafford, A. (2012), 'Wattbox installation in H5a and H6b' (internal communication).

Strengers, Y. (2011), 'Negotiating everyday life: the role of energy and water consumption feedback', *Journal of Consumer Culture*, **11**, 319–338.

Strengers, Y., and Maller, C. (2012), 'Materialising energy and water resources in everyday practices: insights for securing supply systems', *Global Environmental Change*, **22**, 754–763.

Thrift, N. (2005), 'But malice aforethought: cities and the natural history of hatred', *Transactions of the institute of British Geographers*, **30**, 133–150.

15. Climate change and human security: the international governance architectures, policies and instruments
Michael Mason

Other chapters in this volume set out the multiple threats posed by climate change to human security. Without restating these claims here, it is nevertheless useful to remind ourselves that a prominent theme concerns how human security framings recast the idea of climate change as a development-oriented rather than environmental challenge. According to this approach, the dangers of climate change reside less in the incidence and magnitude of (predicted) biophysical events than in their apprehension as threats to human well-being, particularly for the poor and disadvantaged. The human security lens invites us to view climate change in a people-centred way, admitting it as only one of a number of conditions of life which may in practice jeopardize opportunities for safe, dignified and inclusive human development. What can be readily acknowledged is that there are diverse trajectories of climate-related influence on human lives and livelihoods, though it is the severe stress on vulnerable peoples attributed to present and future climate change that has justified its 'human securitization'.

This chapter examines the disparate architectures, policies and instruments drawn on in efforts to conjoin climate change and human security governance. The institutional starting-point is not promising: Held (2010, pp. 185–188) pinpoints the 'paradox' of global governance as the chronic mismatch between, on the one hand, the growing cross-border scope and intensity of collective problems – including transnational security and environmental risks – and, on the other, weak problem-solving capacities at international and regional levels. He ascribes the key shortcomings here to a multilateral order, still bearing the institutional imprint of a mid-twentieth century geopolitical settlement, not yet fit for purpose in our times. As noted below, neither climate change nor security are consensual domains for global governance. Both are dominated by state-centred decision-making which prioritizes national interests over human well-being: the international security system features enduring schisms over conflict prevention and management, while there are also significant disagreements between states concerning the ambition of international action on climate change. Conjoined governance addressed to 'environmental security' or 'climate security' might be expected to compound the institutional indeterminacy here. However, there is also the possibility that such a convergence could instead offer the prospect of coordinated action integrating diverse policy communities (Barnett et al., 2010, pp. 9–10). This, at least, is acknowledged within the United Nations (UN) system, where, as we shall see, explicit governance connections have been made between human security and climate change.

The first half of the chapter reviews those limited global governance policies and instruments recruited to address the human security implications of climate change. This involves both a survey of the emergence of human security concerns within global climate governance – notably the UN Framework Convention on Climate Change (UNFCCC) – and the recognition by some security governance actors that climate-related harm represents a significant threat to the lives and livelihoods of many people. Reference is made to several governance initiatives to determine whether there is anything other than scattered institutional moves to enhance human security against major climate hazards. In the second half of the chapter, I switch to a normative analysis of whether a more integrated governance to this end is justified and, if so, what it might look like. There are systemic obstacles to the clustering of human security and climate change decision-making. Even if operational concerns can be allayed, it would be rash to suggest more than general principles of institutional design which may be appropriate. I argue that, while the global climate regime holds epistemic and governance authority over the management of 'dangerous' climate change, the effective inclusion of climate concerns in human security decision-making is most likely to be achieved by consolidating the legal coherence and force of the latter – especially in relation to the development of human rights and humanitarian norms on the prevention of climate harm. Such a rights-based architecture would also afford human security governance at least some protection from, and critical engagement with, the power-oriented politics of the international security system.

GOVERNING CLIMATE CHANGE AND HUMAN SECURITY: REGIME INDETERMINACY

There is no fixed institutional site for international rule-making on climate change as a threat to human security. What I characterize in this section as 'regime indeterminacy' denotes the absence of an agreed institutional authority or forum for addressing climate change and human security in an integrated manner. As will be shown below, while human security is now acknowledged as a legitimate governance issue within the international climate regime, this has not yet led to any formal policy decision by the UNFCCC to protect or enhance human security in the context of climate change. Similarly, climate change is recognized as a serious threat to well-being within various global initiatives on human security, but its incorporation into these nascent governance mechanisms is ad hoc and uncertain.

Since its signing in 1992, the UNFCCC has served as the key instrument for global rule-making on climate change. Global climate governance is generally cooperative, but away from the UNFCCC, heterogenous and fragmented – encompassing related treaties (e.g. Montreal Protocol), transnational municipal networks, subnational actors, bilateral or club agreements and corporate climate initiatives (Biermann et al., 2010; Keohane and Victor, 2011). The near-universal membership of the climate change convention reveals the shared agreement of states that action is necessary to prevent 'dangerous anthropogenic interference with the climate system' (UNFCCC, Article 2), although the history of the treaty regime attests to enduring differences between Parties over appropriate responsibilities and commitments. To achieve global climate stabilization at safe levels, the UNFCCC has focused on mitigation actions, notably cuts in greenhouse gas emissions. Binding commitments to reduce emissions had to await the 1997 Kyoto Protocol (in force since 2005), which placed differentiated obligations on industrialized countries who ratified the agreement (excluding of course the United States). Even the modest Kyoto targets for 2008–2012, well short of the emissions reductions deemed necessary by the Intergovernmental Panel on Climate Change (IPCC), were only partially met. By 2012, UNFCCC negotiations to secure more comprehensive and deeper reductions had resulted in voluntary commitments from all major emitting countries but no new legally binding agreement. The tortuous progress of the UN climate treaty regime has led to increasing calls for a more decentralized system of global climate decision-making. Taking note of the myriad climate governance initiatives outside the UNFCCC process, the argument is that a loosely coordinated, flexible approach will facilitate innovative, ambitious measures that, tailored to specific sectors or regions, can bypass barriers to global collective action (e.g. Falkner et al., 2010; Hoffmann, 2011; Victor, 2011).

It is not necessary here to engage with the debate on the appropriate architecture for global climate governance, except to note that the idea of human security barely registers in the competing blueprints for institutional renewal or reform. Both the UNFCCC and Kyoto Protocol contain no references to human security, yet if it is conceded that dangerous climate change poses a potentially catastrophic threat to future conditions of human life, then its framing as a serious challenge to human security seems justifiable. For O'Brien et al. (2010, p. 13), this implies a shift of focus from human-induced environmental change to what these changes mean for vulnerable individuals and communities. Indeed, it is through the lens of climate vulnerability and adaptation that the human security concept is gaining currency within the UN climate regime. The UNFCCC requires Parties to take into account the needs of particularly vulnerable developing countries in managing the adverse effects of climate change, while the Kyoto Protocol provides a financial mechanism, the Adaptation Fund, to assist these countries in meeting the costs of climate change adaptation. Responsibility to support funding of adaptation in developing countries is also enshrined in the 2010 Cancun Agreements, including the allocation of substantial new monies through a Green Climate Fund. While these actions are expressed as state obligations and entitlements, UNFCCC assistance to developing countries has included vulnerability assessments centred on human well-being, gauging local coping strategies and adaptive capacity in the context of natural resource-dependent livelihoods (UNFCCC, 2007, pp. 15–16).

UNFCCC deliberations on climate vulnerability have been significantly shaped by the IPCC, especially the contribution of Working Group II (Impacts, Adaptation and Vulnerability) to successive assessment reports and IPCC-sponsored research on adaptation and vulnerability. IPCC framings of vulnerability to climate change have increasingly stressed the socioeconomic and political conditions that affect how individuals and communities cope with the impacts of climate-related change (Adger, 2006; Leary et al., 2008). A concern for the plight of particularly vulnerable populations has invited human security concepts and interpretations. Initially these centred on climate variability and change impacts on food production, recasting the notion of 'food security' as applying to households and individuals rather than national agricultural production (Boko et al., 2007, pp. 454–456). This concern with food security fed into UNFCCC decision-making on adaptation, most recently in the Work Programme on Loss and Damage (Subsidiary Body for Implementation, 2011, p. 47). Since 2010 the IPCC has explicitly embraced ideas of human security, which can be attributed at least in part to the agenda-setting activities of lobbying coalitions (e.g. the Climate Change, Environment and Migration Alliance) and epistemic communities (e.g. the Global Environmental Change and Human Security Project). By the time of the Fifth Assessment Report, IPCC Working Group II devoted a full chapter to human security as a necessary conceptual matrix for understanding the differential vulnerability and adaptive capacity of people exposed to climate change.

The invocation of human security in the forthcoming Fifth Assessment Report of the IPCC may be a necessary step in legitimating the concept for UNFCCC Parties, but as yet it has received no policy endorsement within the climate change treaty regime. Climate vulnerability and adaptation seem the most likely governance domains to receive such a move. In their review of UN funding schemes relevant to climate change adaptation, encompassing UNFCCC-managed or mandated funds, McGlynn and Vidaurre (2011) observe no references to human security in the terms of reference or guidelines in any of these financial instruments. Yet human security analysis has emerged in climate adaptation initiatives funded by the United Nations Development Programme (UNDP), which is assisting poorer countries in adaptation financing and policy-making. In its 2007/2008 Human Development Report, UNDP treated climate change as arguably the greatest challenge facing global poverty reduction and human development efforts (UNDP, 2007). Its emphasis on protecting poor and vulnerable households from climate shocks resonated with IPCC work on climate vulnerability, and the gravity of dangerous climate change as an existential threat justified, for UNDP, a human-centred understanding of climate insecurity.

To be sure, the UNDP interest in climate change emerged from its own influential formulation of human security, broadly defined in the 1994 Human Development Report as 'freedom from fear and freedom from want' then specified as 'safety from such chronic acts as hunger, disease and repression and . . . protection from sudden and hurtful disruptions in the patterns of daily life – whether in homes, jobs or communities' (UNDP, 1994, p. 23). While this human security concept was articulated by a UN body with a human development mandate, the global governance domain addressed, even challenged, by its construction was the international security system. The timing was opportune for its re-definition of security away from state-centred threats and military conflicts: the post-Cold War had opened up discursive space for recognizing multiple, inter-related threats to human well-being. For its proponents (e.g. Commission on Human Security, 2003; Gasper, 2010), the idea of human security exposes the dysfunctionality of global security decision-making, tied to national military capacities and technologies: the person-centred understanding of human security implies instead a universal or cosmopolitan coverage indifferent to state borders. For its critics, by contrast, the concept mistakenly retains 'security' as a desired end goal, which not surprisingly has resulted in its co-option by wielders of political and economic power (e.g. Neocleous, 2008; Turner et al., 2011).

The human security concept has acquired significant currency since the mid-1990s. Not surprisingly, it has appeared in a range of National

Human Development Reports sponsored by UNDP, particularly those covering volatile, conflict-prone countries – including Afghanistan, East Timor, Iraq and Sierra Leone (Jolly and Basu Ray, 2007; UNDP, 2008). Environmental degradation constitutes a separate category of threat to human security in UNDP's 1994 Human Development Report, and features in, for example, the national reports for East Timor and Sierra Leone. UNDP embraces a broad concept of human security, which corresponds to its usage in the UN system and an influential formulation by the Japanese government. However, narrower definitions, which are less amenable to environmental threats, have also attracted policy support. For example, the Canadian and Norwegian governments interpret human security as pertaining only to personal physical security and civil rights (Tadjbakhsh and Chenoy, 2007, pp. 9–38; Gasper, 2010).

Minimal definitions of human security can *exclude* climate change damage due to the lack of intentionality of climate-induced harm to persons (Gasper, 2010, p. 27). There has also been resistance by several UN Security Council members – notably China and Russia – to the inclusion of climate change as relevant to the Council's primary responsibility for maintaining international peace and security (a proposal raised by the UK during its April 2007 presidency of the Security Council). Nevertheless, climate change is now recognized across the UN system as posing major challenges to both human and national security. A report issued by the Secretary-General in 2009 presented climate change as a threat multiplier, exacerbating existing sources of conflict and security (UN Secretary-General, 2009). This statement reflected the preference of many Member States to focus on the security of individuals and communities, and also for a strengthening of efforts to mainstream climate change within the UN (e.g. UN Chief Executives Board for Coordination, 2008). A subsequent report by the Secretary-General on human security repeated the claim that climate change, and its interactions with other insecurities, represents a serious threat to human lives and livelihoods (UN Secretary-General, 2010, p. 12).

That the human security concept has some policy traction in the UN system reflects the recommendations of a global Commission on Human Security, including its broad definition of human security as the protection of fundamental freedoms (Commission on Human Security, 2003). One commission proposal led to the creation of an Advisory Board on Human Security to advise the UN Secretary-General on the promotion and dissemination of the human security concept. Beyond UNDP, human security projects have been undertaken by a wide range of UN agencies, including the Office of the UN High Commissioner for Refugees, the Food and Agriculture Organization (FAO), the World Health Organization and the UN Development Fund for Women (UN Secretary-General, 2010, pp. 2–3). The main dedicated instrument for advancing human security goals is the UN Trust Fund for Human Security established in 1999 and managed, since 2004, by a Human Security Unit in the Office for the Coordination of Humanitarian Affairs (OCHA). However, a review of the Trust Fund issued in 2010 by a UN internal audit body noted limited buy-in on the part of Member States, with a heavy dependence on financial support from the Japanese government. It also expressed concern that there had been no comprehensive evaluation of the Fund's activities in terms of human security outcomes, despite financial commitments by 2009 of $355 million (Office of Internal Oversight Services, 2010, pp. 3–5). The Human Security Unit at OCHA is ambitiously tasked with integrating UN human security activities across all UN activities: obstacles to this integration, according to senior UN staff, are thrown up by divergent financial systems and mandates, as well as a lack of understanding of the concept by UN Country Teams (Advisory Board on Human Security, 2011).

Given this challenging institutional context, it is noteworthy that OCHA is making modest progress across the UN system in highlighting climate change as a human security issue. Recent (2011–) multi-agency projects approved by the UN Trust Fund for Human Security include an FAO-led project to strengthen rural livelihoods severely affected by climate change-induced drought in Lesotho and a UNICEF-led project to enhance community resilience and coping with climate change and natural disasters in Vanuatu (Human Security Unit, 2011, pp. 7–8). More generally, climate change is a thematic priority for OCHA and, since December 2008, the agency has led an advocacy campaign to raise awareness of the humanitarian implications of climate change. As set out in a 2009 speech by the UN Under-Secretary General for Humanitarian Affairs and Emergency Relief Coordinator, climate change threatens to overwhelm a global humanitarian system designed to respond to trigger events (e.g. natural disasters, conflicts) rather than chronic humanitarian needs arising from 'slow onset' climate-induced harm. Addressing effectively the latter requires, it is claimed, new models of prevention, preparedness and response (OCHA, 2009, pp. 9–11).

Ironically, by representing climate change as an autonomous driver of humanitarian impacts, OCHA has neglected the particular trajectories of climate-related harm experienced by civilians in (post)conflict environments – affected communities that it has an express mandate to assist. Populations facing or recovering from conflict are especially vulnerable to climate variability and extremes because of impaired coping options and low adaptive capacity (Barnett, 2006; Mason et al., 2011). Scholarly work on the relationships between determinants of climate change, human security and conflict (e.g. Barnett and Adger, 2007; Mason et al., 2012) has raised questions about the responsibilities and capabilities of institutional actors – including states, international organizations and donors – which have yet to be addressed by UN agencies facing climate vulnerability in war-related environments. There are actors and modalities in the UN system which could in principle accommodate a human security understanding of climate (and other environmental) hazards in (post)conflict areas; for example, the post-conflict needs assessments undertaken by the UN Environment Programme and an emerging concern with environmental stresses in the UN Peacebuilding Commission (Swain and Krampe, 2011, pp. 207–208). As yet, though, there is little institutional movement in this direction.

It can be concluded, therefore, that despite the embrace of human security by international organizations addressing climate change and human development/humanitarian needs, there remains no settled global governance space for enhancing human security against climate change. That the main impetus for promoting this goal has come from UN bodies suggests that the multi-agency funding being facilitated by OCHA's Human Security Unit is the most appropriate instrument for conjoining climate change and human security activities – at least within the UN bureaucracy. The 'mainstreaming' of human security in UN activities is a politically delicate project, deferential to the sovereignty sensitivity of those Member States anxious that the concept could be used to justify interventionist actions (e.g. under the responsibility to protect norm). The Report on Human Security of the Secretary-General asserts that, on the basis of its mandate to address security, development and human rights, human security is central to the work of the UN (UN Secretary-General, 2010, p. 17); yet at the same time, human security is not presented as a strategic priority under the Secretary-General's Five-Year Action Agenda launched in January 2012. This is in contrast to the 'generational impera- tive' of climate change (UN Secretary-General, 2012). Similarly, the wave of collaborative governance with non-state actors fostered since 1998 by the UN Fund for International Partnerships has generated far more funding for environmental (including climate change) projects than those relating to peace, security and human rights; in other words, human security-oriented activities are also perceived as contentious by many private partners (Andanova, 2010, p. 45).

The ambivalence over human security in UN governance practice, com- bined with the uncertain location of human security in the UN climate change convention, invites a normative analysis of what institutional (re) design may be needed to advance more effectively the protection of core human values in the context of harmful climate change. This is the focus for the second half of this chapter.

INTEGRATING HUMAN SECURITY NORMS IN THE GLOBAL GOVERNANCE OF CLIMATE HARM

Understanding climate change as an issue of human security represents an explicitly normative approach which, according to its proponents, rejects the dominant 'environmental' narrative of climate change. By reifying the environment as an independent, naturalized category, governance responses under this mainstream perspective are viewed as favouring techno-managerial fixes which preclude questions about the moral responsibility and political-economic interests of those who profit most from high-carbon development paths. A human security framing is, in contrast, seen as opening space for critical reflection on, and policy engagement with, the structural drivers of climate vulnerability as they interact with other threats to human well-being (Adger, 2010; O'Brien et al., 2010, pp. 11–14). The normative intent is both to reveal how climate harm is implicated in the creation of particular insecurities faced by vulnerable people and how their fundamental freedoms can be protected or even enhanced.

From such a normative perspective, the identification of (potential) harm is pivotal to the justification and governance application of human security ideas. As Mike Hulme (2009, pp. 191–196) argues, the ultimate objective of the UN climate change convention, as articulated by Article 2, is usually abbreviated as the avoidance of 'dangerous climate change'; yet such a notion of danger defies easy quantification, exposing the limits of scientific risk analysis. Interpretations of danger, he claims, are always context-specific and value-laden, because the experience of insecurity only becomes meaningful for individuals or groups in particular situations. This experiential realm of harm perception is a necessary element for the construction of human security. The additional challenge of defining climate danger is that it constitutes an 'un-situated risk', with distant, intangible sources and diffuse, indirect causes (Hulme, 2009, p. 196). Viewing UNFCCC Article 2 through the lens of human security would therefore seem to render even more intractable the global governance efforts to manage climate risk, as the current and projected trajectories of specific harm are too uncertain and imprecise to register in terms of human security.

However, the tendency to portray dangerous climate change as a governance problem *sui generis* overlooks the international legal architecture in which the UNFCCC is embedded, which offers, if not blueprints for collective action, then at least principles to steer institutional coordination. Of cardinal relevance to the Article 2 objective to avoid dangerous anthropogenic interference with the climate system is the *harm prevention principle* recalled in the preamble of the climate change convention:

States have, in accordance with the Charter of the United Nations and the principles of international law, the sovereign right to exploit their resources pursuant to their environmental and developmental policies, and the responsibility to ensure that activities within their jurisdiction or control do not cause damage to the environment of other States or of areas beyond the limits of national jurisdiction.

This paragraph repeats Principle 2 of the 1992 Rio Declaration on Environment and Development, which itself is a slightly amended version of Principle 21 of the 1972 Stockholm Declaration on the Human Environment. Both UN declarative principles codify what is widely acknowledged as a general environmental obligation in international law. The principle is endorsed, beyond the UNFCCC, in a range of multilateral environmental agreements, including treaties addressing air and marine pollution, biodiversity conservation, radioactive contamination and desertification. All these legal regimes share as their governance problem the prevention and/or mitigation of inadvertent environmental harm, usually as a result of behaviour by state or non-state actors otherwise deemed permissible. The issue is that the accumulation of harm for vulnerable entities has, for a given governance authority, approached or crossed a threshold of unacceptable injury.

Again, the core harm prevention obligation within the climate change convention is accorded determinacy by its resonance with a background set of normative expectations in international environmental law. A common corollary of the harm prevention principle is the legal requirement of due diligence, such that states take all necessary measures as may reasonably be expected in all circumstances (Okowa, 2000, p. 81). Due diligence allows a consideration of problem-specific and other contextual factors in the application of harm prevention rules, tailoring obligations with reference, for example, to historical responsibility, state capability and the likelihood or seriousness of danger. The UNFCCC famously differentiates responsibilities for tackling climate change and its adverse effects in relation to the greater culpability and mitigation capacity of developed country Parties, as well as the special needs and circumstances of developing country Parties. These 'common but differentiated responsibilities and respective capabilities' are underpinned by an explicit appeal to 'equity' (Article 3(i)). Harm prevention obligations in the climate change treaty system are also qualified by the adoption of a precautionary principle – that Parties take 'precautionary measures to anticipate, prevent or minimize the causes of climate change and mitigate its effects' (Article 3(iii)). As expressed in Principle 15 of the Rio Declaration and other multilateral environmental treaties, the intent is to prevent a lack of full scientific certainty serving as a justification for postponing cost-effective measures designed to prevent serious or irreversible damage as a consequence, in this case, of climate change. This principle is not immune to criticism (e.g. Sunstein, 2005), but nevertheless is regarded by UNFCCC Parties as a legitimate expression of due diligence in negotiations over commitments to avoid dangerous climate change.

The notion of human security invites a reconceptualization of climate harm prevention from the perspective of human safety, well-being and freedom. Neil Adger provides useful pointers on applying human security to climate change impacts, highlighting, for example, how 'freedom from want' could encompass increased resource scarcity caused by declining water availability or land productivity, and 'freedom from fear' could include risks to health or place of residence as a consequence of climate-induced damage (2010, p. 281). The assessment of climate harm according to human security would, in any given context, need to identify both major climate risks and also preventive mechanisms for risk avoidance or reduction, while at the same time situating these stresses in relation to other sources of human insecurity. Uniting the specification and prioritization of threats is a concern with the vulnerabilities of affected individuals and communities, including their capacity to adapt to threats in a way that builds resilience (Adger, 2010).

If, following Adger, we take as a key criterion of human security the protection of the most vulnerable people, its institutional application in the UN climate change regime is currently constrained by the state-centred formulation of responsibilities and entitlements. As already noted, there are express commitments to address the needs of particularly vulnerable countries, justified by the principle of differentiated treatment; but the measures adopted in practice have not effectively captured unequal vulnerabilities within national populations. For example, under the Clean Development Mechanism (CDM) – one of the flexibility mechanisms of the Kyoto Protocol – there is a problem, first, with the outcomes of disbursements to developing country Parties under the Adaptation Fund (mainly financed by a 2 percent levy on Certified Emissions Reductions under the CDM). Level of vulnerability to climate change is an important criterion for Adaptation Fund support of adaptation projects and programmes in developing countries. Yet McGlynn and Vidaurre (2011) have highlighted the relatively modest financing available in relation to the climate adaptation needs of target countries and the long-term sustainability of a fund tied to the uncertain future of the Kyoto Protocol. Second, in relation to the CDM projects themselves, there is no international supervision of the extent to which these address the conditions of the most vulnerable: authority on technological and project choices is delegated to host countries, who often have recourse to other policy considerations, such as trade and investment interests. Moreover, the prominent role of private sector actors in CDM projects, and carbon markets more generally, has skewed climate financing away from particularly vulnerable groups (Cullet, 2010, pp. 191–192).

Any expectation that the protection of those people most vulnerable to climate change harm would find normative support from the commitment to equity in Article 3 of the UNFCCC is also frustrated by the consistent expression of this moral principle in terms of inter-state relations. To be sure, in this article there is an explicit recognition of inter-generational equity – that the climate system should be protected for the benefit of both present and future generations of humankind. And human security would seem a relevant framework for identifying scenarios that pose catastrophic costs to future generations. Yet the weighting given to the welfare of future generations is typically interpreted through the lens of national interests. The present costs of abatement actions presumed to benefit future generations usually appear in UNFCCC negotiations in terms of disputed burden-sharing between developed and developing countries. Similarly, UNFCCC decision-making on inter-generational equity addresses the costs (and benefits) of mitigation and adaptation actions as received by Parties to the climate change convention and Kyoto Protocol. There are of course differences over what equity means in practice, including its moral kinship to ideas of fairness and justice (see Soltau, 2009), but not over the understanding that it applies above all to an interdependent community of sovereign states.

While public international law is by definition state-centric, there are multilateral instruments which demonstrate that it is indeed possible to prioritize harm protection for vulnerable individuals and groups. Andrew Linklater (2011, pp. 36–41) identifies a growing cluster of 'cosmopolitan harm conventions' designed to protect people from avoidable harm regardless of their national citizenship status. With roots in international agreements covering the welfare of non-combatants during war, a cosmopolitan regard for protecting the vital interests of vulnerable people is now embedded in conventions covering, for example, genocide, apartheid, torture and terrorist bombings. Their preoccupation with safeguarding the bodily integrity and dignity of human beings in situations of extreme danger falls within the moral compass of human security. Recognizing that the damage caused by climate change is, like other instances of transnational environmental harm, distinct from direct forms of violence, its potential for serious injury to people can still justify a cosmopolitan construction of harm prevention (Linklater, 2011, p. 39; Mason, 2005, pp. 69–75). I follow this logic to argue that, given the very limited inroad of human security thinking in the UN climate change regime, there is currently more institutional scope for addressing the human security effects of climate change by applying rules of conduct developed in the fields of (1) human rights governance and (2) global humanitarian governance.

1.   Human Rights Governance

The relationship between climate change and human rights has already been examined at the UN by the Office of the High Commissioner for Human Rights (OHCHR), which was charged in March 2008 by the UN Human Rights Council with undertaking an analytical study of the issue. In its report, OHCHR cautions that climate change effects may not necessarily violate human rights, but that human rights obligations still provide important protection to individuals whose rights are negatively affected by climate change or responses to it. These rights impacts are potentially wide-ranging, though those human rights most directly threatened are judged to include rights to life, food, water, health, housing and self-determination (OHCHR, 2009, pp. 8–15). Importantly, OHCHR recognizes that the effects of climate change will be felt most acutely by vulnerable individuals and groups, noting that states are already legally bound to address such vulnerabilities in accordance with rights instruments promoting equality and non-discrimination (OHCHR, 2009, p. 15). The report stresses existing governmental duties under international human rights law, which encompasses national obligations to realize substantive and procedural rights, as well as international obligations to cooperate in the promotion and protection of human rights. These recommendations have not been ignored: the UN Human Rights Council has cited the report in encouraging its expert groups ('special mandate-holders') to address climate change within their domains of responsibility, and human rights bodies now have justification to consider climate change effects within their monitoring remits (Knox, 2009, p. 477). Insofar as respect for human rights serves as a benchmark for human security, such institutional moves are a necessary step in protecting vulnerable people from various trajectories of serious harm as a direct or indirect consequence of climate change. Of course, they also rely on high levels of human rights protection across the international community, which remains an ambitious expectation.

Human rights governance features standards and instruments for ensuring that those breaching human rights are held to account, including the availability of remedies and redress for victims. This is arguably the most difficult area for institutionalizing the protection of vulnerable individuals and groups from climate-induced injury, in part because of difficulties attributing harm to particular state or private actors. As an application of territorial responsibility, the UNFCCC apportions accountability obligations to governments, but the treaty regime does not provide direct remedies to those people disproportionately affected by climate harm. In this respect, the climate change convention is no different from other multilateral environmental agreements which avoid prescribing state liability for actual environmental damage, reflecting the preference of the international community for private liability systems (Mason, 2005, pp. 116–119). Opportunities are gradually opening up for affected individuals and groups to pursue transnational civil litigation against particular state or corporate actors for climate harm. Vulnerable communities are often reliant on assistance from external actors to bring such actions, which creates an inevitable selectivity in victims represented. Similarly, a reliance on diverse domestic systems of liability also limits the exercise of private law remedies to particular legal jurisdictions, with the US currently serving as a key testbed for such 'climate justice' actions (Grossman, 2009; Abate, 2010). While an imperfect instrument for providing direct remedies to parties facing serious climate harm, tort-based climate litigation does highlight an accountability deficit in global climate governance.

2. Global Humanitarian Governance

Human rights protections are an important element of international assistance dealing with the humanitarian consequences of climate-related emergencies. However, humanitarian rules comprise a distinctive set of governance policies and instruments, which are designed to limit the harm caused by disasters and armed conflict. As such, they prescribe assistance for those experiencing particular patterns of human insecurity. Alongside the climate change work of OCHA mentioned above, there have been efforts to promote global coordination of humanitarian assistance on climate harm through the work of a task force on climate change, migration and displacement reporting to an Inter-agency Standing Committee of UN and non-UN humanitarian agencies. The focus of the task force on climate-induced migration and displacement reveals a high-level humanitarian concern about developing coherent, workable rules to meet the challenge of substantial numbers of people expected to be displaced or even made stateless by climate change. Significantly, the UN High Commissioner for Refugees (UNHCR) has criticized use of the term 'climate refugees' to describe such people, fearing that it 'could potentially undermine the international legal regime for the protection of refugees whose rights and obligations are quite clearly defined and understood' (UNHCR, 2009, p. 9). For the same reason, it has opposed amending the 1951 Convention Relating to the Status of Refugees to include such displaced persons. Nevertheless, there is growing support for a new legal instrument to provide dedicated protection to those suffering involuntary movements due to climate-related changes (e.g. Docherty and Giannini, 2009; McAdam, 2011). No such assistance is set out in the UN climate change treaties, so the most likely domain for developing relevant measures is global humanitarian governance.

Concerned as it is with regulating the conduct of armed conflict, international humanitarian law is a distinct and essential subset of this governance field. Its broad aim to protect individuals and groups from the effects of war finds resonance with those applications of the human security concept to (post)conflict areas, as noted above for UNDP Human Development Reports. The legal and policy authority of international humanitarian law is of course far greater than that of human security, but the latter still has value in situations of armed conflict by highlighting in a holistic way the compounded stresses (war-related and otherwise) faced by vulnerable individuals and groups. This is how a human security perspective could identify the direct and indirect relations between conditions of warfare and climate vulnerability. Existing humanitarian law includes provisions prohibiting extreme and disproportionate damage to the environment by combatants, but is imprecise regarding the specific protection and assistance available to civilians whose vulnerability to climate harm may be exacerbated by military actors (e.g. severe restrictions in livelihood options and personal insecurity). The emerging subdiscipline of 'warfare ecology' has opened up areas of enquiry looking at the combined biophysical and human effects arising from war-related conditions (Machlis et al., 2011). Should this research delineate how hostile military forces are accentuating the vulnerability of populations to serious climate harm, there will be evidence for arguing that such actions may breach the general duty on combatants to protect civilian populations. At the moment, climate change tends to be represented as an external variable shaping the long-term environmental conditions for people in post-conflict environments (UNEP, 2009, p. 11). A recognition, instead, that combatants can intensify the short-term vulnerability of civilians to climate (and other environmental) stresses would bring climate-related harm more squarely within the scope of international humanitarian rule-making and enforcement.

For global governance concerned with human rights and humanitarian assistance, there are existing rules of conduct which, in principle, cover certain expressions or experiences of climate harm. This has been acknowledged by lead UN agencies on human rights and humanitarian affairs, even if there are also uncertainties about the appropriate governance instruments to assist those facing a high risk or incidence of climate-related damage affecting their vital freedoms. A fundamental challenge here is a structural mismatch between the UN climate change regime and the rights-based global regimes directed at the protection of vulnerable human beings. Humphreys captures neatly the systemic differences between the two governance domains:

> One [climate change] is a regime of flexibility, compromise, soft principles and differential treatment; the other [human rights] of judiciaries, policing, formal equality and universal truths. Faced with injustice, one regime tends to negotiation, the other to prosecution. (2010, pp. 316–317)

The search within human rights and humanitarian governance for remedies available to people seriously threatened by climate harm reflects a major accountability deficit in the UNFCCC architecture. It is not surprising, therefore, that human security has received support as a mediating or bridging concept between these disconnected areas of governance – one that highlights the predicament of those most vulnerable to 'dangerous' climate change.

At the same time, there are also potential pitfalls with such a move. Including climate change threats within the purview of human security risks institutional overstretch in what is still a nascent, uncertain domain of governance. The expansive take on harm prevention provided by human security has unsettled critics of this approach on account of its apparent open-endedness (see Jolly and Basu Ray, 2007, pp. 465–466; Gasper, 2010, p. 40): according policy attention to diffuse climate change effects is unlikely to allay such concerns. In operational terms, much will depend, therefore, on ascertaining climate risks or harm with enough precision to factor them in when prioritizing serious threats to human well-being. Here the epistemic authority of the IPCC is likely to be crucial in validating the type of climate information deemed credible enough to feature in substantive assessments of human security, which would also need to incorporate the bottom-up, experiential accounts of affected groups. Such assessments may enable the identification of prima facie breaches of human rights or humanitarian rules caused by climate-related harm largely attributable to the actions or omissions of responsible parties. Reaching this (appropriately) high threshold of proof should trigger further investigation by relevant monitoring and enforcement bodies within global human rights and humanitarian governance. Of course, human security goals are also served by preventative policy actions – for example climate adaptation planning to increase the resilience of the most vulnerable (Adger and Nelson, 2010). However, mechanisms providing rights-based accountability for individuals and groups facing serous climate harm are still rudimentary, warranting greater political attention and policy development.

CONCLUSION

The idea of human security now has significant currency in a range of global policy discourses and has received high-level endorsement within the UN system. It is also recognized as relevant to the global governance of climate change, most notably in recent deliberations by the IPCC and, implicitly, in UNFCCC commitments to prioritize the needs of those judged to be most vulnerable to climate-related harm. As noted in this chapter, though, there has been no formal declaration on human security by the UNFCCC, so the principal international architecture for regulating climate change does not (yet) feature dedicated measures for advancing human security as a regime goal. Climate change as a threat to human security has instead emerged as a key concern within the human development reporting process overseen by UNDP – the agency credited with popularizing the concept – while the UN has given its lead humanitarian office the strategic responsibility for managing climate change projects supported by a human security trust fund. Despite statements by the UN Secretary General that climate-induced harm is a major source of human insecurity, there remains no settled governance authority for enhancing human security against 'dangerous' climate change – a situation I labelled 'regime indeterminacy'.

In the second half of the chapter I argued that greater governance coherence for this area finds normative support from cosmopolitan notions of harm prevention, which universally accord moral priority to the protection of people from unnecessary injury. Cosmopolitan harm conventions represent the institutionalization of this perspective: they resonate with ideas of human security, but hold greater political and legal force. Interpreting climate threats along cosmopolitan lines therefore allows an integration with pre-existing rules of conduct relating to the prevention and mitigation of harm for vulnerable individuals and groups. At the moment, the clearest expression of such rules is found in international regimes concerned with human rights and humanitarian protection. I noted above that these regimes feature instruments that could address particular instances of climate-related harm. While their coverage is necessarily selective, and may not capture all the possible human security effects of climate change, the added value of these regimes to global climate governance is both a recognition of serious climate-related threats infringing on vital human interests and also adjudicatory mechanisms offering rights-based means of accountability. The climate harm prevention value of these governance regimes is, ironically, more likely to be effective by remaining separate from the UNFCCC until such time that its member states commit to compatible human security goals.

REFERENCES

Abate, Russell S. (2010), 'Public nuisance suits for the climate justice movement: the right thing and the right time', *Washington Law Review*, **85** (1), 197–252.

Adger, W. Neil (2006), 'Vulnerability', *Global Environmental Change*, **16** (3), 268–281.

Adger, W. Neil (2010), 'Climate change, human well-being and insecurity', *New Political Economy*, **15** (2), 275–292.

Adger, W. Neil and Donald R. Nelson (2010), 'Fair decision making in a new climate of risk', in Karen O'Brien, Asuncion St. Clair and Berit Kristoffersen (eds), *Climate Change, Ethics and Human Security*, Cambridge, UK: Cambridge University Press, pp. 83–94.

Advisory Board on Human Security (2011), *Minutes of the Meeting with Heads of UN Agencies, Funds and Programmes*, New York, USA: UN Trust Fund for Human Security.

Andanova, Liliana B. (2010), 'Public-private partnerships for the earth: politics and patterns of hybrid authority in the multilateral system', *Global Environmental Politics*, **10** (2), 25–53.

Barnett, Jon (2006), 'Climate change, insecurity and injustice', in W. Neil Adger, Jouni Paavola, Saleemul Huq and M.J. Mace (eds), *Fairness in Adaptation to Climate Change*, Cambridge, USA: MIT Press, pp. 115–129.

Barnett, Jon and W. Neil Adger (2007), 'Climate change, human security and violent conflict', *Political Geography*, **26** (6), 639–655.

Barnett, Jon, Richard A. Matthew and Karen L. O'Brien (2010), 'Global environmental change and human security: an introduction', in Richard A. Matthew, Jon Barnett, Bryan McDonald and Karen O'Brien (eds), *Global Environmental Change and Human Security*, Cambridge, USA: MIT Press, pp. 3–32.

Biermann, Frank, Fariborz Zelli, Philipp Pattberg and Harro van Asselt (2010), 'The architecture of global climate governance: setting the stage', in Frank Biermann, Philipp Pattberg and Fariborz Zelli (eds), *Global Climate Governance Beyond 2012: Architecture, Agency and Adaptation*, Cambridge, UK: Cambridge University Press, pp. 15–24.

Boko, Michel, Isabelle Niang, Anthony Nyong, Coleen Vogel, Andrew Githeko, Mahmoud Medany, Balgis Osman-Elasha, Ramadjita Tabo and Pins Yanda (2007), 'Africa', in Martin L. Parry, Osvaldo F. Canziani, Jean P. Palutikof, Paul J. van der Linden and Clair E. Hanson (eds), *Climate Change 2007: Impacts, Adaptation and Vulnerability. Contribution of Working Group II to the Fourth Assessment Report of the Intergovernmental Panel on Climate Change*, Cambridge, UK: Cambridge University Press, pp. 433–467.

Commission on Human Security (2003), *Human Security Now*, New York, USA: UN Secretary-General's Commission on Human Security.

Cullet, Philippe (2010), 'The Kyoto Protocol and vulnerability: human rights and equity dimensions', in Stephen Humphreys (ed.), *Human Rights and Climate Change*, Cambridge, UK: Cambridge University Press, pp. 183–206.

Docherty, Bonnie and Tyler Giannini (2009), 'Confronting a rising tide: a proposal for a convention on climate change refugees', *Harvard Environmental Law Review*, **33** (2), 349–403.

Falkner, Robert, Hannes Stephan and John Vogler (2010), 'International climate policy after Copenhagen: towards a "building blocks approach"', *Global Policy*, **1** (3), 252–262.

Gasper, Des (2010), 'The idea of human security', in Karen O'Brien, Asuncion St. Clair and Berit Kristoffersen (eds), *Climate Change, Ethics and Human Security*, Cambridge, UK: Cambridge University Press, pp. 23–46.

Grossman, David (2009), 'Tort-based climate litigation', in William C.G. Burns and Hari M. Osofsky (eds), *Adjudicating Climate Change: State, National and International Approaches*, Cambridge, UK: Cambridge University Press, pp. 193–229.

Held, David (2010), *Cosmopolitanism: Ideals and Practices*, Cambridge, UK: Polity.

Hoffmann, Matthew J. (2011), *Climate Governance at the Crossroads: Experimenting with a Global Response after Kyoto*, New York, USA: Oxford University Press.

Hulme, Mike (2009), *Why We Disagree About Climate Change*, Cambridge, UK: Cambridge University Press.

Human Security Unit (2011), *Progress Report to the Advisory Board on Human Security: September 2010 to September 2011*, New York, USA: Human Security Unit/OCHA.

Humphreys, Stephen (2010), 'Conceiving justice: articulating common causes as distinct regimes', in Stephen Humphreys (ed.), *Human Rights and Climate Change*, Cambridge, UK: Cambridge University Press, pp. 299–319.

Jolly, Richard and Deepayan Basu Ray (2007), 'Human security – national perspectives and global agendas: insights from national human development reports', *Journal of Human Development*, **19** (4), 457–472.

Keohane, Robert O. and David G. Victor (2011), 'The regime complex for climate change', *Perspectives in Politics*, **9** (1), 7–23.

Knox, John H. (2009), 'Linking human rights and climate change at the United Nations', *Harvard Environmental Law Review*, **33** (2), 477–498.

Leary, Neil, Cecilia Conde, Jyoti Kulkarni, Anthony Nyong and Juan Pulhin (eds) (2008), *Climate Change and Vulnerability*, London, UK: Earthscan.

Linklater, Andrew (2011), *The Problem of Harm in World Politics: Theoretical Investigations*, Cambridge, UK: Cambridge University Press.

Machlis, Gary, Thor Hansen, Zdravko Špirić and Jean E. McKendry (eds) (2011), *Warfare Ecology: A New Synthesis for Peace and Security*, Dordrecht, The Netherlands: Springer.

Mason, Michael (2005), *The New Accountability: Environmental Responsibility Across Borders*, London, UK: Earthscan.

Mason, Michael, Mark Zeitoun and Rebhy el Sheikh (2011), 'Conflict and social vulnerability to climate change: lessons from Gaza', *Climate and Development*, **3** (4), 285–297.

Mason, Michael, Mark Zeitoun and Ziad Mimi (2012), 'Compounding vulnerability: impacts of climate change on Palestinians in Gaza and the West Bank', *Journal of Palestine Studies*, **41** (3), 38–55.

McAdam, Jane (2011), *Climate Change Development and International Law: Complementary Protection Standards*, PPLA/2011/03, Geneva: Division of International Protection, UN High Commissioner for Refugees.

McGlynn, Emily and Rodrigo Vidaurre (2011), *UN Funding Schemes Relevant to Climate Change Adaptation*, Berlin, Germany: Ecologic Institute.

Neocleous, Mark (2008), *Critique of Security*, Edinburgh: Edinburgh University Press.

O'Brien, Karen, Asunción St. Clair and Berit Kristoffersen (2010), 'The framing of climate change: why it matters', in Karen O'Brien, Asunción St. Clair and Berit Kristoffersen (eds), *Climate Change, Ethics and Human Security*, Cambridge, UK: Cambridge University Press, pp. 3–22.

OCHA (2009), *Climate Change: Campaign Toolkit*, New York, USA: Office for the Coordination of Humanitarian Affairs.

Office of Internal Oversight Services (2010), *Audit Report: Management of the United Nations Trust Fund for Human Security*, New York, USA: OIOS, UN.

OHCHR (2009), *Report of the Office of the High Commissioner for Human Rights on the Relationship Between Climate Change and Human Rights*, A/HRC/10/61, New York, USA: UN.

Okowa, Phoebe N. (2000), *State Responsibility for Transboundary Air Pollution in International Law*, Oxford, UK: Oxford University Press.

Soltau, Friedrich (2009), *Fairness in International Climate Change Law and Policy*, Cambridge, UK: Cambridge University Press.

Subsidiary Body for Implementation (2011), *Synthesis Report on Views and Implementation on the Thematic Areas in the Implementation of the Work Programme*, FCCC/SBI/2011/INF.13, Bonn, Germany: UNFCCC Secretariat.

Sunstein, Cass R. (2005), *Laws of Fear: Beyond the Precautionary Principle*, Cambridge, UK: Cambridge University Press.

Swain, Ashok and Florian Krampe (2011), 'Stability and sustainability in peace building: priority areas for warfare ecology', in Gary Machlis, Thor Hansen, Zdravko Špirić and Jean E. McKendry (eds), *Warfare Ecology: A New Synthesis for Peace and Security*, Dordrecht, The Netherlands: Springer, pp. 199–210.

Tadjbakhsh, Shahrbanou and Anuradha M. Chenoy (2007), *Human Security: Concepts and Implications*, London, UK: Routledge.

Turner, Mandy, Neil Cooper and Michael Pugh (2011), 'Institutionalised and co-opted: why human security has lost its way', in David Chandler and Nik Hynek (eds), *Critical Perspectives in Human Security: Rethinking Emancipation and Power in International Politics*, London, UK: Routledge, pp. 83–96.

UN Chief Executives Board for Coordination (2008), *Acting on Climate Change: the UN System Delivering as One*, New York, USA: UNCEB.

UN Secretary-General (2009), *Climate Change and its Possible Security Implications*, A/64/350, New York, USA: UN General Assembly.

UN Secretary-General (2010), *Human Security*, A/64/701, New York: UN General Assembly.

UN Secretary-General (2012), 'The Secretary General's Five-Year Action Agenda', 25 January, New York: UN. http://www.un.org/sg/priorities/index.shtml.

UNDP (1994), *Human Development Report 1994: New Dimensions of Human Security*, New York, USA: United Nations Development Programme.

UNDP (2007), *Human Development Report 2007/2008: Fighting Climate Change: Human Solidarity in a Divided World*, New York, USA: United Nations Development Programme.

UNDP (2008), *Iraq: National Report on the Status of Human Development*, New York, USA: United Nations Development Programme.

UNEP (2009), *Integrating Environment in Post-conflict Needs Assessments*, Geneva, Switzerland: United Nations Environment Programme.

UNFCCC (2007), *Climate Change: Impacts, Vulnerabilities and Adaptation in Developing Countries*, Bonn, Germany: UNFCCC Secretariat.

UNHCR (2009), *Climate Change, Natural Disasters and Human Displacement: A UNHCR Perspective*, Geneva, Switzerland: UN High Commissioner for Refugees.

Victor, David G. (2011), *Global Warming Gridlock: Creating More Effective Strategies for Protecting the Planet*, Cambridge, UK: Cambridge University Press.

16. A human rights-based approach from strengthening human security against climate change
Steve Vanderheiden

Those harmed as the result of anthropogenic climate change suffer from a chain of causes and effects that is expected to visit its damage upon persons directly through extreme weather events like storms, floods, or heat waves, as well as indirectly through the heightened scarcity of food or water and impaired ecological capacity (IPCC 2007). These expected climate-related impacts threaten human security as they undermine ecological, agricultural, economic, and social systems in manifold harmful ways. While none of the harmful events is unique to a world of rising atmospheric concentrations of greenhouse gases like carbon dioxide, defying efforts to conclusively trace any specific impacts to human-induced changes to the planet's climate system, the frequency and intensity of such harmful weather events are expected to increase as the result of human activities, threatening to increase the human harm that is suffered as a result. Regardless of how its harmful human impacts are experienced, the manner in which climate change is known to be caused—through actions that increase net emissions of greenhouse gases into the atmosphere, including fossil fuel combustion and degradation of natural carbon sinks—has aptly been described as a moral wrong (Gardiner 2011), because avoidable harm is visited upon innocent victims as the foreseeable if unintentional result of the various activities and policies that drive current greenhouse gas emissions rates.

Whether this wrong is best understood through the lens of individual ethics, whereby morally culpable acts by some are linked to morally unacceptable consequences for others, or that of international and intergenerational justice, in which policies and institutions rather than individual actions are ultimately faulted for the harmful effects noted above, these impacts can usefully be understood in terms of moral rights. The quintessential moral right not to be harmed is seen as being violated by the effects of climate change, whether the normative analysis of its cause is rooted in what has come to be known as climate ethics or climate justice. Such normative analyses need not employ rights-based approaches, but the moral case for minimizing further human emissions of heat-trapping gasses and protecting those vulnerable to climate-related change can

perhaps most cogently be expressed through the kind of categorical prohi-
bition against wrongful conduct that rights provide. Since they call for the
strongest kinds of protections for the interests they safeguard, overriding
countervailing claims other than those also based in rights, rights capture
the wrongfulness of imposing harm upon others and motivate the measures necessary for preventing harm from occurring as well as providing a
remedy for its imposition when it does occur.

One might quibble with the notion of moral rights, or their application to the harm associated with climate change, since for reasons to be
considered below this phenomenon takes a form that challenges conventional rights analysis. For present purposes, however, I shall assume a
moral right not to be harmed by the avoidable actions of others, along
with a demonstrable link between human activities, climate change, and
increased human harm. In this chapter, mine is the more specific question
of whether imperatives surrounding climate change might usefully be cast
in the language and politics of *human* rights, which comprise a subset of
the moral rights that provide their ethical foundation. Framing political
efforts to minimize ongoing contributions to climate change and to shield
vulnerable peoples from its effects as a human rights issue, rather than
one of individual ethics or global justice, narrows the purview of its normative inquiry by focusing upon only a small range of expected climate
impacts and expands the resources that might be wielded on its behalf. As
legal constructs, human rights provide normative resources and political
tools that promise unique advantages, and as theoretical ones may offer a
reply to objections that have been levelled against person-affecting ethical
approaches to capturing the wrongness of climate change, especially if
rights against environmental harm can be formalized among recognized
human rights. But these approaches have baggage of their own, including
the political impotence of current human rights regimes when attempting
to discipline powerful states and the normative indeterminacy of many
right claims when it comes to identifying culpable parties or providing a
remedy.

RIGHTS-BASED APPROACHES

In general, moral rights against climate-related harm vest those protected
by them with *prima facie* claims to injunctive relief against those responsible for causing climate change, as well as to compensation for injuries
that they are made to suffer. Recognizing this right against being harmed
provides a very strong, if not necessarily conclusive, reason for others to
cease their involvement in harmful causal processes and to rectify any

experienced harm for which they are responsible. In Hohfeldian terms, this involves the claim-right that those causing the injury refrain from performing harmful actions, through a passive right that governs the actions of would-be polluters (Hohfeld 1919). In principle, the duties associated with this right are binding upon all, although the aggregative and indirect way in which one person's rights might be violated by the acts of others presents a problem for conventional rights theories, to be further discussed below. The moral right, then, is not against harm itself, but rather only harm resulting from responsible human agency. To illustrate: being killed in a routine flood is unfortunate but not wrongful and violates no rights, but being killed in a flood for which a responsible party can be identified is a wrongful violation of one's negative rights. It may not necessarily be wrong to violate someone's rights: diverting a raging flash flood so that it is concentrated upon one home and kills its owner rather than inundating an entire town and killing its residents may not be wrongful, but violates the single victim's rights nonetheless. Her right against being harmed may not require that she be spared at the expense of the larger group, but it may call for her survivors to be compensated for her injury and their loss.

The question of what counts as responsible human agency, without which rights are not violated, requires more discussion than is possible here. As I have elsewhere argued (Vanderheiden 2008, ch. 6), the appropriate standard for establishing culpability in standard cases of rights against harm is *contributory fault*, which as Feinberg notes of standard legal liability has three components:

> First, it must be true that the responsible individual did the harmful thing in question, or at least that his action or omission made a substantial causal contribution to it. Second, the causally contributory conduct must have been in some way *faulty*. Finally, if the harmful conduct was truly "his fault," the requisite causal connection must have been directly between the faulty aspect of his conduct and the outcome. It is not sufficient to have caused harm *and* to have been at fault if the fault was irrelevant to the causing (Feinberg 1970a, p. 222).

Given its reliance upon relatively direct causal links between culpable actions and resultant harm, contributory fault is challenged by environmental harm like that with climate change, for reasons to be further discussed below. For now, though, consider the importance of the rights-violating act or omission being a faulty one. By this standard, environmental harm that results from excusable ignorance (Vanderheiden 2004) may bring about morally bad outcomes but does not violate rights against harm, since agents cannot be faulted for causing outcomes that they could not have reasonably foreseen and thus avoided. Because moral rights involve claims against others, which can only be made against their culpable contributions to harm, they cannot be violated by purely accidental injuries, which are similar in structure to those resulting from natural causes. Moral fault is thus additional to causal responsibility, but can be coupled with it under conditions that change during the course of an ongoing series of contributory acts. If an agent is contributing to some environmental harm as the result of excusable ignorance, for example, but is then informed of this contribution, refusing to curtail it or initiate appropriate remedial measures initiates culpability for subsequent harm, which begins to violate the right of an injured party in a way that mere injury did not.

The right against being harmed as the result of anthropogenic environmental change, especially when such harm involves serious suffering or death, can be regarded as among the most basic of rights, and in the context of climate change bridges the categories of security and subsistence rights, as extreme weather events can threaten human security directly while climate-related scarcity can threaten subsistence indirectly (Shue 1996). For example, climate change threatens basic security interests including territoriality with sea level rise and health with changing disease vectors, as well as driving crop failures through droughts and floods. As such, it trumps lesser rights claims, such as those to the use of property or to the exercise of political sovereignty, as well as to non-rights based imperatives, such as that to accumulate wealth, which is frequently held out as a trumping reason against taking regulatory actions (Dworkin 1977). Without yet considering the specifics of how the actions of some can harm others through climate change, few would deny that it is wrong to cause serious and avoidable harm to others, and that if any rights exist, those against being avoidably harmed by others are basic among them.

Human rights are legal and political constructs that are built upon the foundation of moral rights, which provide key aspects of their justification. Because human rights are held by all *qua* human, regardless of national origin or citizenship, they extend the protections that are granted through constitutional rights beyond national borders, doing so through treaties and other multilateral agreements by which states pledge their protection internally and their promotion abroad, in principle offering their protections even to persons and peoples residing within states that reject human rights conventions. First declared through the 1948 Universal Declaration of Human Rights, and later codified, expanded, and clarified through a series of subsequent conventions, human rights prescribe norms of conduct in world politics, provide a mechanism for evaluating regimes and measuring social progress, and may in some cases be used to justify humanitarian interventions within territories of rights-abusing states on their behalf. Since they are universal in scope, human rights are presumed to protect interests that are common to all as well as being vulnerable to interference by others, making culturally relative interests as well as those threatened only by uncontrollable natural disasters or the vagaries of luck poor subjects of human rights. In other words, they protect against *wrongs*, and not merely bad outcomes, typically through the combination of normative force designed to prevent such wrongs from occurring and remedial processes designed to intervene when they do occur. Rights regimes, then, disseminate the norms that rights protect and enforce their prescriptive implications where necessary, utilizing various power resources in so doing.

Moral rights provide a necessary but insufficient condition for protecting an interest as a human right, which provides additional force to the prescriptions that it issues. The case for wielding human rights on behalf of climate change, then, turns largely upon the value of this additional force, which some view as useful given the recent inability of the international community to adopt sufficient regulatory measures for preventing dangerous climate change, and in doing so protecting those rights that climate change threatens (Caney 2008; Shue 2011). If *new* legal or political measures designed to protect the vulnerable against climate-related harm cannot be adopted, one might surmise, perhaps *existing* international imperatives could be wielded in the service of similar ends. The United States, for example, is not bound by the terms of the 1997 Kyoto Protocol, having refused to ratify the treaty and thus to accept the limits on national emissions that it prescribes, but it *is* party to human rights conventions that might be used to accomplish similar goals, as some have suggested. As constitutional rights provide alternative points of access to policy systems for the politically disadvantaged, to whom legislatures are often unresponsive, human rights may allow for an alternative means for compelling states to take action on climate change, even if they manage to avoid taking on any commitments through the quasi-legislative UN Framework Convention on Climate Change (or UNFCCC) processes that were designed to develop international climate policy.

The clearest rationale for wielding human rights on behalf of international action on climate change, then, is pragmatic: human rights discourse and regimes offer one potential solution to the current impasse over the development of effective international climate policy. In principle, if it could be shown that climate change violates one or more recognized human rights *and* that some nation-state party was responsible for that violation, that demonstration could pave the way for a remedy being ordered as a matter of international law, or at least build support for stronger action among parties whose concern for human rights exceeds their current interest in environmental protection. Of course, international humanitarian law is at present too weak to compel powerful states to change their internal environmental policies, and the United States in particular has come to actively resist the application of international law to its internal affairs, typically citing threats to national sovereignty as its rationale for multiple recent efforts to undermine international humanitarian law. Finding that climate change threatens human rights and that a powerful state like the U.S. is culpable for its contributions to climate change would not *necessarily* bring about domestic policy change, nor would international courts likely succeed in ordering compensation for victims, as most suggesting this strategy are surely aware. Indeed, a serious risk of the human rights approach concerns the fallout for human rights if the U.S. was officially found to be violating human rights, but opted to ignore the remedy ordered under international law. Why, then, pursue a strategy with such likely downside risks and unlikely upside benefits?

AGAINST HUMAN RIGHTS APPROACHES

Before considering reasons for viewing climate change through the lens of human rights, several reasons against doing so must be noted, one of which concerns the relatively modest aims of rights-based approaches. One general contrast with distributive justice-based analyses of climate change is that the rights rest on sufficientarian principles, where rights protection requires that all have access only to that minimal set of resources necessary for their rights not to be violated, whereas justice approaches often rest on egalitarian principles, where equal rather than sufficient resources is the default position from which principled exceptions are granted. As this contrast is typically cast, rights are concerned with reducing absolute but not relative deprivation, focusing only upon the worst off, whereas justice is concerned with narrowing the gap between the best and worst off. Depending upon where thresholds for rights violations are set, human rights approaches to climate change can leave in place significant injustice, provided that these do not yet violate the rights of the affected. With regard to the distribution of or access to resources relevant to climate change, whether shares of atmospheric space (Vanderheiden 2009) or basic necessities like food and water, approaches based in rights may tolerate relatively more inequality and demand less remedial action be taken. As such, a human rights strategy may compromise climate justice goals, allowing the world's affluent to continue to contribute toward climate change up to the point where rights are violated—or in what may for reasons to be discussed below be a much lower bar, where they can be conclusively shown to have been violated by culpable acts—offering no reasons for preventing bad outcomes that are not rights-violating.

For example, climate change may violate the human right to food, which according to the Universal Declaration of Human Rights (UDHR) requires each person to have access to enough "food, clothing, housing and medical care and necessary social services" to allow for a "standard of living adequate for the health and well-being of himself and of his family" (Article 25). Climate-related floods and droughts can adversely affect crop yields in famine-prone regions, threatening this subsistence right in a way that might be linked to the polluting actions that are altering the climate system. Food is here grouped together with other basic necessities, and the adequacy standard invoked by the UDHR might be viewed as requiring only that persons not be deprived of a baseline threshold of calories or nutrients below which further deprivation would violate their subsistence rights, with climate change responsible for any additional deprivation to which it can be shown to be causally linked. Significant climate-related crop losses could be compatible with this right being satisfied, especially with transnational food trade or aid, so long as some basic level of subsistence is maintained. While the human right to food or to subsistence more generally might if protected alleviate a key vulnerability associated with poverty, there is no human right against poverty itself, and indeed rights find nothing wrong with adequately-fed but otherwise impoverished masses existing alongside extreme wealth. If the goal is taken to be some form of egalitarian justice, as analyses of climate justice claim (Caney 2005; Hayward 2007; Vanderheiden 2011), wide inequality in access to the resources that enable food security rather than absolute deprivation would impugned as the problem, with more ambitious remedial measures the remedy to it.

In the context of climate change, claiming violations of the human right to food may, as Caney has argued, identify a moral threshold below which no one is allowed to fall, and thus trigger remedial action once that threshold has been crossed, but the remedial bar would be set at just above that threshold again, at which point rights would not be violated (Caney 2010). Those vulnerable to deprivation-related suffering would remain vulnerable, if not quite deprived, with only those deprived of the objects of subsistence rights having a claim to aid, and then only just enough to lift them above the threshold. With increasing climatic effects resulting from rising global temperatures, the disaster response orientation of sufficientarian human rights approaches may never catch up to the threat that climate change poses to subsistence, at least if understood as justifying *post hoc* remedies to those falling below moral thresholds rather than underscoring the need for proactive efforts to enhance resilience and thus decrease

vulnerability to environmental change. But even if human rights imperatives were construed in this more proactive manner, they would still yield
more modest goals than those prescribed by climate justice approaches,
making them narrower in scope and less ambitious in the kinds of impact
to which they apply, if for such reasons also making them more politically
feasible.

Caney, who continues to defend a comprehensive account of climate
justice that requires a more robust international policy response to the
problem than would be available under the human rights approach that
he also endorses (Caney 2009), has recently begun to argue that climate
change could be held to violate human rights to life, health, and subsistence. Citing the 1976 International Covenant on Civil and Political
Rights, he notes the right of each not to be "arbitrarily deprived of his
life," which he interprets as impugning those deaths caused by climate-
related events like floods, storms, and heat waves as examples of arbitrary
deprivations (Caney 2010, pp. 166–67). As Caney argues, this negative formulation of the right requires that no one may "act so as to deprive other
people arbitrarily of their lives," but less clear are the causal links between
human actions and the loss of life he associates with climate change. Since
climate change only raises the frequency and intensity of extreme weather
events that result in loss of life, and so might be probabilistically linked to
higher mortality rates but not conclusively linked to any particular event
or consequent death, it would in any instance of weather-related deaths
be impossible to say whether anyone's rights were violated in the way that
he supposes. Distinguishing between the wrongful and the merely unfortunate would in such cases simply not be possible, given the causal links
between human activity, climate change, extreme weather events, and
human fatalities. So long as the right to life is treated as a negative right
against interference, requiring correlative negative duties not to be complicit in arbitrary deprivation but no positive duties to assist the vulnerable when they are placed at risk, regardless of cause, it would be difficult
to establish climate change as having violated anyone's rights, and not
merely intensified natural phenomena that threaten human lives.

Moreover, even if some death from an extreme weather event could
conclusively be linked to anthropogenic climate change rather than naturally occurring events, there would still be difficulty in identifying culpable
parties that violated the rights of the victim, failing in their negative duties
to avoid causing arbitrary deprivation of life. Due to the fragmented
agency and the dispersed and aggregative causality associated with climate
change, no one's emissions on their own produce discernible impacts on
global climate, let alone any harmful weather event, so they cannot be said
to have caused the arbitrary deprivation in question (Sinnott-Armstrong

2005). Even very large groups of relatively profligate greenhouse polluters, such as Americans or residents of developed countries, cannot be said to have caused climate change on their own, since the phenomenon results from what humanity has in total done to alter atmospheric concentrations of heat-trapping gasses. It is, to paraphrase the UNFCCC, our common (if differentiated) responsibility, with harmful effects resulting from what humans together have done to increase atmospheric concentrations of greenhouse gases and failed to do to increase resilience to the consequent climatic effects. Moreover, climatic effects are themselves the product of aggregate polluting activity, not the result of even large partial quantities of emitted pollution. If not for the actions of non-members (non-Americans or residents of developing countries), neither of these groups would have produced the climatic effects that would identify them as culpable for violating anyone's human right to life. In short, rights violations, in contrast to adverse impacts upon the interests that rights protect, may require more direct and clearly demonstrable causal links than are available between culpable acts or policies, environmental change, and its specific impacts upon rights-protected interests.

Caney also argues that climate change could be seen as violating the human right to health, which he formulates as a right "that other people do not act so as to create serious threats to their health" (Caney 2010, p. 167). He points to scientific estimates concerning the increased frequency of diseases, including gastrointestinal and cardio-respiratory distress, dengue, and malaria, but again the indirect causality of climate-induced impacts prevent the clear and direct association of any particular episodes of ill health with any culpable human actions. As with the human right to life, the practical impossibility of distinguishing between disease or other injuries that result from human agency and those which would have occurred without climate change renders contributory fault elusive, and thus alleged rights violations tendentious. To claim otherwise would be to jettison the individualistic focus of rights (Waldron 1987), either transforming the right to health into a collective interest rather than an individual one, claiming evidence of rights violations from statistical change in health outcomes within a population, or conflating bad outcomes for persons with violations of their rights, absent a causal link to any faulty act. One might formulate health rights positively rather than negatively, requiring that remedial action be taken to protect against threats to human health, even if not anthropogenic, but this would not serve the strategy of linking climate change to human rights, and would raise a variety of other objections about the scope of correlative duties associated with such a right.

The third human right that Caney identifies as threatened by climate change concerns subsistence, relying upon the UDHR-declared right to food and other basic necessities noted above. He again phrases the right negatively, such that all have a right "that other people do not act so as to deprive them of the means of subsistence" (Caney 2010, p. 168), protecting persons from culpable actions that can be causally linked to deprivation, but rejecting a positive right to the means of subsistence. He notes climate-related threats to food security from droughts, floods, and other extreme weather events, citing estimates that 45–55 million additional persons may be put at risk of hunger as the result of a modest 2.5°C of warming by 2080. Again, the key for establishing that culpable acts leading to deprivation violated human rights to subsistence is a demonstration of contributory fault: not only were those whose rights were violated deprived of the good around which the right in question is based, but this deprivation resulted from the faulty acts of others in contributing to climate change. For reasons already noted, adding this third category of human rights violations adds little to the ethical analysis, other than to more comprehensively document climate-related impacts. It shares previously noted problems in linking climate change to human rights violations with the other two rights discussed above.

SALVAGING HUMAN RIGHTS APPROACHES?

Perhaps one could follow Caney's strategy in pointing to scientific estimates of total human impacts of climate change, arguing that x additional deaths from more frequent and intense weather events means x additional arbitrary deprivations of life, violating the human rights of those affected. Even if we cannot distinguish the anthropogenic drivers of particular weather events from otherwise similar events that would have occurred at lower atmospheric concentrations of greenhouse gases, one might reason, we can identify links between human action and the loss of life, which after all is the object of a human right to life. Although we could not say with any certainty *whose* life was lost as the result of the polluting actions that cause climate change, we could estimate the number of rights violations that a given level of warming was likely to cause, showing climate change (and, by extension, the actions and policies that cause climate change) to be responsible for those additional losses of life.

The problem with this approach is twofold. First, apart from several collective rights like those to self-determination, territory, and culture, most human rights protect individuals against threats to important human interests. The right to life is one such individual right, and it is violated when, as Caney notes, any person is arbitrarily deprived of their life. It cannot protect groups against deaths beyond normal mortality rates, such as those resulting from conflict or environmental change, while retaining its structure as an individual right. Indeed, rights against harm are inherently individualistic, rejecting the exercises in aggregation used above, which as Rawls notes of utilitarianism fail to take seriously the separateness of persons (Rawls 1971, p. 24). While higher group mortality also entails additional individual deaths, the human right to life cannot be said to be violated by statistical deviation from normal group mortality rates. Even if climate change could be identified as the cause of x additional deaths within a group, this would not violate any right held by the group. Unless some person can be shown to have lost their life as the result of anthropogenic climate change, *they* cannot be said to have had their human right to life violated. If no identifiable persons can be shown to have lost their lives as the result of climate change, then climate change cannot be responsible for violating any identifiable person's right to life, where such identification would be crucial for granting standing or assessing damages to sufferers of climate-related harm.

Second, by engaging in this sort of aggregative analysis, the human rights approach does not appear to add clarity or scope to standard climate ethics analyses of wrongful climate-related harm. Deontological approaches resting on categorical prohibitions against causing avoidable harm to innocent victims, or those resting on moral rights, already capture the same diagnosis, and suffer from the same problems of distinguishing human-caused from naturally occurring deaths. Consequentialist ones, for reasons suggested above, do better in capturing the wrongness of causing additional deaths, but reject the inviolability of persons characteristic of rights theories. Distributive justice-based approaches to climate change likewise enjoy an advantage over rights theories, since they focus on allocations of resource shares or the manner in which climate change exacerbates existing inequalities, both of which are more amenable to the aggregative impact strategies noted above. Compared to showing that some set of like acts or policies caused some bad outcome, it is relatively easy to show that some person or group has emitted more than their just share of greenhouse gases, since this analysis need not link those excessive emissions to any particular effects. If the goal is to link offending actions or policies to the criteria that some normative theory uses to identify departures from their expressed ideal, rights-based approaches appear to be least well-suited for applications to climate change, with human rights offering little improvement upon the already murky links provided by its analysis through moral rights.

Another strategy for shoring up the causal links between the human activities and public policies that cause climate change and the objects of human rights involves the official recognition of a kind of penumbra right, implied by human rights against harm, specifically against environmental harm. Recognizing as among protected human rights the right to a safe environment, as it is often formulated, substantiates the role that environmental hazards can play in threatening existing human rights. In this sense, it builds upon human interests that have already been protected by rights to life, health, or subsistence, formalizing the need for protection against environmental threats as instrumental to those interests. By focusing upon the human impacts of environmental change, it does not assign rights to the natural world or its resources, as do calls for granting rights to nonhuman animals or natural objects. Rather, like rights against severe poverty or to due process, the right to a safe environment would add a second-order instrumental protection to the list of legal rights as a measure designed to call attention to the instrumental relationship that environment plays in human welfare. It would thus expand the current list of human rights, but not that of the objects of such rights, as the unsafe environmental conditions it protects against also threaten other rights. The legal protections it would offer if formalized are thus already implied within the penumbra of other rights, and its formal recognition would thus merely make explicit this implication.

One influential formulation of such an environmental human right can be found in the Stockholm Declaration, from the 1972 UN Conference on the Human Environment, which declared as its first Principle that:

> Man has the fundamental right to freedom, equality and adequate conditions of life, in an environment of a quality that permits a life of dignity and well-being, and he bears a solemn responsibility to protect and improve the environment for present and future generations.

Elevating the interest in a safe or adequate environment to the status of a human right would confer several benefits. As Hayward notes of constitutionally protected environmental rights, but applicable to human rights insofar as they obtain the status of legal rights, such protection

> entrenches a recognition of the importance of environmental protection; it offers the possibility of unifying principles for legislation and regulation; it secures these principles against the vicissitudes of routine politics, while at the same time enhancing possibilities of democratic participation in environmental decision-making processes. (Hayward 2005, p. 7)

Other human rights approaches are reductionist in valuing environmental quality only to the extent that it is instrumental to the protection of other interests that through their status as protected are deemed to be inherently valuable, but guaranteeing a safe environment by right obviates the contingency associated with this instrumental relationship. Moreover, such legal recognition would in principle extend standing to NGOs and other parties in legal challenges to the policies or practices that degrade the environment and so threaten rights, since the rights to a safe or adequate environment would be held by all, and so need not depend upon demonstrations of harm to particular persons or require that aggrieved parties demonstrate existing injuries. By contrast, approaches that turn on linking environmental degradation to the loss of right-protected aspects of well-being require more particular threats or existing injuries combined with the willingness on the part of those threatened or injured to press their claims through human rights processes. Finally, the text of this right as expressed here pairs the declaration of the right with a charge of responsibility for protecting it, refusing to wait for violations of the right to occur before ordering remedial action, which in cases of some environmental problems would come too late for such action to be effective in protecting the interests around which it is designed.

This approach, however, may not fully address objections that are made against the reliance of human rights discourses to express imperatives of environmental protection. As with Caney's application of existing rights to environmental threats, the right to a "safe" or "adequate" environment requires only that some threshold of sufficiency not be crossed, and so offers a poor substitute for claims of environmental justice in ensuring equity in access to environmental goods and services. Even more explicitly environmental formulations of this right encounter difficulty in specifying how much of the planet's ecological goods and services each might be entitled to before their deprivation constitutes an unsafe or inadequate amount, since their focus is upon the impact of deprivation on other interests rather than taking any baseline of environmental quality as good in itself. With global environmental hazards like climate change, where links between pollution or resource depletion and threats to human dignity or well-being can more conclusively be drawn, there remain the problems of associating any culpable actions with the harmful phenomenon, for reasons noted above. If greenhouse polluting actions or policies cannot be shown to violate existing human rights like those to life or subsistence, it would be unclear how recognizing a right against climate change would improve the ability of rights discourses to target responsible parties or actions. If they can establish this connection between agents and harm through the causing of unsafe environmental conditions, on the other hand, it is unclear what would be added by expanding the list of rights that are violated by such actions, as all would seem to stand or fall together.

What, then, is gained by identifying three separate human rights that are allegedly violated by climate change, rather than simply resting the case for remedial action on the most serious among them (the right to life)? If rights approaches depended upon passing some threshold of badness or moral tipping point, chronicling these additional human rights impacts beyond loss of life might be used to weigh against concerns that presently prevent action from being taken. In human rights politics, of course, such thresholds often do matter, as remedies like sanctions or the use of force are typically thought to be justified only when rights are violated on a sufficiently wide scale, but this pragmatic consideration defies the essential structure of rights theories, which do not depend upon such impact comparisons. In principle, at least, remedial action would be warranted if *any* rights were being violated—this, indeed, is why Caney and others find rights approaches to be so attractive against stubborn threats like climate change—and countervailing considerations would be limited to more serious rights violations, not mere quantitative comparisons. Unlike criminal charges, where sentences might be extended by adding together multiple violations, the remedy that would be required of culpable parties if they were found to be violating human rights by contributing to climate change would first be to stop contributing to it, then perhaps to compensate its victims. Identifying more rights impacts and thus more victims might justify this expansion of human rights analyses, but recall the elusive links between climate change and specific harm that make this consideration dubious.

After all this, we might again ask: what does this kind of human rights-based analysis add to normative approaches to climate change based either in distributive justice or ethics? Philosophically, the answer might be: not much. All three of the rights identified by Caney as being threatened by climate change could be captured by straightforward moral injunctions against harming, so human rights merely replicate charges that climate ethics has already brought against climate change, along with the difficulties in establishing them. These objections are not unique to rights-based approaches, however, and should dissuade us from invoking the human right to life on behalf of efforts to compel carbon polluters to take on greater climate change mitigation commitments if other reasons for taking a rights approach can be identified. There is little doubt that the failure to prevent avoidable deaths or deprivation-related suffering is morally wrong, even if the standard model of contributory fault prevents these morally bad outcomes from being definitively established as rights violations, and wielding human rights against climate change might yet be vindicated if it could in practice improve the chance of such outcomes being avoided.

DEFENDING HUMAN RIGHTS APPROACHES

Why not simply reject the application of human rights to climate change, then? One reply is political rather than philosophical. Legal rights provide institutional means of redress when persons are wronged by having their rights violated, whereas moral injunctions are only able to invoke the lesser executive power of normative force. With robustly protected constitutional rights, for example, victims can pursue redress through the courts, often with the assistance of counsel funded by states or human rights NGOs, and if successful have the backing of other state actors in obtaining their ordered remedies. Human rights rely upon less robust "soft" law, but nonetheless are in principle protected by the force of state coercion against significant threats, whether through quasi-judicial bodies like the International Criminal Court or through the multilateral protection of states pledged with the UDHR and reaffirmed through the Responsibility to Protect (R2P) doctrine, as well as being protected under domestic law by signatory states to the various human rights conventions, which assign primary responsibility for rights protection within each state to its recognized government. While less robust than constitutional rights, human rights promise stronger protections than ethical proscriptions and prescriptions, which carry no sanction and thus are unenforceable, and at least attempt to bring practical enforcement to the protection against moral wrongs, aligning political power with the ends that render it legitimate.

Beyond the legal and political power that they might potentially invoke, human rights are important sources of norms in international politics and governance, affirming the objects protected by them as universally held and crucial to human welfare, heralding those rights as the basis for evaluation of regimes and international institutions or agreements, and calling attention to humanity's most urgent threats. Even if not justiciable sources of legal power, the recognition of human rights against environmental harm may confer significant discursive benefits for the development of international climate policy, linking climate change with other human rights imperatives. By treating climate change as a human rights issue rather than an economic one, as Caney argues, the case might more effectively be made for strong mitigation efforts even if these come at high costs, as rights protection is not subject to cost-benefit analysis. Human rights impacts, rather than economic costs, might more effectively be used to measure both mitigation measures and adaptation programs, and to compare the relative efficacy of both kinds of programs as they compete for scarce resources. Expressing climate impacts in human rights terms would also shift the focus from emissions, atmospheric concentrations of greenhouse gases, or temperature increase goals—which have thus far been used to define the goals of international climate policy efforts, but which may be too abstract for members of the public—to the human impacts of climate change (McInerney-Lankford et al. 2011). While there remains a danger that linking climate change and human rights could weaken the latter rather than bolstering the former, a human rights approach could help to consolidate support for climate action within the human rights community, not only by drawing upon its central normative aim but also through the implicit pledge that climate policy development would make human rights promotion central to its mission, reducing worries about competition for scarce aid and development resources being diverted to nascent international climate efforts.

In addition, approaches that invoke quasi-legal rights may be more personally or politically empowering for those suffering from climate-related harm than are those based in ethics, which may involve moral rights but give no practical form to them, given their characterization of that harm as infringing upon entitlements that can be *claimed* rather than merely victimizing those affected by it and relying upon those who are culpable also being conscientious. As Feinberg suggests, "having rights allows us to 'stand up like men,' to look others in the eye, and to feel in some fundamental sense the equal of anyone" (Feinberg 1970b, p. 252), for rights enable persons to make valid claims of entitlement when they are wronged by others, rather than limiting them to moral suasion or depending upon the sympathy or charity of others. Merely being in possession of rights, even without having to invoke them before the state (where claims may be dismissed), might therefore confer some salutary benefits upon would-be sufferers of climate-related harm. This personal empowerment may be partly dependent upon the legal or political empowerment that protected rights also provide, when states recognize rights holders as having valid claims against each other as well as against the state itself, but issues also partly from the formulation of claims in terms of rights, and persists even as such claims are officially denied. Being in the position to demand a remedy, as opposed to merely pointing out a morally bad outcome in which one is involved, implies an equal moral status that may embolden those accustomed to disadvantage to act on behalf of their interests rather than accepting harm visited upon them as inevitable or irresistible.

Official public recognition as a rights holder undergirds the personal empowerment that Feinberg describes, transforming what might otherwise be an unenforceable moral claim into an enforceable political one, bolstering the confidence of those vulnerable to climate-related harm to demand the objects or their rights by granting them the power that is necessary for attaining it. Where rights exist on paper only, and are not supported through the legal and political enforcement mechanisms that make rights regimes function, moral rights denied a basis in positive law cannot fully vest those who are only nominally protected by them with this dignity and sense of equality, but the aspirational nature of human rights might still inspire potential claimants to the status that they promise, in the project of gaining their ultimate legal recognition (Donnelly 2003, pp. 191–92). Recognition of those affected by climate change as bearers of rights, rather than as innocent victims seeking beneficent adaptation aid, may strengthen the cause of climate justice even if rights claims are denied by relevant political authorities, as they were for many years in nearly every successful rights-based social movement.

This legal and political empowerment carries with it several practical advantages in motivating strong climate change mitigation policies along with these more psychological or sociological benefits. Perhaps most obviously, recognizing that climate change violates some legal right grants the injured access to the mechanisms of government, whether through the court system that hears their case and has the power and obligation to redress their grievances by protecting those rights or through the legislative and executive functions of governments that are legally committed to protecting rights and can be ordered to do so when they fail. While no international body protects human rights in the way that domestic courts are able to protect legal rights, and human rights jurisprudence creates no precedent for domestic law, a legal declaration that some party has violated the human rights of another could potentially have significant agenda-setting and opinion-mobilizing effects. Even if not legally binding, human rights may have sufficient normative weight to condemn in popular opinion those parties or policies responsible for violating such rights, and governments that are found to be complicit in rights violations would face the burden of demonstrating that they adhere to the system of rights to which they declare their allegiance.

While not relying directly upon human rights, a demonstration of the link between the acts and policies that cause climate change and threats to domestically protected rights to life, health, or subsistence could empower those threatened by climate change and protected by these domestic legal rights, if not yet all those now putatively protected by human rights. Official recognition of such a link would turn on similar analyses used on behalf of human rights claims, bolstering the normative force of the human rights appeal. Courts have thus far been reluctant to grant legal standing to actual or potential victims of climate-related harm by recognizing rights that are threatened by the predicted impacts of climate change, although the U.S. Supreme Court did in *Massachusetts vs. Environmental Protection Agency* (2007) allow a consortium of cities and states to sue the EPA for failing to regulate carbon dioxide under the statutory authority of the Clean Air Act, granting states legal standing by virtue of the harm to their "capacity of quasi-sovereign" in endangering the public health and welfare of their citizens. In requiring the EPA to promulgate emission controls for automobiles, the Court affirmed no individual rights against impacts associated with climate change, but nonetheless granted legitimacy and urgency to regulatory efforts carried out under existing law at a time when Congress was unwilling to pass any new statutory protections to do the same. Had the Court granted standing to individual persons, legal precedent would allow for much broader challenge to the current national climate policy intransigence, including the expansion of regulatory power beyond automobile tailpipe emissions and potentially also the power to order adaptation financing as a form of injunctive relief for vulnerable people and compensation for those that cannot be insulated from climate-related harm. Human rights claims might emanate from domestic legal rights cases, expanding their protections outward.

Conversely, human rights challenges under international law to the policies of carbon polluting states could affect domestic climate politics from the other direction, mobilizing support for better domestic rights protections by faulting domestic policies for human rights violations abroad. For example, a 2005 challenge by the Inuit before the Inter-American Commission on Human Rights (IACHR) alleged that the United States was violating their human rights through its contribution to climate change, which has had some of its most palpable effects in the arctic regions of Canada, Alaska, Greenland, and Russia inhabited by Inuit people. According to the petition, climate change "caused by the acts and omissions of the United States" violated Inuit human rights "to the benefits of culture, to property, to the preservation of health, life, physical integrity, security, and a means of subsistence, and to residence, movement, and inviolability of the home" (Watt-Cloutier 2005, p. 5). The petition was denied without prejudice in November 2006, but had it been accepted it would have marked a significant victory for the Inuit as well as others vulnerable to climate-related harm. Included within the petition and within the IACHR's authority to order were demands that the U.S. "adopt mandatory measures to limit its emissions of greenhouse gases and cooperate in efforts of the community of nations" to mitigate anthropogenic climate change, that it assess and consider impacts of domestic emissions on Inuit people "before approving all major government actions," and that it develop and finance an adaptation a plan for Inuit people to adapt to changing climatic conditions (Watt-Cloutier 2005, pp. 7–8). Again, the petition's ultimate failure might give pause concerning the strategy, but the potential for agenda setting and mobilizing support through human rights discourse ought also to be considered.

CONCLUSIONS

Human rights are typically viewed as aspiration rather than connoting any positive law, and indeed are described as such in the Universal Declaration of Human Rights, which calls upon signatory states to "strive by teaching and education to promote respect for these rights and freedoms and by progressive measures, national and international, to secure their universal and effective recognition and observance." While subsequent declarations of human rights through multilateral treaties are legally binding, in principle giving them the status of legal rights and requiring enforcement by signatory states no matter where rights violations occur, in practice they enjoy a significantly weaker status than domestic legal rights. Some of this relative weakness issues from institutional deficits at the global level, where no court system equivalent to those designed to protect domestic constitutional rights exists, leaving the enforcement of human rights largely to the discretion of states rather than impartial legal authorities. Other sources of their relative weakness are legal, as for example with the U.S. reservation to the International Covenant on Civil and Political Rights declaring its provisions not to be self-executing, which denies petitioners access to U.S. courts and rejects normal treaty requirements that its provisions also be made a part of domestic law.

While human rights law could potentially require state parties to human rights treaties and conventions to take on more strenuous carbon abatement efforts as injunctive relief for current rights violations or in the interest of rights protection, thereby strengthening human security against climate-related harm, it is only one of several sources of international law that already require large carbon polluters like the United States to take meaningful action to mitigate climate change. Although the U.S. avoided legally binding greenhouse emissions caps under the 1997 Kyoto Protocol by refusing to ratify the treaty, it is a signatory to the 1990 Rio Declaration, through which it committed to freezing its emissions at 1990 levels pending further policy actions, along with being party to the 1979 Geneva Convention on Long-range Transboundary Air Pollution, which offers another international basis in international law for requiring greenhouse emissions controls. Linking recognized human rights to climate change mitigation and adaptation efforts, or recognizing penumbra rights like the right to a safe or adequate environment, could build upon existing international law in further defining national obligations in and goals of international regulatory regimes.

Although both the seriousness of climate change as a global policy concern and the urgency of action to prevent climate-related suffering suggest the connection to human rights, which are properly reserved for humanity's greatest moral and political challenges, the upside of invoking such rights on behalf of climate change mitigation might best be seen as political rather than philosophical, and of the political upshots the primary benefits may reside in the recognition and empowerment of current and potential sufferers of climate-related harm rather than the legal mobilization of recognized political authorities. To be sure, there remain downside risks of a human rights approach, including potential damage to support for other human rights imperatives from linking it to a politically unpopular if urgent policy issue, and compromise to the more ambitious egalitarian climate justice imperatives that the rights approach can only partly fulfil. Whether or not to make such a case in terms of climate justice, climate ethics, or human rights must be regarded in strategic rather than analytical terms, as the value or risks of various approaches turns less upon their ability to clarify the moral stakes in the human response to climate change and more upon their propensity for mobilizing an effective response, but these considerations may warrant at least some further work on human rights approaches, perhaps seeking to offer them as more directly normative than either analytic or authoritative, connecting recognized ethical commitments with nascent efforts to build those into the way the world confronts this looming environmental threat.

REFERENCES

Caney, Simon (2005), 'Cosmopolitan justice, responsibility and global climate change', *Leiden Journal of International Law*, **18** (4), 747–75.

Caney, Simon (2008), 'Human rights, climate change, and discounting', *Environmental Politics*, **17** (4), 536–55.

Caney, Simon (2009), 'Justice and the distribution of greenhouse gas emissions', *Journal of Global Ethics*, **5** (2), 125–46.

Caney, Simon (2010), 'Climate change, human rights, and moral thresholds', in S. Gardiner, S. Caney, D. Jamieson, and H. Shue (eds), *Climate Ethics: Essential Readings*, Oxford, UK and New York, US: Oxford University Press, pp. 163–77.

Donnelly, Jack (2003), *Universal Human Rights in Theory and Practice*, Ithaca, US: Cornell University Press.

Dworkin, Ronald (1977), *Taking Rights Seriously*, Cambridge, US: Harvard University Press.

Feinberg, Joel (1970a), *Doing and Deserving*, Princeton, US: Princeton University Press.

Feinberg, Joel (1970b), 'The nature and value of rights', *Journal of Value Inquiry*, **4**, 243–57.

Gardiner, Stephen M. (2011), *A Perfect Moral Storm: The Ethical Tragedy of Climate Change*, Oxford, UK and New York, US: Oxford University Press.

Hayward, Tim (2005), *Constitutional Environmental Rights*, Oxford, UK and New York, US: Oxford University Press.

Hayward, Tim (2007), 'Human rights versus emissions rights: climate justice and the equitable distribution of ecological space', *Ethics & International Affairs*, **21** (4), 431–50.

Hohfeld, Wesley N. (1919), *Fundamental Legal Conceptions*, New Haven, US: Yale University Press.

Intergovernmental Panel on Climate Change (IPCC) (2007), *Climate Change 2007: Impacts, Adaptation, and Vulnerability*, Contribution of Working Group II to the Fourth Assessment Report of the Intergovernmental Panel on Climate Change, ed. by M.L. Parry, O.F. Canziani, J.P. Palutikof, P.J. van der Linden and C.E. Hanson, Oxford, UK and New York, US: Oxford University Press.

McInerney-Lankford, Siobhán, Mac Darrow, and Lavanya Rajamani (2011), *Human Rights and Climate Change: A Review of the International Legal Dimensions*, Washington, US: The World Bank. siteresources.worldbank.org/INTLAWJUSTICE/Resources/HumanRightsAndClimateChange.pdf.

Rawls, John (1971), *A Theory of Justice*, Cambridge, US: Belknap Press.

Shue, Henry (1996), *Basic Rights: Subsistence, Affluence, and U.S. Foreign Policy*, Princeton, US: Princeton University Press.

Shue, Henry (2011), 'Human rights, climate change, and the trillionth ton', in Douglas Arnold (ed.), *The Ethics of Global Climate Change*, Cambridge, UK and New York, US: Cambridge University Press, pp. 292–314.

Sinnott-Armstrong, Walter (2005), 'It's not my fault', in W. Sinnott-Armstrong and R.B. Howarth (eds), *Perspectives on Climate Change: Science, Economics, Politics, Ethics*, San Diego, US: Elsevier, pp. 285–306.

Vanderheiden, Steve (2004), 'Knowledge, uncertainty, and responsibility: responding to climate change', *Public Affairs Quarterly*, **18** (2), 141–58.

Vanderheiden, Steve (2008), *Atmospheric Justice: A Political Theory of Climate Change*, Oxford, UK and New York, US: Oxford University Press.

Vanderheiden, Steve (2009), 'Allocating ecological space', *Journal of Social Philosophy*, **40** (2), 257–75.

Vanderheiden, Steve (2011), 'Globalizing responsibility for climate change', *Ethics and International Affairs*, **25** (1), 65–84.

Waldron, Jeremy (1987), 'Can communal goods be human rights?', *European Journal of Sociology*, **28** (2), 296–322.

Watt-Cloutier (2005), *Petition to the Inter-American Commission on Human Rights Seeking Relief from Violations Resulting from Global Warming Caused by Acts and Omissions of the United States*, with support from the Inuit Circumpolar Conference, 7 December, earthjustice.org/sites/default/files/library/legal_docs/petition-to-the-inter-american-commission-on-human-rights-on-behalf-of-the-inuit-circumpolar-conference.pdf.

Index